FOREWORD

GEOTRAP, the OECD/NEA Project on Radionuclide Migration in Geologic, Heterogeneous Media, is devoted to the exchange of information and in-depth discussions on present approaches to acquiring field data, and testing and modelling flow and transport of radionuclides in actual (and therefore heterogeneous) geologic formations for the purpose of site characterisation and evaluation and safety assessment of deep repository systems of long-lived radioactive waste. The project comprises a series of structured, forum-style workshops on the following themes:

(i) Field tracer transport experiments: design, modelling, interpretation and role in the prediction of radionuclide migration.

(ii) The basis for modelling the effects of spatial variability on radionuclide migration.

(iii) The characterisation of water-conducting features and their representation in models of radionuclide migration.

(iv) Confidence in models of radionuclide transport for site-specific performance assessment.

(v) Geological evidence and geological bases for radionuclide retention processes in heterogeneous media.

The workshops, which comprise technical presentations, each discussion periods, poster sessions and *ad hoc* working group discussions, are designed to enable national waste management agencies, regulatory authorities and scientists to interact and contribute to the advancement of the state of the art in these areas.

The fourth workshop in the series was held in Carlsbad, New Mexico (United States) on 14-17 June 1999, and hosted by the US Department of Energy/Carlsbad Area Office (United States).

The workshop discussed the approaches that have been taken within national waste management programmes to evaluate, enhance and communicate confidence in the geosphere-transport models that are used in site-specific performance assessments (i.e. exercises carried out for a real or hypothetical repository, but based on a real – not hypothetical – geologic location). Forty-five delegates, and five observers, from twelve NEA Member countries attended the workshop. They represented waste management organisations (implementing agencies and regulatory authorities), nuclear research institutes, the academic community and scientific consulting companies. The workshop provided an overview of the state of the art in this technical field, both in national waste management programmes and the scientific community, as well as a forum for discussing more general aspects of confidence building in the framework of repository development.

The present publication is a synthesis of the materials that were presented (oral and poster presentations), the discussions that took place, and the conclusions and recommendations drawn, notably during the working group sessions. The synthesis also puts these conclusions and recommendations into perspective within the scope of the GEOTRAP project. The opinions, conclusions and recommendations expressed are those of the authors only, and do not necessarily reflect the view of any OECD Member country or international organisation.

ACKNOWLEDGEMENTS

On behalf of all participants, the NEA wishes to express its gratitude to the US Department of Energy/Carlsbad Area Office – USDOE/CAO (United States) which hosted the workshop in Carlsbad, New Mexico, and conducted technical visits to the Waste Isolation Pilot Plant (WIPP), the USDOE's deep geologic repository for transuranic nuclear waste, and to the Flow Visualisation and Processes Laboratory at Sandia National Laboratories.

Special thanks are also due to:

- M. Matthews, USDOE/CAO, S. Altman, K. Hart, and P. Comiskey, SNL, (United States) for their help in the technical and practical preparation of the workshop;

- the members of the Programme Committee who structured and conducted the workshop: S. Altman, SNL (United States), J. Hadermann, PSI (Switzerland), P. Lalieux (NEA), P. Lebon, ANDRA (France), M. Matthews, USDOE/CAO (United States), S. Norris, Nirex (United Kingdom), R. Patterson, USDOE/YMP (United States), P. Smith, SAM Ltd. (United Kingdom) and R. Storck, GRS (Germany);

- P. Smith, SAM Ltd. (United Kingdom) and S. Altman, SNL (United States) who helped the Secretariat in drafting the workshop synthesis;

- the working-group chairmen: G. de Marsily, University Paris VI (France), N. Eisenberg, NRC (United States), R. Storck, GRS (Germany) and R. Beauheim, SNL (United States), who led and summarised the debates that took place in the four working groups;

- the speakers and the authors of posters for their interesting and stimulating presentations; and

- the participants for their active and constructive contribution.

P. Lalieux and C. Pescatore from the Radiation Protection and Waste Management Division of the OECD Nuclear Energy Agency were responsible for the GEOTRAP Project's scientific secretariat. P. Lalieux (now ONDRAF/NIRAS, Belgium) was also a member of the workshop Programme Committee.

TABLE OF CONTENTS

PART A: SYNTHESIS OF THE WORKSHOP

PART B: WORKSHOP PROCEEDINGS

SESSION I

CONFIDENCE IN TRANSPORT MODELS IN THE OVERALL CONTEXT OF REPOSITORY DEVELOPMENT

Chairs: M. Matthews (USDOE/CAO, United States) and P. Lalieux (NEA)

SESSION II

DEVELOPMENT OF, AND CONFIDENCE IN, GEOSPHERE-TRANSPORT MODELS USED IN PERFORMANCE ASSESSMENT

Chairs: J. Hadermann (PSI, Switzerland) and P. Lebon (ANDRA, France)

SESSION III

APPROACHES TO CONFIDENCE IN MODELS THAT SUPPORT GEOSHPERE-TRANSPORT MODELS USED IN PERFORMANCE ASSESSMENT
Chair: R. Patterson (DOE/YMP, United States)

SESSION IV

EXPECTED IMPACT OF FUTURE SITE INVESTIGATION STRATEGIES AND APPROACHES ON CONFIDENCE IN TRANSPORT MODELS
Chairs: S. Norris (Nirex, United Kingdom) and W. Weart (SNL, United States)

PART A:

SYNTHESIS OF THE WORKSHOP

1. INTRODUCTION

1.1 Background, aims and scope of the 4th GEOTRAP workshop

Confidence in the long-term radiological safety of underground repositories for the disposal of radioactive waste is required to support decision making in the step-wise process of repository planning and development. Confidence is required in the appropriateness of the site and design for their intended use, and in the methodologies, models and databases used to assess their performance. This includes the assessment of the geosphere as a barrier to the transport to the biosphere of radionuclides released from the repository. The nature and scope of the confidence-building process is still under discussion by many organisations. Discussions are complicated by the subjective nature of the term confidence, and of the term uncertainty, to which confidence is closely related. Both the NEA[1] and NRC/SKI[2] have recently published documents on these topics.

Although the NEA document[1] presents a logical framework for confidence evaluation, enhancement and communication, there is still much room for development in defining the practical means that are required to achieve these goals. The aims of the 4th workshop of the GEOTRAP project were:

- to examine the approaches that have been taken within national waste-management programmes to evaluate, enhance and communicate confidence in models of radio-nuclide transport for site-specific performance assessment;

- to prepare the present synthesis report, that reviews and summarises the lessons learned at the workshop, putting them into perspective within the scope of the GEOTRAP project and the state of the art in the technical field.

The scope of the workshop was limited to "technical confidence", which includes confidence in the methodologies, models and data used in performance assessment, although the wider aspects of the confidence required for decision making in repository planning and development were recognised in the workshop discussions.

1.2 Structure of the present synthesis

The findings of the workshop are summarised in Section 2. These include an assessment of the achievements of the workshop, observations arising from the presentations and discussions, and recommendations. Sections 3-6 discuss the topics covered by the workshop in more detail. Section 3 discusses the evolution of confidence in the course of repository planning and development and Section 4 discusses general aspects of confidence in performance assessment. These set the scene for a

---

1. "Confidence in the Long-term Safety of Deep Repositories – its Development and Communication", OECD/NEA, 1999.

2. "Regulatory Perspective on Model Validation in High-Level Radioactive Waste Management Programs: A Joint NRC/SKI White Paper", NUREG-1636, 1999.

discussion of aspects of confidence more specific to models of radionuclide transport for site-specific performance assessment in Section 5. Some concluding remarks are made in Section 6.

Annex 1 details the workshop structure and lists the key questions that were established by the Programme Committee for each session and working group. Annexes 2 to 5 present the main points made during the discussions that took place within the *ad hoc* working groups. These annexes consist of reports prepared by the working group chairmen, without further editing.

2. FINDINGS OF THE WORKSHOP

2.1 Achievements and observations

The workshop, which began with a presentation of the NEA document on confidence building[1], demonstrated that the concepts developed and explained in this document are applicable to many national programmes, and are likely to assist programmes in placing their approaches to confidence building in a structured and logical framework.

Although discussed at GEOTRAP in the context of geosphere-transport modelling, many aspects of confidence building are generic in nature, and apply to repository development and performance assessment in general. In particular, the workshop demonstrated a realisation, on the part of the waste-disposal community, that "absolute confidence" in a model (or "absolute validation") is unnecessary in the context of repository development and is, in fact, unattainable. Rather, in the course of a repository programme, models will be developed and refined, and confidence in the models will evolve, as increasing scientific understanding of the system is acquired. Model results are often used to support decision making, and confidence must be *adequate to support the decision at hand*.

Regarding the development of confidence in the course of repository planning and development, the workshop highlighted:

- the adoption of a step-wise approach to repository planning and development in many programmes, with models and databases for performance assessment refined iteratively, as information and experience are acquired;

- a recognition that confidence in the long-term safety provided by a particular site and design can both increase and decrease in the course of a project, as understanding is developed, with the discovery of unexpected and potentially negative features being a common experience in several programmes;

- a recognition that the level of confidence that is required for decision-making in each stage of a repository programme, from the initial planning stages to licensing and implementation, depends on the decision at hand. The existence of open issues does not necessarily preclude a positive decision for a specific stage, particularly in the earlier stages of repository planning.

These aspects are discussed in more detail in Section 3.

The workshop discussed the evaluation, enhancement and communication of confidence in performance-assessment models in general (Section 4), and models of radionuclide transport for site-specific performance assessment in particular (Section 5). Important points of consensus between workshop participants emerged, both on general aspects of confidence in performance assessment, such as:

- the importance of traceable and transparent documentation, and of peer review, as measures to enhance and communicate confidence;

- the benefits of publicising both favourable and unfavourable findings and aspects of an assessment in order to enhance the credibility of the organisations carrying out performance assessments, both in the eyes of the wider scientific community and in the eyes of the general public;

- the widespread view that performance assessment is to be judged as an entity, so that specific regulatory requirements regarding confidence in geosphere-transport models are not expected;

- the usefulness, for example, of natural analogues and experimental visualisation as methods to build public confidence.

and on aspects more specific to models of radionuclide transport for site-specific performance assessment, namely:

- the recognition that "the geosphere is the key component of any geological disposal system" and that the host rock and overlying strata may serve several safety functions in addition to those of transport barriers;

- the need to develop a consistent understanding of a site, that takes account of a wide range of observations and, in particular, the usefulness of observations from nature to support understanding and build confidence that the full range of safety-relevant features, events and processes has been taken into account;

- the recognition by implementers and regulators that the degree of confidence in performance-assessment results is a function of the modelled time scale, and that studying the past evolution of a system is a valuable means of understanding the behaviour of a system over the long time-scales of relevance to performance assessment;

- the need for site characterisation and research and development to narrow, where possible, parameter uncertainty and the range of conceptual model uncertainty, since the existence of several alternative models that are consistent with observations is not conducive to confidence;

- the need to maintain site characterisation and research and development efforts that are guided by, but not exclusively driven by, specific performance-assessment models in order to create and maintain confidence.

As an indicator of the confidence that can be attached to current geosphere-transport models, it is noted that, in the last few years, no fundamentally new processes and events have been identified in the course of site characterisation and research and development that would challenge the applicability of current models. Methods exist to identify sources of uncertainty and analyse their significance with respect to evaluated performance and, in some cases, performance assessment has provided the motivation for the development of new experimental techniques. In response to (and in anticipation of) the availability of more site-specific data, many programmes have refined their models to permit, for example, the representation of geosphere heterogeneity in a more realistic manner, as discussed at the third GEOTRAP workshop. The current challenges are to develop confidence in site-specific understanding of features such as channelling, dominant pathways and diffusion-accessible porosity, to better understand temporal changes and to improve methods of deriving effective parameters from modelled parameters.

In spite of recent refinements, the complexity of many geosphere-transport models is still less than the state of the art would permit (although the data for more complex models may be unavailable and the transparency and reproducibility of the results of highly complex models are questionable). It was agreed that:

- confidence requires that models incorporate the essential features of the system, based on a reasonable understanding of the processes being represented, and on site understanding and conceptual models that are consistent with wide-ranging observations and evidence, even though all such information cannot be incorporated directly into the models and databases;

- the use of (conservative) simplifications is inevitable, and the degree to which it is acceptable, without undermining confidence, depends on the purpose to which the model is applied; and

- a degree of simplification that leads to a clear loss of realism should be avoided.

There was, however, some lack of consensus at the workshop as to the degree of complexity that is appropriate for performance-assessment models, which may reflect problems of acceptance of simplified models in the wider scientific community, and, more generally, reflects that the evaluation of confidence is a qualitative judgement. Model simplification and its implications for confidence in model results are discussed further in Section 6.

It was noted that efforts to achieve consensus between technical experts may be an important step towards gaining public confidence.

2.2 Recommendations

> **(i)** The workshop identified a number of technical areas where more work could be particularly beneficial to confidence in performance–assessment models in general, and geosphere–transport models in particular.

Natural analogues and site-specific evidence (e.g. paleohydrogeological arguments):

Although some examples were presented at the workshop, it was generally agreed that chains of argument based on natural analogues and site-specific evidence are underused. There is scope for better and more direct use of natural analogues and paleohydrogeological arguments in process identification, in discrimination between alternative conceptual models, in parameter estimation and in the making of safety cases.

Performance-assessments should more directly take advantage of observations at repository sites, or at other sites that show similar features and processes. More quantitative comparisons of model predictions with observations from analogues, and the better integration of natural analogues into performance-assessments, are seen as areas warranting further development. The difficulties arising, for example, from the highly uncertain initial conditions in the evolution of natural analogues are, however, recognised.

Long-term experiments:

More consideration could be given to experiments performed over the entire development and operational period of a site, such as natural-gradient tracer tests. The primary purpose of such tests would be to confirm predictions made for larger time and spatial scales than could be investigated prior to a construction licence application. The tests could be integrated with any monitoring system that is set up around a repository, and should have only modest personnel and equipment costs. Such tests would contribute to confidence in a site for as long as the site was operational or the monitoring programme active.

Additional areas:

Additional areas where future work would be beneficial to most, if not all, national programmes include:

- the use of complementary performance measures (e.g. containment by particular barriers) for regulatory purposes,

- methods for deriving effective parameters from measured parameters, and

- methods for propagating uncertainties related to the formulation of conceptual models and to performance-assessment model abstraction.

(ii)	**Confidence building is a process that needs to be inclusive in order to ensure a wide acceptance of the methods, models and data used in performance-assessment.**

Within the project team of an implementing body:

There is a need to involve site-characterisation staff in model development, in order to build confidence in the resulting models throughout the project team. The workshop identified a formal process of parameter elicitation as a method of promoting good, traceable documentation and of fostering good co-operation between experimentalists and modellers.

Between implementers and regulators:

Confidence building is a process that needs to include both implementers and regulators, in order, for example:

- to identify any differences in views regarding the fundamental principles of the approach to the evaluation of safety (such as the degree to which highly simplified geosphere-transport models are acceptable, as discussed at length at the workshop), and

- to develop consensus on specific model concepts (e.g. at Konrad in Germany, the need for a model in which fractures are explicitly modelled within the porous continuum).

In this way, any potential differences in view between the two bodies are identified and resolved as early as possible.

Within the wider community of scientists, politicians and the general public:

Consensus between the implementer and the regulator responsible for a particular programme, and also consensus between different programmes internationally, are not necessarily sufficient to ensure the adequacy of an approach in the view, for example, of the wider community of scientists, politicians and the general public. Although different arguments, and different emphases, may be appropriate in different fora, these latter groups also need to be included in the confidence-building process, in order:

- to avoid unrealistic expectations from performance-assessment modelling – i.e. to communicate the notion that the evolution of the repository and its environment need not, and cannot, be described completely in performance-assessment;

- to build the credibility, in the view of these wider groups, of the implementing and regulatory bodies, through their openness with respect to unresolved issues and their willingness to address non-technical concerns.

(iii) **To instil the confidence necessary for decision making in the early stages of a repository project, when many unresolved issues exist and before the necessary data to resolve the issues have been collected, a step-by-step plan showing how and when the necessary data will be collected should be created.**

In the absence of particular data, confidence is created by a clear plan that recognises data requirements and provides a strategy for data acquisition. The plan must, however, be flexible, with detailed planning proceeding iteratively as information and experience are acquired. The plan should focus site characterisation and research and development efforts on the reduction of parameter uncertainty and the discrimination between alternative conceptual models, even when models have similar consequences for compliance in terms of system performance.

(iv) **More efforts should be made by the radioactive-waste community to look outside the radioactive-waste field when planning site characterisation, and for peer review.**

Particular lessons may be learned from the oil industry, where great efforts have been made in recent years to develop and apply techniques for the integration of geophysical data and other "soft data" in site models. This includes the definition of correlations among different types of well logs, core measurements and field measurements. These correlations are used to condition geostatistical reservoir models.

In order to make use of as wide a range of experience and expertise as possible, peer review groups should also draw on individuals from outside the radioactive-waste field.

3. THE EVOLUTION OF CONFIDENCE IN THE COURSE OF REPOSITORY PLANNING AND DEVELOPMENT

The workshop highlighted the step-wise approach to repository planning and development adopted by many programmes, with models and databases for **performance-assessment** *refined iteratively, as information and experience are acquired.*

Site characterisation and performance-assessment activities run in parallel to the step-wise planning and implementation of a repository. Generic understanding of factors relevant to long-term safety, as well as site- and concept-specific models and data, are initially limited, with many unresolved issues. Indeed, the sparseness of site-specific data may lead to discrepancies between model simulations and field observations (e.g. in the SKI SITE 94 project it was found that a careful investigation of regional scale hydrology and hydrochemistry was necessary to account for discrepancies between model simulations and observed data). Thus, early studies in the course of a repository project often emphasise the robustness of the site and design, with respect, for example to the effects of geological evolution. An example is the case of the Waste Isolation Pilot Plant (WIPP), located near Carlsbad, New Mexico (United States), where early investigations focused on salt dissolution and transport and used highly simplifying assumptions in conceptualising transport, depending more on geological than hydrological evaluations. Performance-assessment models and databases often deal with the need to illustrate safety in the presence of the initially high degree of uncertainty through the use of highly conservative model assumptions and parameter selection.

At later stages of a repository project, confidence building tends to focus more on site characterisation (the quality of the information on the repository site and design) and on more realistic models and databases, which are better supported by a wide range of information. This is especially true for models and databases for near-field rock, which is more amenable to detailed characterisation than the less accessible parts of the geosphere more distant from the repository.

The evaluated confidence in the long-term safety provided by a particular site and design can both increase and decrease in the course of a project.

New information can increase confidence in the evaluated performance if it provides additional support for the parameter values and the conceptual models that are selected. Furthermore, data that can be used only qualitatively in the early stages of a project (e.g. as "independent indicators") may be incorporated directly into model parameters at a later stage. The GEOTRAP presentations, however, highlighted the fact that confidence in the long-term safety provided by a particular site and design can decrease, as well as increase, in the course of a project. As noted at previous GEOTRAP workshops, there is, for example, a tendency for site characterisation to reveal more variability, with its associated uncertainties, than originally expected (e.g. when a site-characterisation programme moves from surface-based investigations to the underground). Indeed, programmes noted that periodic "confidence crises" can occur, when a new observation or finding is made that challenges the existing level of confidence. This was illustrated in the case of observations of brine seepage into the repository at the WIPP site. Prior to the observation, the prevailing conceptual model for the WIPP was that there was no intercrystalline brine. The observation of seepage during mining led to a long series of investigations and a complete overhaul of the conceptual model of the repository behaviour.

The arguments that support decision making, including arguments that give confidence in performance-assessment findings, need to be adequate to support the decision at hand. The existence of "open issues" does not necessarily preclude a positive decision, particularly in the earlier stages of repository planning.

The sufficiency of confidence depends on:

- the stage of repository planning and development and the purpose of the performance-assessment within the step-wise development process, i.e. the level of confidence required will evolve during the course of a repository project, according, for example, to the consequences of a decision proving to be incorrect – these tend to be most severe at later stages, when performance-assessments are used to support license applications,

- the intended audience (as discussed by Working Group C, the sufficiency of confidence will differ depending on whether the audience is technical/scientific, regulatory/official, or public/political), and

- in relation to the modelling of specific processes, the relative importance of these processes in the overall demonstration of safety.

The existence of open issues, which may be defined as those issues that must eventually be resolved for the final licensing of a repository, may be acceptable at earlier stages, but there must be reasonable assurance that their resolution can be achieved at later stages. This was a topic discussed by Working Group D.

It was pointed out by Working Group D that the sufficiency of confidence in data (and other aspects of performance-assessment modelling) will be judged according to the scientific standards ("state of the art") of the day, and not the standards that applied when the data were collected (or when the methods/models were developed). As the state of the art advances, performance-assessment methods, models and data must keep pace.

4. GENERAL ASPECTS OF CONFIDENCE IN PERFORMANCE ASSESSMENT

It is widely acknowledged that the degree of confidence in **performance-assessment** *results is a function of the modelled time-scale.*

Uncertainties in the evaluated performance of a repository generally become greater as the time-scale over which performance is evaluated is increased. As a result, the required degree of rigour in the performance-assessment may become less at later times, with confidence in long-term safety built increasingly on qualitative and semi-quantitative arguments, rather than quantitative evaluation of risk or dose. In this context, the use of complementary measures of performance (e.g. containment by particular barriers) may be valuable, although more work is needed before these can be applied in practice for regulatory purposes.

Traceability of documentation, avoiding undocumented analytical procedures and missing data, is important in the communication of confidence in that it allows the possibility of independent verification of performance-assessment results.

The communication of confidence, through, for example, traceable and transparent documentation, was discussed at length by Working Group C. In the NEA document on confidence building[1], it was pointed out that such documentation can itself provide a method of confidence enhancement:

> "... a carefully laid out strategy for the refinement (of performance-assessment methods, models and databases) through successive development stages, and the traceable and transparent documentation of the process of refinement, should foster the view that the detrimental effects of uncertainties have been reduced to the maximum reasonable extent. This should promote confidence in the quality of the safety case".

It may be necessary to communicate confidence in the findings of performance-assessment modelling to multiple audiences, ranging from technical specialists to the general public. Due to the potentially broad audience, documentation needs to be transparent, with scientific explanations that are clear and not obscured by details. Multiple levels of documentation, including, for example non-technical summaries for a general audience, complemented by more technically detailed documents for specialist audiences can also be useful. It is important that there is coherence between these different documents. In all cases the documents must be traceable.

The importance of peer review in the evaluation of confidence, and as a major confidence-enhancing measure, is noted for all programmes.

Peer review can highlight topics where more confidence is required and indicate priorities for future research and development and site characterisation. Peer review is most effective if a wide range of expertise is drawn upon. External peer review can work to counteract the potential for false confidence within the disposal community, which may arise where several organisations use the same approach.

The features, events and processes that are identified as important in a **performance-assessment** *can depend on the acceptance guidelines with which results will be compared.*

For example, in the USA, the adoption of a dose-based regulatory standard has led, at Yucca Mountain, to more importance being attached to the saturated zone.

Confidence in the findings of **performance-assessment** *needs to be complemented by wider confidence in the credibility of the organisation(s) that have carried out the work in order to support decision making in repository development.*

The credibility of the organisation carrying out an assessment is favoured by openness of publication and, in particular, willingness to publicise both favourable and unfavourable findings and aspects of an assessment. The inevitable limitations of modelling and model testing should be acknowledged. As discussed by Working Group D, these include:

- the difficulty/impossibility of performing field tests with the radionuclides of most relevance to long-term safety;

- the need for upscaling of measurements in order to apply them to performance-assessments and the consequent uncertainties that this introduces;

- more generally, the various sources of uncertainty associated with an assessment;

- the external constraints, including finite financial resources and externally imposed timetables, within which a repository programme must operate.

Communication of confidence to the general public is seen as an area of increasing importance. Methods have been identified that have the potential to communicate confidence effectively to the public, but the development of additional methods, and the refinement of existing methods, are priorities for many programmes.

Many of the concepts relevant to confidence that are used by radioactive waste disposal specialists, such as conservatism and robustness, are difficult to communicate effectively to the general public. Indeed, these terms may have different meanings, even within the disposal community. It is, therefore, difficult to communicate the confidence of technical specialists to the general public on this basis. Potentially more effective methods that have been identified to facilitate communications with the general public, and to build public confidence, include:

- assisting the public to understand risk as a safety indicator by comparing the evaluated risk from a disposal project with those associated with other activities with which the public is familiar;

- use of natural analogues to demonstrate the possibility of isolating radioactive materials over very long time scales (though clear explanations of the limitations of such analogues are necessary, and the differences between analogues and engineered repositories must be acknowledged);

- experimental visualisation methods for transport and other processes: the "power of pretty pictures" to explain complex processes and how they affect a potential repository should not be underestimated.

Working Group C also recognised that disagreements within the scientific community may lead to a loss in public confidence. Efforts to resolve such disagreements and achieve consensus between technical experts are therefore important to gaining public confidence.

5. CONFIDENCE IN MODELS OF RADIONUCLIDE TRANSPORT FOR SITE-SPECIFIC PERFORMANCE-ASSESSMENT

5.1 The role of the geosphere within the overall disposal system

A widespread view among both regulators and implementers is that a disposal system should be judged as an entity, and that specific regulatory requirements regarding confidence in the **geosphere-transport** *barrier, and in* **geosphere-transport** *models, are not to be expected. There is also a recognition that "the geosphere is the key component of any geological disposal system" and that the host rock and overlying strata may serve several safety functions in addition to that of transport barriers.*

An important aspect of the geologic disposal concept is that it is based on multiple barrier functions, including those provided by the repository host rock, to ensure that (unforeseen) deficiencies in any particular barrier function will not jeopardise the safety of the entire system. In some programmes, there has been a trend towards the adoption of robust engineered systems in order to offset some of the difficulties inherent in assessing the performance of complex and heterogeneous natural systems, and thus increase overall system confidence. The current emphasis on enhanced assurance through engineering does not, however, detract from the fact that the geosphere remains the key component of any geological disposal system.

It was pointed out at the workshop that, although the focus of discussion was on the role of the host rock in providing a barrier to radionuclide transport, the host rock has additional barrier functions, i.e. protecting the engineered barriers and providing a suitable (e.g. mechanical, geochemical, hydrogeological) environment for them to operate. Indeed, the most important consideration in regulatory review is likely to be that the implementer has convincingly demonstrated that stable and favourable conditions will be maintained over the time scale of interest, such that the barriers will function as intended. This view is reflected in the approaches adopted in many programmes to the demonstration of long-term safety.

5.2 Confidence in the completeness of phenomena considered in geosphere performance-assessment and their conceptualisation

Confidence in the completeness of phenomena analysed in **performance-assessment** *does not come from performance-assessment modelling itself. Rather, it comes from the development of a consistent understanding of the disposal system (including the site), taking into account general scientific understanding and a wide range of measurements and observations from the field and laboratory.*

The workshop distinguished between investigations and observations that are an integral part of the database on which a numerical model is calibrated, and "independent indicators", which contribute to confidence in the completeness of the phenomena analysed, as well as to distinguishing between alternative conceptual models. Independent indicators can be divided into:

- indicators of the present behaviour of the system, and
- indicators of the geological/hydrogeological/geochemical history of the system over long periods.

In either case, relevant observations can originate from an identified repository site, or, especially in the early stages of a repository programme, from analogous sites. Working Group A further divided indicators according to the type of measurement or observation used, namely:

- geochemical,
- thermal,
- geomechanical and structural, and
- biological.

The working group report provides a number of specific examples of the use of each type of indicator.

Studying the past evolution of a natural system is a valuable means of understanding the behaviour of the system over the long time-scales of relevance to **performance-assessment.** *Observation from nature can enhance confidence that the full range of safety-relevant events and processes have been considered in* **performance-assessment** *and can lend support to conceptual models.*

The risk that phenomena such as infrequent geological events and slow geological processes will perturb a disposal system unfavourably in the long term needs to be considered in performance-assessment. There is thus a need to draw on observations from nature in order to build confidence that the full range of safety-relevant events and processes have been considered in performance-assessment, and that the potential effects of these events and processes are understood. Quoting from the first presentation at the GEOTRAP workshop:

"Performance-assessment without nature observation is empty, nature observation without performance-assessment is blind".

The following types of observations from nature were described and discussed at the workshop:

(i)   Observations at repository sites and natural analogues

Observations at repository sites, include those leading to an understanding of the origin and evolution of groundwater, examples of which were provided at the workshop for the Konrad site in Germany and the Wellenberg site in Switzerland. A further example is the observation of the slow rate of salt dissolution at WIPP, which gives confidence in the longevity of the host formation. An example of the use of natural analogues is provided by the uranium/thorium deposit at Palmottu, Finland, which has been used to assess the disturbance of groundwater chemistry by glacial effects. The use of these observations tends to be limited to building confidence in the suitability of a site and design to provide long-term safety. Their value lies in the fact that it is often possible to understand past behaviour (e.g. the paleo-hydrogeology of a repository site) over time-scales similar to, or exceeding, those relevant to performance-assessment. Assuming that the site is not significantly perturbed by the presence of a repository, this understanding can then be used to put bounds on possible future behaviour. Even though it cannot be excluded that the future evolution of the system may be different from the past, the demonstrated ability to explain and model past evolution provides confidence in modelling future evolution.

Natural analogues have been successfully used to test some performance-assessment databases quantitatively (e.g. testing by the US DOE at the Yucca Mountain site of the data used to model rock-water interactions). The analysis of observations at repository sites and natural analogues can also be used to build confidence in the models used in performance-assessment, providing a qualitative understanding of some key processes and building confidence that all relevant features and processes are included. Natural analogues illustrate, for example, the concentration of actinides at geochemical fronts and the consequent importance of considering the possibility of such fronts, and the most appropriate way to represent them, in performance-assessment. It should be possible to build confidence in models of radionuclide migration by testing whether the processes observed in national analogues are reflected in model results when similar conditions apply at a potential repository site. The workshop participants were, however, of the view that the studies of natural analogues have not, so far, contributed greatly to the quantitative testing of performance-assessment models or the process models that support them.

(ii)   Observations of anthropogenic analogues

Examples presented at GEOTRAP include studies of tailings from uranium mining and of the nuclear weapons tests at the Mururoa and Fangataufa Atolls. The analysis of the Mururoa and Fangataufa Atolls is possibly unique in that it provided an opportunity to develop models of groundwater flow and radionuclide transport through a geological medium, apply them over length scales (though not time scales) that are relevant to repository performance-assessment, and test them against observations. Confidence in the groundwater flow model was, for example, favoured by its consistency with temperature profiles that provided independent indicators. The contribution to confidence in transport models is, however, limited, particularly due to the severe limitations in the available data and in the resources that were available to carry out the study.

Some **performance-assessment** *models, or their underlying concepts, are amenable to testing on a field scale. Such testing, especially if model predictions are made in advance of an experiment, contributes to confidence in the model and its underlying concepts. Renewed interest is noted in large-scale field tracer transport experiments.*

The particular value of "blind testing" was pointed out at the workshop, in the context of field tracer transport tests in the unsaturated zone at Yucca Mountain in the USA. The value of blind testing was also discussed at length at the first GEOTRAP workshop. Other examples of such tracer tests, and their relevance to confidence building, include:

- the Tracer Retention and Understanding Experiments – TRUE – at the Äspö Hard Rock Laboratory, which are used to address the question of whether laboratory-derived retention parameters are applicable at the field scale,

- multi-well and single-well tracer tests at the WIPP site, which have been used (along with laboratory visualisation experiments) to support a multi-rate diffusion model for transport in the Culebra dolomite, and

- tracer tests at the planned ANDRA underground rock laboratory, which will be used to obtain defensible values for diffusion coefficients, and to address up-scaling issues and field measurement difficulties.

5.3 Enhancing confidence through site characterisation and research and development

Methods exist whereby the range of conceptual models that are consistent with current understanding of a disposal system can be explored.

Nirex gave an example of how a matrix diagram (derived from a Master Directed Diagram) can be used to provide an audit trail for the compilation of a database of features, events and processes and a full set of conceptual models and their interactions. The matrix is used to define a set of conceptual models and identify those that have the most influence on the disposal system, thus producing a reduced set of conceptual models.

It was pointed out at the workshop that the participation in a project of more than one modelling group, independently developing and applying models and codes, can build confidence that the range of model uncertainty has been explored.

An analysis of the ranges of conceptual models and parameter values may not provide sufficient confidence for decision-making purposes. A focus of site characterisation and research and development efforts should be the reduction of uncertainty and the discrimination between alternative conceptual models, even when models have similar consequences for compliance in terms of system performance.

As pointed out by Working Group D, a position of not knowing which of two or more models is incorrect does not instil confidence, and experiments should be designed to discriminate between these alternatives. Furthermore, as noted in previous GEOTRAP workshops, the possibility of new, alternative conceptual models, and the operation of phenomena that are not included in existing models, must be borne in mind when planning and interpreting site characterisation and research and development activities. Site characterisation and research and development should thus aim to enhance confidence in the completeness of the phenomena considered in performance-assessment and to reduce conceptual model uncertainty.

An example of such an exercise is the unsaturated-zone tracer tests at Yucca Mountain, designed for the purpose of discriminating between conceptual models of fracture-matrix interaction used in performance-assessment. Another example is the natural and artificial tracer experiments being run at the Tournemire site (France), which aim to determine the transport properties of fracture zones. Specifically, the investigators at that site are interested in understanding the relative importance of transport by diffusion through intact claystones versus transport through fracture zones. As discussed by Working Group D, for site understanding in general and geosphere-transport in particular, any single type of evidence alone may not be sufficient to distinguish among alternative conceptual models, but discrimination may be possible when different lines of evidence are combined (in this context, natural analogues and paleohydrogeological evidence were considered by the GEOTRAP participants to be currently underused).

Even after extensive site characterisation, multiple conceptual models may still need to be considered. For example, several different hydrogeological and numerical models were used in modelling the Konrad site in Germany. Furthermore, radionuclide transport paths from a repository through the geosphere may depend on the details of repository layout (e.g. the orientation of tunnels relative to the hydraulic gradient and fracture zones), which may not be fixed until a late stage in a repository planning and development programme. Posiva/VTT, in particular, noted the difficulty in identifying potential transport paths, and the need to consider alternative possibilities.

Maintaining site characterisation and research and development efforts that are guided by, but not exclusively driven by, specific performance-assessment models is important in creating and maintaining confidence. Although there is some scope for improvement, methods exist to identify various sources of uncertainty and analyse their effects on evaluated performance. In some cases, **performance-assessment** *has provided the motivation for the development of new experimental techniques.*

While they do not, in themselves, identify new features, events and processes, the development and application of performance-assessment methods and models can reveal those uncertainties that have the greatest impact on performance, and hence the greatest impact on confidence in the evaluated long-term safety of the repository. Site characterisation and research and development efforts can then be focused on reducing these uncertainties. Examples of performance-assessment driven experiments are:

- the unsaturated zone test at Yucca Mountain, USA,

- the Canadian *in situ* diffusion tests, and

- the WIPP tracer tests, USA.

Examples of such methods that can be used to assess the impact of particular uncertainties are:

- geostatistical methods (as used, for example, by GRS to model the three-dimensional spatial distribution of low-permeability clay layers and to assess the impact of associated uncertainties with respect to performance measures, such as groundwater travel times and the times of maximum flux),

- parameter sensitivity analyses, and

- "insight models" that analyse in detail a subset of features, events and processes.

- In addition, a qualitative consideration of the equations governing radionuclide transport can identify key parameters. In particular, some programmes have identified the ratio of

24

the groundwater flow rate to the flow-wetted surface (alternatively referred to in the Swiss programme as the "channel transmissivity") as the most important nuclide-independent parameter in geosphere-transport modelling.

Determining the ratio of the groundwater flow rate to the flow-wetted surface requires an understanding of groundwater movement on a small scale. In Finland, this has led to the development of a new flow meter that measures groundwater velocities under natural gradients directly.

The workshop discussed other advances in the techniques available for site characterisation and noted significant current activity in the refinement and application of:

- laboratory techniques to examine interactions between solutes and mineral surfaces (e.g. of infrared spectrometry to analyse surface interactions);

- *in situ* ("downhole") geochemical experiments (e.g. the CHEMLAB experiments at Äspö).

Geophysical techniques for the remote sensing of geological features, and the correlation of data from different geophysical techniques, have also advanced considerably, principally due to developments in the oil industry. Their application in the field of radioactive-waste disposal has, however, been relatively limited and of a mainly qualitative nature. The application of these techniques in performance-assessment should, perhaps, be more widely considered.

The workshop also noted (with the exception of the Finnish flow meter, above) relatively little progress in field hydrogeological characterisation techniques, especially for fractured, saturated media. This may partly indicate the maturity of experimental approaches to the hydrogeological characterisation of these media, and, in the case of field investigations for radioactive waste disposal, the limits in the possibilities for (or impossibility of) *in situ* experimentation using radionuclides.

*In evaluating the current level of confidence in site-specific **geosphere-transport** models, the workshop identified the need for data to characterise heterogeneous geological media, the need for a better understanding of temporal changes and the need for improved methods of deriving effective parameters from measured parameters as issues with the potential to undermine confidence.*

In recent years, considerable work has been carried out to acquire site-specific data to support performance-assessments. This includes the characterisation of the heterogeneity of geological media, which was a topic of the third GEOTRAP workshop. In response to (and in anticipation of) the availability of more site-specific data, many programmes have refined their models to permit, for example, the representation of geosphere heterogeneity in a more realistic manner. Nevertheless, and although opinions were divided (as might be expected given the diverse backgrounds of the participants), many participants felt that the heterogeneity of potential host rocks, and the resultant uncertainty in models and data, continues to be a factor with the potential to undermine confidence. Specific areas where, in many cases, additional data would enhance confidence include (Working Group D):

- the characterisation of channelling within fractured media;

- the identification of dominant transport pathways;

- the determination of diffusion-accessible porosity.

Other issues identified as having the potential to undermine confidence were temporal variability[3] and upscaling. As discussed by Working Group D, parameters that are measured in field and laboratory experiments must generally be processed (upscaled) in some way in order to provide input parameters for geosphere-transport models. Confidence in upscaling approaches is therefore critical to confidence in transport model parameters. Due mainly to the uncertainties associated with the upscaling of parameters measured at a smaller scale than their application in performance-assessment, there is a trend towards block-scale characterisation of flow and transport parameters, especially for crystalline (or tuff) rocks. Although the block scale is still smaller than the scales relevant to performance-assessment, it represents a significant advance over laboratory-scale experiments and the single-borehole scale. SKB is specifically designing experiments in the Äspö Hard Rock Laboratory in order to provide information for upscaling from laboratory scale to borehole scale experiments and further upscaling to block scale experiments.

5.4 Model simplifications and its implications for confidence in model results

The discussions of model simplification illustrated how the evaluation of confidence is a qualitative judgement. Even within the scientific and technical community, the degree of confidence that is placed in particular methodologies, models and data sets may vary widely.

The complexity of performance-assessment models is often less than the state of the art would, in principle, permit. Working Group B addressed the issue of simplification of models in performance-assessment (this was also a topic discussed at the third GEOTRAP workshop). It was concluded that simplification is inevitable in performance-assessment, and that the degree of simplification depends on the purpose of the assessment (e.g. design optimisation or compliance with regulatory guidelines). Working Group B listed the following reasons for simplifications:

- the inevitably limited information and data, resulting in uncertainties;
- limited resources for an assessment (e.g. available time, budget, etc.); and
- the desire to produce results that can be explained clearly (transparency and reproducibility of calculations).

As well as their use in identifying critical uncertainties to be addressed, for example, by site characterisation, "insight models" and sensitivity studies can provide useful tools for identifying processes that do not have a strong impact on performance, and thus provide support for model simplification. A relevant example presented at the GEOTRAP workshop was the use of insight models to investigate the importance of Onsager processes to radionuclide transport through argillaceous media (e.g. Opalinus Clay). These processes are normally not included in performance-assessment models.

The workshop discussed model simplification and its implications for confidence in model results. The workshop participants agreed that:

- the waste-disposal community needs to be aware of the state of the art in transport modelling, even if such models may be either unnecessary for safety demonstration or inapplicable in a performance-assessment context;

---

3. The importance of processes (principally climate-induced changes and geological evolution) that lead to time-dependent conditions in a geological medium was discussed at length in the third GEOTRAP workshop.

- performance-assessment models should incorporate essential features of the system based on a reasonable understanding of the processes being represented, and a degree of simplification that leads to a clear loss of realism is to be avoided;

- although it is generally impractical to incorporate all relevant observations and evidence explicitly in numerical performance-assessment models, confidence in the modelling approach requires that all such information should be consistent with the underlying conceptual model, or models, of the system;

- even if complex models are not applied directly in a near-field/geosphere/biosphere model chain, they may play a role in ensuring the conservatism of simpler models, and in estimating the degree of conservatism in performance-assessment model results;

- confidence in a numerical performance-assessment model, and in the conceptual models on which it is based, is favoured by its consistency with many diverse and independent observations and sources of evidence (i.e. multiple lines of reasoning), i.e. if a particular uncertainty is perceived to cast doubt on an observation, that uncertainty may not affect other independent observations or sources of evidence;

- confidence is also favoured by the testability of a model; i.e. its ability to withstand counter-arguments successfully; and

- simplifications need to be well documented, and supported, e.g. by showing they are conservative and/or do not have a strong impact on performance.

There was, however, some divergence of views regarding the extent to which simplification is acceptable without undermining confidence in assessment results.

Some participants, especially those from the wider scientific community, were of the opinion that the use of excessively simple models in performance-assessment (vis-à-vis the state of the art in the academic world) could undermine their confidence in the findings of the assessment. For example, treating the retardation of actinides as a "sorption" process, modelled using the K_d approach, may be inappropriate (and possibly non-conservative, although, more probably, pessimistic, depending on the way in which K_ds are selected) if, in reality, retardation is, at least in part, due to precipitation or other processes for which K_d provides a poor representation. Furthermore, the development of more complete models is desirable in order to have confidence that all interactions between phenomena have been taken into account.

On the other hand, it is widely accepted within implementing and regulatory bodies that numerical models that can be demonstrated to be bounding or conservative can yield results that are suitable for decision making in repository development. Furthermore, the consensus view within these bodies is that the incorporation of all identified phenomena into a single, all-inclusive computer model that aims at "realism" does not necessarily lead to increased confidence. Rather, use of an excessively complex model, with many parameter values that are poorly supported by site-characterisation and other evidence, would yield non-transparent results, and may give a false impression of confidence. In this view, the adding of new processes and couplings in performance-assessment models is warranted only if they are well supported by observations and relevant data. Posiva/VTT gave an example of a simple, conservative, steady-state geosphere-transport model of groundwater flow, which implied high and probably unrealistic flow rates of saline water, but was adequate for the purposes of the assessment.

6. CONCLUDING REMARKS

The wide range of experience and backgrounds of the workshop participants, and the attendance of two scientists from the broader academic community, played a major role in developing, discussing and challenging the findings that were drawn. It should be noted that, in contrast to previous GEOTRAP workshops, the majority of implementing organisations and regulatory bodies sent their own delegates to the 4th workshop, instead of being represented by technical consulting companies.

As in previous GEOTRAP workshops, the establishment, in advance, of a series of key questions to be addressed in each session and in each working group proved to be a very effective way of focusing the discussions and reaching practical conclusions and recommendations. The workshop achieved lively interaction among the participants and many points of consensus emerged, although differences in views on the topic of model simplification highlighted the need for waste-management specialists to communicate the reasons for selecting particular models (and for the use of simplified models) to their scientific peers, who may tend to favour more complex, "all inclusive" models.

The workshop focused on "technical confidence", which includes confidence in the methodologies, models and data used in performance-assessment. As an indicator of the confidence that can be attached to current geosphere-transport models, it was noted that, in the last few years, no fundamentally new processes and events have been identified in the course of site characterisation and research and development that would challenge the applicability of current models. The current challenge is to develop confidence in site-specific understanding of features such as channelling, dominant pathways and diffusion-accessible porosity, and to improve methods of deriving effective parameters from modelled parameters.

"Technical confidence" is only one component of the confidence requirements for decision-making in repository development. Other components include ethical, economical and political aspects of the appropriateness of underground disposal and confidence in the organisation structures and legal and regulatory framework for repository development. These aspects were defined as lying outside the scope of the workshop, as were methods for communicating confidence to the general public. The importance of such communication was, however, mentioned several times in the course of the workshop. Indeed, efforts to achieve consensus between technical experts are considered an important step towards gaining public confidence.

STRUCTURE AND PROGRAMME OF THE WORKSHOP

1. Introduction

The workshop comprised sixteen technical presentations, each allocated a discussion period, and a poster session with eight posters that provided a good complement to the oral presentations. The guiding theme for structuring the presentations and discussions was the way in which confidence in performance assessment requires confidence in geosphere-transport models that, in turn, requires confidence in supporting (i.e. process) models and, ultimately, in the general scientific understanding of the site. This led to a "top-down" structure to the workshop (Sessions I-IV):

I: Confidence in transport models in the overall context of repository development.

II: Development of, and confidence in, geosphere-transport models used in performance-assessment.

III: Development of, and confidence in, (process) models that support geosphere-transport models.

IV: Expected impact of future site investigation strategies and approaches on confidence in transport models.

A key part of the workshop (Session V) consisted of focused discussions within small, *ad hoc* working groups, on specific themes, namely:

A: The role of independent indicators to support transport models.
B: Implications of simplification in geosphere-transport modelling in performance-assessment.
C: Aspects and sufficiency of confidence in supporting (process) modelling.
D: Strategies for increasing confidence in transport models through data acquisition.

These discussions were each introduced by a presentation by the working group chairmen. The outcomes of the working group discussions provided the basis for a plenary, concluding discussion (Session VI) and for the present synthesis.

For each session and for each working group, the Programme Committee had established a series of key questions to be addressed. As in previous GEOTRAP workshops, this proved to be a very effective way of focusing the discussions and reaching practical conclusions and recommendations. The main elements of the workshop programme are reproduced in the following section.

2. Workshop programme

SESSION I

CONFIDENCE IN TRANSPORT MODELS IN THE OVERALL CONTEXT OF REPOSITORY DEVELOPMENT

Chairs: M. Matthews (USDOE/CAO, United States) and P. Lalieux (NEA)

This introductory session should set the scene for the whole workshop. By providing a common background to all participants to the workshop, it should also help reduce the difficulties encountered due to the differences in terminology in the various disciplines covered and waste management programmes represented.

Confidence in the Long-term Safety of Deep Geological Repositories – Approach Developed under the Auspices of the NEA
H. Röthemeyer (BfS, Germany)

Evolution of Confidence in Transport Models for Performance Assessment Purposes: the WIPP Case
W. Weart (SNL, United States)

Confidence in Geosphere Performance Assessment – the Canadian Nuclear Fuel Waste Disposal Programme 1980-1999: a Retrospective
M. Jensen (Ontario Power Generation, Canada) and B. Goodwin (GEAC, Canada)

Radionuclide Transport-related Aspects of the Radiological Evaluation of the Mururoa and Fangataufa Atolls
G. de Marsily (Univ. Paris VI, France) and J. Hadermann (PSI, Switzerland)

Regulatory Views and Experiences Regarding Confidence Building in Geosphere-transport Models
B. Dverstorp and B. Strömberg (SKI, Sweden), and J. Geier (Oregon State Univ., United States)

SESSION II

DEVELOPMENT OF, AND CONFIDENCE IN, GEOSPHERE-TRANSPORT MODELS USED IN PERFORMANCE-ASSESSMENT

Chairs: J. Hadermann (PSI, Switzerland) and P. Lebon (ANDRA, France)

This session is concerned with specific aspects of the evaluation, enhancement and communication of confidence in geosphere-transport models that are directly used in performance-assessment exercises.

A geosphere-transport model is defined as the component of the performance-assessment model chain that accounts typically for (i) transport processes in the geosphere (e.g. advection, dispersion, diffusion, retention), (ii) radioactive decay and nuclide build-up during geosphere-transport and (iii) dilution in groundwater. This model is typically supplied with output from a near-field model, and provides input to a biosphere model. It is also typically supported by more detailed modelling of specific processes (to be addressed in Session III) that are treated in a simplified (usually conservative) manner in the geosphere-transport model itself. The level of detail to be included in a geosphere-transport model will depend, for example, upon the purpose of the performance-assessment exercise, upon the confidence in the underlying process-models and in site understanding, and upon the confidence in the abstraction process from underlying process-models to the geosphere-transport model.

Confidence Building in the Nagra Approach to Geosphere-transport Modelling in Performance Assessment
P. Smith (SAM Ltd, United Kingdom), M. Mazurek (Univ. Bern, Switzerland), and A. Gautschi, J.W. Schneider and P. Zuidema (Nagra, Switzerland)

The Posiva/VTT Approach to Simplification for Geosphere-transport Models, and the Role and Assessment of Conservatism
T. Vieno, H. Nordman and A. Poteri (VTT Energy, Finland), A. Hautojärvi and J. Vira (Posiva, Finland)

The Transport Model for the Safety Case of the Konrad Repository and Supporting Investigations
E. Fein and R. Storck (GRS, Germany), H. Klinge and K. Schelkes (BGR, Germany) and J. Wollrath (BfS, Germany)

Regulatory Experience with Confidence-building in Geosphere-transport Models
N. Eisenberg, L. Hamdan, T. Nicholson, B. Sagar and R. Codel (NRC, United States)

SESSION III

APPROACHES TO CONFIDENCE IN MODELS THAT SUPPORT GEOSPHERE-TRANSPORT MODELS USED IN PERFORMANCE-ASSESSMENT

Chairs: R. Patterson (DOE/YMP, United States)

Session III addresses the specific aspects of confidence evaluation, enhancement and communication of the models that are used to support more general (and often simplified) geosphere-transport models that are used in performance-assessment. Such supporting models, sometimes named process-models, address e.g. groundwater flow, geochemistry, geometry/structure, matrix diffusion. They tend to aim at realism (which may not be the case for geosphere-transport models used in performance-assessment). Geometrical/structural and groundwater flow models have been extensively treated in previous GEOTRAP workshops; they are considered to be outside the remit of the present workshop. Focus is therefore on other supporting (process-) models. It should be noted, however, that dispersion, which is largely the result of flow in heterogeneous media, will be addressed in this session.

Session III is also aimed at providing a link to the topics and issues to be addressed in the framework of the 5th workshop of the GEOTRAP project, i.e. *"Geologic Evidence and Theoretical Bases for Radionuclide Retention Processes in Heterogeneous Media"*.

The Role of Matrix Diffusion in Transport Modelling in a Site-specific Performance-assessment: Nirex-97

S. Norris and J.L. Knight (Nirex, United Kingdom)
The Effects of Onsager Processes on Radionuclide Transport in the Opalinus Clay
J. Soler (PSI, Switzerland)

Uranium Migration in Glaciated Terrain: Implications of the Palmottu Study, Southern Finland
D. Read (Enterpris, Univ. of Reading, United Kingdom), R. Blomqvist and T. Ruskeeniem (GTK, Finland) and K. Rasilainen (VTT Energy, Finland)

SESSION IV

EXPECTED IMPACT OF FUTURE SITE INVESTIGATION STRATEGIES AND APPROACHES ON CONFIDENCE IN TRANSPORT MODELS

Chair: S. Norris (Nirex, United Kingdom) and W. Weart (SNL, United States)

On the basis of existing performance-assessment exercises and reviews (from regulators and scientific peers), several organisations have established, or are in the process of establishing, new strategies and approaches to site characterisation in order to increase confidence in the geosphere-transport barrier. This session will deal particularly with the definition and justification of these strategies and approaches, and their expected impacts on confidence.

An Unsaturated Zone Transport Field Test in Fractured Tuff
G. Y. Bussod (LANL, United States)

The Impact of Performance Assessment on the Experimental Studies at Äspö HRL and on Future Site Characterisation and Evaluation in Sweden
P. Wikberg, A. Ström and J.-O. Selroos (SKB, Sweden)

Future Prospects for Site Characterisation and Underground Experiments Related to Transport Based on the H-12 Performance Assessment
Y. Ijiri, A. Sawada, M. Uchida, K. Ishiguro, H. Umeki (JNC, Japan) and E. Webb (SNL, United States)

Current Understanding of Transport at the East Site (Clays) and Relevant Research and Development Programme in the Planned Underground Research Laboratory
P. Lebon (ANDRA, France)

SESSION V

IN-DEPTH DISCUSSIONS BY WORKING GROUPS

Working Group A:

ROLE OF INDEPENDENT INDICATORS TO SUPPORT TRANSPORT MODELS
Chair: G. de Marsily (Univ. Paris VI, France)

Independent indicators are taken to be sources of information and databases that are not drawn upon in the formulation of models, but can be compared to the results of models in order to enhance confidence.

> What types of indicators are available and how can they/are they being used to support transport?
>
> Use of independent indicators to support extrapolation to long-term and larger spatial scales.
>
> How independent are independent indicators?

Working Group B:

IMPLICATIONS OF SIMPLIFICATION IN GEOSPHERE-TRANSPORT MODELLING IN PERFORMANCE-ASSESSMENT
Chair: R. Storck (GRS, Germany)

> Appropriate level of detail in performance-assessment models for geosphere-transport with respect to supporting (process-) models (e.g. complexity of flow model).
>
> Confidence in the geosphere-transport barrier in the presence of (over-) conservatism (e.g. sorption represented by K_d, matrix diffusion represented in a 1D homogeneous matrix).
>
> How to evaluate conservatism?
>
> Implication of geometrical simplifications (e.g. of a three dimensional transport scheme to a one dimensional stream tube) *vis-à-vis* the conservatism of the approach.

Working Group C:

ASPECTS AND SUFFICIENCY OF CONFIDENCE IN SUPPORTING (PROCESS-) MODELLING
Chair: N. Eisenberg (NRC, United States)

> Spatial and temporal extrapolation
> Level and sufficiency of confidence (confidence evaluation)
> Grounds on which to base the evaluation of confidence
> Approaches to the identification of open issues
> Progress over the last 10 years (in particular *vis-à-vis* the INTRAVAL project)
> Communication and documentation of confidence (statement of confidence) both within individual programmes and towards external technical audiences (regulators, peer reviewers)

Working Group D:

STRATEGIES FOR INCREASING CONFIDENCE IN TRANSPORT MODELS THROUGH DATA ACQUISITION
Chair: R. Beauheim (SNL, United States)

> What open issues are expected to be resolved
> How to focus future research and data acquisition to deal with these open issues (lab, in situ, URL, ...)
> Impact of programme deadlines on the scope of future research

SESSION VI

CONCLUSIONS AND RECOMMENDATIONS
Chairs: G. de Marsily (Univ. Paris VI, France) and H. Röthemeyer (BfS, Germany)

POSTER SESSION

The Use of the Nirex Matrix Diagram in Modelling Coupled Transport Processes in Performance-assessment
L.E.F. Bailey and S. Norris, (Nirex, United Kingdom)

From Site Data to Models : Understanding the Influence of Fractures on Transport Properties through Natural and Artificial Field Tracer Experiments (the Tournemire Site)
G. Bruno, L. De Windt, Y. Moreau Le Golvan, A. Genty and J. Cabrera (IPSN, France)

Crushed-Rock Column Transport Experiments Using Sorbing and Nonsorbing Tracers to Determine Parameters for a Multirate Mass Transfer Model for the Culebra Dolomite at the WIPP Site, NM
C.R. Bryan and M.D. Siegel. (SNL, USA)

Analyses Concerning the Estimation of the Release Behaviour of Heaps of the Former Mining Activities and Uranium Ore Mining
K. Fischer-Appelt and J. Larue, (GRS, Germany), G. Henze, (BfS, Germany) and J. Pinka (GEOS, Germany)

Dual Porosity Approaches for Fractured Media: from the Single Fracture Scale to Block Scale
E. Mouche and C. Grenier (CEA, France)

Anisotropy in Diffusion in Soft Clays
M. Put, G. Volckaert and J. Marivoet (SCK/CEN, Belgium)

Evaluation of Transport Model Uncertainties Using Geostatistical Methods: A Case Study Based on Borehole Data from the Gorleben Site
K.-J. Röhlig and B. Pöltl, (GRS, Germany)

Coupled Phenomena Potentially Affecting Transport by Diffusion in Siltrocks: Evaluation Approach and First Results Regarding Thermodiffusion
E. Tevissen (ANDRA, France), M. Rosanne, N. Koudina and B. Prunet-Foch (LPTM-CNRS, France) and P.M. Adler (IPG, France)

CONCLUSIONS OF WORKING GROUP A

ROLE OF INDEPENDENT INDICATORS TO SUPPORT TRANSPORT MODELS

Chairman: Ghislain de MARSILY (University Paris VI, France)

Members: *Lokesh CHATURVEDI (EEG, United States), Richard CODELL (US NRC, United States), Eckhard FEIN (GRS, Germany), Peter FLAVELLE (AECB, Canada), Jörg HADERMANN (PSI, Switzerland), Augustin MEDINA (UPC, Spain), Emmanuel MOUCHE (CEA, France), Russell PATTERSON (DOE/YMP, United States), Helmut RÖTHEMEYER (BfS, Germany), Peter WIKBERG (SKB, Sweden)*

1.   DEFINITION

The Group reached a definition of what could be an "indicator" for supporting transport models.

An **Indicator** is an information on the status or functioning of a natural system **that has not been used for developing or calibrating a model of the system under consideration**, and can therefore be used for an independent assessment of the quality of the model by comparing the model predictions or the model concepts with the indicator.

The indicator may be directly or indirectly correlated to the quantity of interest in the models (e.g. when a co-kriging approach is used).

As soon as the indicator becomes an integral part of the data base on which the model is calibrated or validated, it is no longer an indicator, but an input data.

An indictor which is consistent with the model results increases confidence in the transport model. An indicator which is inconsistent is an anomaly. At an early stage of model development, indicators are often used to validate/falsify a conceptual model of the system (see examples below). Any anomaly which is not consistent with the model assumptions or results is also an indicator. This anomaly should be resolved by additional site characterization or through modification of the transport model. Alternatively, the cause of the anomaly could be explained and could be ignored if it can be proven that the model results are conservative for performance-assessment.

An indicator can sometimes be taken from observations made at another site, assumed to behave or have behaved in a similar manner.

- Two categories of indicators have been found, which may however overlap :
- indicators of the present behavior of the system
- indicators of the geological/hydrological/geochemical history of the system over long periods of time.

2. INDEPENDENCE OF INDICATORS

Some indicators depend on a model for their evaluation. The use of such indicators may thus need a separate model different from the one to be tested. Since transport is affected by different processes, only a subset of which is included in any model, care should be taken that both models include the same relevant processes and are mutually consistent.

Example: Evaluation of water transit time in a fractured medium, based on isotope measurements including, or not, matrix diffusion. Use of a 2-D flow field to evaluate the dispersivity in a 1-D model.

However, if it can be shown that the transport model used in PA is conservative, then the two models may be different.

Example: Neglecting density due to salinity in performance-assessment in a purely diffusive regime, leading to predominantly advective mass transport. See the Konrad example in the "Examples" section below.

Consistency of models is a necessary condition, but might not be sufficient: Indicators can be consistent when different conceptual models are used. One should therefore verify that alternative conceptual models are not also "validated" by the same indicator, thus making it impossible to discriminate between them.

Example: ^{14}C measurements interpreted as age of water or as mixing of waters of different ages.

3. GROUPING OF INDICATORS

Indicators can be grouped according to the type of behavior they reveal (e.g. isolation, flow, transport…) or to the effect they have on the system (e.g. changes in the physical properties of the system such as fracturing). The Group decided to use a classification of indicators according to the type of measurements/observations used to determine the indicator:

geochemical
thermal
geomechanical and structural
biological

Potential indicators belonging to each of these categories are listed below, and examples of their use in supporting transport models are given. These lists are by no means exhaustive, but serve as illustrations of the use of indicators in transport models for performance-assessment.

3.1 Geochemical

Radionuclides:
^{3}H, ^{14}C, ^{36}Cl, U series…

Stable isotopes:
^{18}O, ^{2}H, $^{87}Sr/^{86}Sr$, $^{37}Cl/^{35}Cl$, $^{10}B/^{11}B$ …

Aqueous Chemistry:
- Major elements, minor elements and traces
- Fluid inclusions and solutions
- Salinity

3.2   Mineralogical

- Alteration products
- Infillings in fractures
- Microscale thermodynamics for estimating Kds

3.3   Thermal

- Temperature
- Hot springs
- Basaltic dikes
- Temperature of mineral formation (for thermal history)
- Biomarkers

3.4   Geomechanical and structural

- Fault movement
- Direction or density of fractures and joints
- Stress field
- Differences in lithology
- ^{4}He to detect open fractures
- Movement of air and moisture along faults

3.5   Biological

Populations of micro-organisms as indicators of different types of water (e.g. halophilic bacteria, sulfate reducing bacteria)

Pack-rat middens, pollen records as indicators of climate change.

EXAMPLES OF INDICATORS

1. GEOCHEMICAL

1.1 Unstable radionuclides

Yucca Mountain: Bomb-pulse Chlorine-36 to indicate the presence of fast paths

In the early days of site characterization, flow through the unsaturated zone (UZ) at Yucca Mountain, Nevada, was hypothesized to be very slow to non-existent. Free-draining pathways were expected to exist, however, and were considered a positive attribute of the site, because they would facilitate drainage through the repository and prevent accumulation of water in the repository. Due to the low precipitation rate in the area, net infiltration was expected to be very small (Montazar & Wilson, 1984). The preliminary UZ numerical model used measured saturations in the drill-cores to calibrate the model. An average percolation rate of 0.01 mm/year through the potential repository horizon was calculated by this model and corroborated by age dating of matrix pore water which indicated that the ages of the pore water exceeded 100,000 years.

Selected samples from the drill-cores were analyzed for bomb-pulse Chlorine-36 and Tritium to determine at what rate and by which pathways water might enter the mountain. Fast pathways deep inside the UZ were not expected. The first indications of a fast path was the detection, in early 1995, of tritium in borehole UZ-16, which is located approximately 1 km East of the repository, at a depth of around 350 m below the soil surface. This borehole is located in an area called the "imbricate fault zone" that contains several small offset faults. Since this was only one sample with no corroborating evidence and tritium can be transported as water vapor, not a lot of attention was paid to it.

As the Exploratory Studies Facility (ESF) consisting of a 7.8 km long tunnel into Yucca Mountain was constructed in 1996-97, both systematic (every 200 meters) and feature-based samples were collected for bomb-pulse Chlorine-36 analysis.

Some of the feature-based samples indicated the presence of bomb-pulse Chlorine-36. This is considered by the project to be an indication of a "fast path" from the land surface to the potential repository horizon (300 m). Both the conceptual and the numerical model of unsaturated zone flow and transport were revised based on this data.

The conceptual model presently assumes that fast paths are present. Based on observations of the occurrence of bomb-pulse Chlorine-36, a hypothesis was developed, assuming that certain factors must be present if bomb-pulse Chlorine-36 were to reach the repository host rock horizon. These factors are: (i) adequate precipitation in areas where the bedrock is covered by less than 2 meters of soil, (ii) a fault or series of fractures that totally penetrate the non-welded paintbrush tuff, and (iii) adequate net infiltration to carry the bomb-pulse Chlorine-36 into the depths.

The numerical model was revised to allow Chlorine-36 to travel to the ESF depths in the 50 year time, increasing the effective percolation flux to 5-10 mm/year. This percolation flux rate is being corroborated by further on-going testing in the ESF.

Ref.: Murphy, W.M. Commentary on studies of ^{36}Cl in the exploratory studies facility at Yucca Mountain, Nevada. MRS Symp.Proc.506, 407-414, 1998.

1.2 Stable Isotopes

Wellenberg: An example of stable isotopes

The proposed site for low- and intermediate-level waste repository in Switzerland is Wellenberg, a hill in the Alps. The host formation is a marl with adjacent limestone. The site has been investigated by deep drillings. The hydrological model developed in a first step was based on measurements of transmissivities and heads in these deep drillings and a number of piezometers. The main feature was an underpressure zone in which the repository would be hosted.

Analysis of **stable isotopes** ($\delta^2 H$, $\delta^{18} O$) in the saline (NaCl) water of the units overlying and underlying the underpressure zone at the Wellenberg site indicated that formation water is mixed with younger water. This observation is corroborated by the aqueous chemistry, described below in section 1.3. These waters lie below the global meteoric water line (GMWL) on the $\delta^2 H$ vs. $\delta^{18} O$ plot, indicating long-time water-rock interaction, on the order of millions of years.

Evidently, the long water-residence times induce a slow radionuclide transport in the zones that have been investigated. In addition, on the basis of these and other water constituent measurements, certain water/radionuclide flow paths could be excluded.

Ref.:   Nagra. Geosynthese Wellenberg 1996 – Ergebnisse der Untersuchungsphasen I und II. Technical Report NTB 96-01, Wettingen, Switzerland, p. 226 ff, 1997.

Yucca Mountain: Stable isotopes

^{18}O and deuterium in the unsaturated zone. Preliminary conceptual models of transport in the fractured part of the unsaturated zone took into account matrix diffusion above the regional water table, as a retardation barrier to radionuclide transport. However, major ions, dissolved solids, and stable isotopes such as ^{18}O and deuterium were much more concentrated in the rock pore water than in the perched water below. Since the perched water had to pass through the fractures, exchange between the fractures and the pore water ought to be minimal and doubt was cast on the relevance of the matrix diffusion mechanism at this site.

Ref.:   Yang, C., P. Yu, G. Rattray, J. Ferarese, and J. Ryan "Hydrochemical investigations in characterizing the unsaturated zone at Yucca Mountain, Nevada", U.S. Geological Survey, WR Investigations Report 98-4132, Denver, Colorado (USA), 1998.

1.3 Aqueous Chemistry

Konrad: an example of salinity indicator

The spatial distribution of water salinity can be used as an indicator of flow and transport. At the Konrad mine in Germany, the site of a proposed low-level nuclear waste repository, weakly mineralized fresh water is found down to a depth of 150 m. Below this depth, the salt concentration increases linearly. This vertical linear increase in concentration indicates that diffusion is the dominant mechanism of solute transport from the salt at the bottom to the soil surface. Hence, the distribution of salt shows that the use of a fresh-water model is extremely conservative.

Ref.:   Klinge, H., Vogel, P. Schelkes, K. Chemical composition and origin of saline formation waters from the Konrad Mine, Germany. In Water Rock Interaction Khamara, M.K., Maest, A.S. (Eds), 7th International Symposium on Water-Rock Interaction – WRI-7, Park City, Utah, USA, 1-18 July 1992.

          Schelkes, K., Klinge, H., Vogel, P., Wollrath, J. Aspects of the Use and Importance of Hydrochemical Data for Groundwater Flow Modelling at Radioactive Waste Disposal Sites in Germany. In Proceedings of the SEDE Workshop on "Use of Hydrogeochemical Data Information in Testing Groundwater Flow Models". Borgholm, Sweden, 1-3 September 1997. OECD, Paris 1999.

Wellenberg: An example of groundwater composition

Water samples in the marl showed that the underpressure zone had NaCl-water originating from the time of the formation of the rock (the rock in the infiltration region is limestone). The Cl content of the water in the marl is much higher (~0.5 times that of seawater) than in the overlying and underlying units and in the adjacent limestone. It was estimated that there has been an exchange of less than one pore volume of water in the past 20 million years. The first hydrological model for Wellenberg had predicted much smaller water-residence times. As a consequence, the hydraulic conductivity distribution had to be changed to lower values to be consistent with the observations.

Ref. :   Nagra Geosynthese Wellenberg 1996 – Ergebnisse der Untersuchungsphasen I und II. Technical Report NTB 96-01, Wettingen, Switzerland, p.277f and p.448, 1997.

Yucca Mountain: Hydrochemistry

Hydrochemistry was useful in validating the conceptual models of saturated-zone flow and transport at the proposed Yucca Mountain High-Level Waste repository. Investigators used a combination of hydraulic heads, stable isotopes, radioactive isotopes, major and minor elements, and temperature sampled from sparse wells and springs to estimate the origin of water in the region of the repository. Preliminary studies indicate that the saturated zone at Yucca Mountain is recharged in mountains 100-150 km north of the site, and probably discharges to the south at playas in Death Valley (Wittmeyer and Turner, 1995). More recent studies combine these measurements with highly detailed 3-D flow and transport models to more clearly define the flow boundaries in this highly complex system (Armstrong et al, 1999).

Ref.:   Wittmeyer, G., and D. Turner. Conceptual and Mathematical Models of the Death Valley Regional Groundwater Flow System, Center for Nuclear Waste Regulatory Analyses, San Antonio Texas (USA), September, 1995.

          Armstrong, A., S. Painter, S. Jones, D. Ferrill, N. Coleman. Hydrochemical Inferences from Thermal and Chemical Measurements at Yucca Mountain, Nevada, American Geophysical Union, May, 1999.

Spain: Modeling of the uranium infiltration in an old uranium processing plant

The ore arrived from a nearby mine and the uranium was separated from the other minerals; the tailings were accumulated near the building. Some time later, due to rainfall, uranium was detected in some wells. Numerical modeling was used to analyze the impact of the uranium on the aquifer and on a nearby river. Hydraulic-head and uranium-concentration data from several boreholes were used in an inverse modeling procedure to estimate the parameters controlling groundwater flow and uranium concentration. At this stage, uranium data were not an independent indicator. Three years later, this model was used to predict uranium concentration and to compare the results with the new data. Uranium data were thus used as an independent indicator.

Ref.:   Samper, J., J. Carrera, A. Medina and G. Galarza. Remedial Actions for a Uranium Mill Tailings in Andújar, Spain: Comparison of Different Sources of Prediction Uncertainty. International Conference on Groundwater Quality Management. Tallinn, Estonia, 1993.

WIPP: Regional water chemistry may be an indicator of flow and transport direction

Chapman (1988) pointed out that the total dissolved solids (TDS) in the Culebra dolomite at WIPP decrease by one order of magnitude in the postulated direction of flow and transport. The hydrochemical facies also changed from NaCl to $CaSO_4$. The decreasing salinity in the direction of flow and transport could not be accounted for while the flow was assumed to be strata-bound and without recharge downstream. Chapman (1988) postulated recharge from the surface through karstic features south of the WIPP site.

The issue was resolved by groundwater basin modeling performed by Corbet and Wallace (1993). Although it did not model solute transport processes, this study provided flow fields that can be used to develop the following concepts that help to explain the observed hydrogeochemical facies.

1.   Vertical leakage, which may carry solutes and/or fresher water into the Culebra.
2.   Lateral fluxes in the Culebra.
3.   Effects of climate change.

Ref.:   Chapman, J. B. Chemical and Radiochemical Characteristics of Groundwater in the Culebra Dolomite, Southeastern NM. EEG-39 (DOE/AL/10752-39). Environmental Evaluation Group, Albuquerque, NM, 1988.

Corbet, T. F., and Wallace, M.G. Post-Pleistocene Patterns of Shallow Groundwater Flow in the Delaware Basin, Southeastern New Mexico and West Texas. In *Carlsbad Region, New Mexico and West Texas*, D. W. Love, J. W. Hawley, B. S. Kues, J. W. Adams, G. S. Austin, and J. M. Barker, eds. Forty-fourth Annual Field Conference, October 6-9, 1993, SAND93-13187J, pp.321-351. New Mexico Geological Society, Roswell, NM. WPO 28643, 1993.

1. 4   Mineralogical

Rock mineralogy may provide an estimate of the Kd values, which is often used as an input parameter in the transport models, when cores have been drilled and used in sorption experiments.

For instance, secondary minerals provide a signature of the water chemistry, flow regime or water residence time and temperature. This signature is evaluated by means of geochemical codes.

Known examples are basaltic alteration products (See Mururoa, Icelandic geothermal systems or repreciptation materials in fractures at Äspö).

More precisely, under the assumptions of instantaneous equilibrium, and given the *a priori* mineral selection and the temperature, geochemical codes may determine the water chemistry and the water residence time that fit the observed secondary mineralogy.

2. GEOMECHANICAL AND STRUCTURAL

Yucca Mountains: Flow in the saturated zone

The hydraulic gradient in the saturated zone of the Yucca Mountain site would indicate a flow toward the southeast, then southwards toward the "critical group" of the population, assumed to exist approximately 20 km south of the site. According to this pattern, flow would be mostly through the alluvial valley infilling materials which may greatly retard radionuclide migration. Analysis of the direction, length and width of fractures and faults in the stress field gives an indication of the anisotropy of the hydraulic conductivity in the saturated zone. The anisotropy suggests a different flow direction and path length from the repository to the critical group, than would be inferred from the hydraulic gradient alone. Flow direction and path length, which may be indicated by the stress fields, would mostly be through the volcanic tuffs, with faster travel times and less chance of retardation of radionuclides.

Ref. :   Ferrill, D.A., J. Winterle, G. Wittmeyer, D. Sims, S. Colton, A. Armstrong "Stressed rock strains groundwater at Yucca Mountain, Nevada", GSA Today, Volume 9, no 5, May 1999.

WIPP: Geomechanical features, such as fracture density, faults, joints, etc. may be used as independent indicators for flow and transport models

On the basis of field transmissivity data and with a strata-bound flow and transport model, the Culebra water-bearing zone was postulated to contain an anomalously high transmissivity zone in the southeastern part of the WIPP site. Geophysical investigation was used to identify and confirm the reason for the high transmissivity. These attempts resulted in the identification of a low electric resistivity zone, which was interpreted as an indirect indication of high transmissivity in the area south of borehole H-11 in the southeastern part of WIPP.

Another example of a geomechanical indicator also comes from the WIPP project. During the early period of site characterization, the Salado Formation fluids were considered to be trapped in the formation. This conceptual model was based on the chemical composition of fluid inclusions in the salt crystals. When the fluid was observed to accumulate in the boreholes drilled in the floor of the repository, an investigation was carried out to assess the potential for flow and transport through the Salado Formation. *In situ* measurements of permeability beyond the disturbed rock zone resulted in an understanding of the behavior of flow from the Salado salt into the interbeds of clay and anhydrite. These in situ measurements confirmed Darcian flow through the Salado although at very low permeabilities. The observation of incipient fractures in marker bed 139 provided an independent indicator of the modeled flow and transport processes through the Salado formation.

Ref.:   Bartel, L.C. Results from electromagnetic surface surveys to characterize the Culebra aquifer at the WIPP site: Sandia National Laboratories Report, SAND87-1246, 1987.

Bartel, L.C. Results from a controlled-source audiofrequency magnetotelluric survey to characterize an aquifer: in Geotechnical and Environmental Geophysics Vol. II Environmental and Groundwater, S.H. Ward, ed., Society of Exploration Geophysicists, 216-233, 1990.

Sweden: Present-day stress field indicating anisotropy in hydraulic properties

The dominating horizontal stress direction is presently in the NW-SE direction in the South East part of Sweden. This was expected to result in an anisotropic hydraulic transmissivity with decreasing transmissivity in the NE-SW direction and an increase in the NW-SE direction. Such anisotropy was seen in the hydraulic properties in the boreholes drilled in different directions from the Äspö HRL tunnel system.

Ref. :   Rhen I (ed), Bäckblom (ed), Gustafson G, Stanfors R, Wikberg P. Äspö HRL – Geoscientific evaluation 1997/2. Results from pre-investigations and detailed site characterization. Summary report. SKB TR 97-03, 1997.

Canada: Helium 4 as a Flow and Transport Indicator

The surface expression of deeply penetrating fractures and faults can sometimes be indicated by groundwater discharges that create temperature, salinity or solute concentration anomalies.

One such indicator is ^{4}He, which can accumulate in the groundwater during long exposure to the •-decay of the uranium series. Upon discharge, this excess ^{4}He can degas from the groundwater. If the discharge is into a soil cover, a ^{4}He anomaly develops in the soil gas.

Soil gas ^{4}He was used by AECL as an independent confirmation of their geosphere model for the Whiteshell Research Area. After developing the model, AECL performed a soil gas survey of the Boggy Creek area, where they postulated one of the major low-dipping fault zones subcrops. They measured a ^{4}He anomaly in the soil gas. This provided the independent confirmation of the extent and continuity of the fault zone for postulated contaminant transport.

Ref. :   Gascoyne, M. *et al.* The Helium Anomaly in Soil Gases at the Boggy Creek Site, Lac du Bonnet, Manitoba. Atomic Energy of Canada Limited Technical Record TR-583 (Rev. 1), COG-92-376 (Unpublished report available from SDDO, AECL Research, Chalk River Laboratories, Chalk River, Ontario K0J 1J0, Canada), 1994.

3.   THERMAL

Temperature : Mururoa and Fangataufa

Temperature anomalies are good indicators of "anomalous" heat transport, e.g. transport of heat by convection rather than by simple conduction.

For example, in the Mururoa and Fangataufa study (see abstract by G. de Marsily and J. Hadermann, this volume), an initial possible conceptual model of fluid flow in the atolls could have been that the water is immobile, given the assumed absence of head gradients due to the absence of recharge in such a hydrostatic environment. The temperatures profiles in several boreholes clearly indicates that the system cannot be purely conductive, and that advection is necessary to explain the behavior of the system.

As soon as the temperature data are used to calibrate a density flow model, they are no longer indicators, but part of the input data of the model.

Ref.:   IAEA The Radiological Situation at the Atolls of Mururoa and Fangataufa, Main Report and Technical Vols 1-6, Vienna, 1998.

IGC: C. Fairhurst, E.T. Brown, E. Detournay, G. de Marsily, V. Nikolaevskiy, J.R.A. Pearson, L.Townley, Stability and Hydrology Issues Related to Underground Nuclear Testing in French Polynesia, 2 Vols, La Documentation Française, Paris, 1999.

Biomarkers: The Northern Illinois MVT District

Biomarkers are natural organic molecules that undergo a change of their physical properties (e.g. the angle of deviation of polarized light) as a function of temperature and time. Indeed, the kinetics of the reaction are a function of temperature, therefore the final stage of the biomarker today records a complex integral of the temperature history of the sediment since it was first formed. When a model of the long-term behavior of a sedimentary system is built, including the thermal history of the basin to reconstruct geochemical behavior (secondary mineral formation, for instance), then biomarkers can be used to determine if the simulated temperature history of the basin is consistent with the maturation index of the biomarkers.

Other such thermal indicators are vitrinite reflectance, fission tracks in apatite, etc.

Ref.   Rowan, E.L., Goldhaber, M.B. Duration of mineralization and fluid-flow history of the Upper Mississippi Valley zinc-lead district. Geology, Vol. 23, p. 609-612, 1995.

Rowan, E.L., Goldhaber, M.B. Fluid Inclusions and Biomarkers in the Upper Mississippi Valley Zinc-Lead District – Implications for the Fluid-Flow and Thermal History of the Illinois Basin. In: Evolution of Sedimentary Basins – Illinois Basin. J.L. Ridgley, Project Coordinator. U.S. Geological Survey Bulletin 2094-F, 34 p, 1996.

4.   MICROBIAL

Sweden: Microbial Sulfate reduction as an indicator for present-day groundwater flow

The groundwater in the tunnel section below the Baltic Sea at Äspö HRL has an enhanced bicarbonate contrast (up to 1000 mg/l) and sulfate contrast (50 mg/l). This is interpreted as the result of massive microbial sulfate reduction (high content of microbes).

Interpretation: Baltic Sea water percolates through the bottom sediments and brings down the organic matter into the rock. Sulfate reduction is occurring now and is likely to have occurred prior to the construction of the tunnel.

Under the island of Äspö, there are no enhanced bicarbonate and depleted sulfate concentrations. Water flow is not passing through the sea-bottom sediments into the tunnel system under the Äspö island. This accords with the flow model.

Ref .:   Laaksoharju M (ed). Sulphate reduction in the Äspö HRL tunnel. SKB TR 95-25, 1995.

5.   MULTI-INDICATOR CASE

Gorleben: Geochemical, geological/hydrological, thermal history of the site

(i) Normal and glacially influenced subrosion processes; the presence of the highly soluble potash salt seams with fluid inclusions and solution products (Gebirgslösungen) provides evidence of the depth to which the Gorleben salt dome has been affected by surface-induced processes.

Contribution to performance-assessment: Modeling of an exponential mass change using global data (mass half life: $2.x10^8$ years), and glacial processes indicates an isolation potential of the disposal system of millions of years. The independent indicators provide proof of a past isolation period at the disposal level of $2.5x10^8$ years, the age of the geologic formation. The transport of radionuclides to the accessible environment is now assumed possible only through disruptive events.

(ii) Natural analogue studies revealed the effects of high temperature on rock salt at a depth of 700-800 m. Basaltic melts with temperatures of around 1 150 °C intruded with mobile constituents into evaporites of Zechstein I of the Werra-Fulda mining district 15 to 25 million years ago. The mineral reactions and observed material transport can be attributed to fluid phases. They extended a few cm into the rock salt and up to and over 10 m into the K-Mg mineral association of the potash salt seams.

Contribution to performance-assessment: Even stresses of high temperatures and concentrated salt solutions lead neither to an extensive decomposition of the entire silicate rock, nor to a mobilization of elements, e.g. lanthanides in vitrified components of the basalt or in insoluble silicate compounds of the rock salt.

Ref.:   Herrmann, A.G. & Röthemeyer, H. Langfristig sichere Deponien – Situation, Grundlagen, Realisierung. Springer, Berlin Heidelberg New York 1998.

CONCLUSIONS OF WORKING GROUP B

IMPLICATIONS OF SIMPLIFICATION IN GEOSPHERE-TRANSPORT MODELLING IN PERFORMANCE-ASSESSMENT

Chairman: Richard STORCK, (GRS, Germany)

Members*: Javier RODRIGUEZ AREVALO (CSN, Spain), Gérard BRUNO (IPSN, France), Martin MAZUREK (University of Bern, Switzerland), Simon NORRIS (U.K Nirex Ltd, United Kingdom), Martin PUT (SCK/CEN, Belgium), Klaus-Jürgen RÖHLIG, (GRS, Germany), Paul SMITH (Safety Assessment Management Ltd, United Kingdom), Timo VIENO (VTT Energy, Finland)*

Extensive use is commonly made of simplified models in performance-assessments. This is due to the large number and complexity of the processes, that affect the performance of a final disposal system, the large spatial extension of a repository site, the geometrical complexity of the geological environment and the long time frames to be considered. Limited computer power has also been a factor.

With increased scientific information and understanding, site-characterisation data and computer power, there is a tendency towards reducing the simplifications in performance-assessments and to approach more closely a realistic representation of the disposal system and its evolution. On the other hand, if the aim is to show compliance with regulatory limits, the use of simplified models is still widely considered acceptable, provided the conservatism of the simplifications can be demonstrated.

The working group identified several questions concerning the use of simplifications and came to substantial agreement, as outlined in the following:

WHY DO WE NEED SIMPLIFICATIONS?

The models and databases for performance-assessment are developed iteratively throughout the process of repository planning and development. Performance-assessment need to fulfil different requirements, depending, for example, on the stage reached in this process. Simplifications are, however, needed at all stages because:

- information and data are inevitably limited and, in particular, a complete characterisation of a site is impossible;

- the resources for an assessment, in terms of time, computer capacity, manpower and money are limited;

- it is arguably more easy to understand results presented from the use of simplified models.

During the first stage of an iterative procedure of performance-assessment, when site-specific data and information are necessarily limited, simplified models are also used:

- to start the performance-assessment modelling process;
- to perform scoping calculations.

Even during later stages, when compliance has to be shown and the safety case has to be compiled, the need for simplification still exists. Furthermore, it is likely that scoping calculations can provide valuable insights at all stages.

HOW DO WE JUSTIFY SIMPLIFICATIONS?

Simplifications have to and can be justified by:

- demonstrating conservatism; or by
- demonstrating the minor importance of the simplification under consideration.

In both cases the means of demonstration is of crucial importance.

Simplified models can also be justified if the required additional information for more realistic modelling is not yet available, but further and more detailed assessments are foreseen at later stages. Such simplified models are generally applied using wide parameter variations to cover the range of uncertainty. Note, however, that complete characterisation of a site is impossible.

HOW DO WE EVALUATE SIMPLIFICATIONS?

For demonstrating conservatism or the minor importance of a process considered in simplified models, the simplifications have to be evaluated. This can be achieved, for example, by logical argument, or by comparing the results of simplified models with those of more detailed models or with independent indicators. Note, however, that due to restrictions of time and resources, such detailed models would, by necessity, be restricted to individual components of the system.

WHAT IS THE APPROPRIATE LEVEL OF DETAIL?

The degree of simplification of performance-assessment models, or their appropriate level of detail, have to be defined on the basis of the actual disposal concept and the site under consideration. The appropriate level of detail also depends on the purpose of the assessment.

- For design and optimisation purposes, all essential features have to be included into the modelling.

- To show compliance with regulatory requirements only the essential and potentially detrimental features have to be included in the modelling.

HOW DO SIMPLIFICATIONS AFFECT CONFIDENCE?

Neglecting features, events and processes as derived from site characterisation and related activities in performance-assessment models may affect confidence in the results of the assessment in both positive and negative ways.

- By simplifying the performance-assessment models through neglecting insignificant features, events and processes, the transparency of the results may increase and hence confidence may also increase.

- However, since neglected features, events and processes may have the potential to influence the results of performance-assessment modelling in an uncertain manner, confidence may also be decreased by the use of simplified models.

- If simplifications result in a clear loss of realism in the performance-assessment models, confidence may be decreased still further.

- The degree to which confidence is increased or decreased by simplifications is strongly dependent on the aims of the assessment, and on the audience to whom the performance-assessment models and results are presented.

WHAT TYPES OF SIMPLIFICATIONS DO WE HAVE?

There are a variety of possible simplifications in the models used in a performance-assessment. As a way forward, the working group suggested that they can be grouped into three classes: examples are shown in Table 1. Although examples for each class are listed in this table, no claim is made for comprehensiveness.

Table 1: **Classification of simplifications**

Examples of processes that might be excluded from an assessment	Examples of processes that might be considered by use of simple models	Examples of data that might be considered at simplified levels of detail
<ul><li>gas production</li><li>colloidal transport</li><li>climatic changes</li><li>groundwater density</li><li>matrix diffusion</li><li>complexing agents</li><li>thermal effects</li><li>mechanical effects</li><li>reactive transport</li><li>mineralisation</li><li>precipitation</li></ul>	<ul><li>sorption</li><li>dispersion</li><li>matrix diffusion</li><li>steady state advection</li><li>dilution</li><li>use of effective parameters</li><li>neglecting correlations</li></ul>	<ul><li>geological settings</li><li>geometrical data</li><li>fracture pathways</li><li>homogeneous properties of geological media</li><li>geological dimensions</li><li>averaging procedures</li><li>parallel plate structure</li></ul>

WHAT ARE THE ADVANTAGES AND DISADVANTAGES OF SIMPLE AND COMPLEX MODELS?

Simplified models have advantages and disadvantages with respect to more complex models. Examples of these are given in Table 2. These advantages and disadvantages were considered during the discussion of the working group and have been taken into account in these conclusions. Again, no claim is made for comprehensiveness.

Table 2: **Advantages and disadvantages of complex and simple models**

COMPLEX MODELS	
Advantages	**Disadvantages**
• Strive for realistic representation, therefore could assist in confidence building • Avoid "dubious" simplification procedures which might be difficult to explain • Make full use of available information • Have greater capacity to include uncertainty or soft data, or their effects • Better guide to use in determining future site characterisation activities	• May be very complex, difficult to understand what happens (black box): audience perception difficulties • Database often not fully available • Complexity is difficult to convey to non-specialists (e.g. PDFs) • Expensive (e.g. in time and money) • More difficult to verify
SIMPLE MODELS	
Advantages	**Disadvantages**
• Availability • Quick tools • Good learning tool, easy to trace what happens • Good for sensibility analyses • Flexible, applicable to different disposal concepts, sites and types • Data requirements are limited • Simple to verify	• Not realistic, therefore not good for confidence building • Simplification procedures must be shown to be consistent with the full model, or to be conservative

Annex 4

CONCLUSIONS OF WORKING GROUP C

ASPECTS AND SUFFICIENCY OF CONFIDENCE IN SUPPORTING (PROCESS) MODELLING

Chair: N. EISENBERG (NRC, United States)

Members: *Susan J. ALTMAN (SNL, United States), Julio ASTUDILLO (ENRESA, Spain), Gilles BUSSOD (LANL, United States), Erik FRANK (HSK, Switzerland), Chul-Hyung KANG (KAERI Korea), Patrick LEBON (ANDRA, France), David READ (Enterpris, United Kingdom), Juhani VIRA (POSIVA Oy, Finland), Len WATTS (BNFL, United Kingdom), Jürgen WOLLRATH (BfS, Germany)*

Sufficiency of Confidence.

After some general discussion, the Working Group (WG) decided that an essential initial task was to determine what was meant (in this context) by sufficient confidence. Without devising a precise definition of sufficient confidence, the WG determined that two important aspects of sufficient confidence were:

1. To determine whether the confidence in a model is sufficient, the function of the model must be specified.

2. The function of a process model is to help describe a sufficient understanding of physicochemical processes in order to substantiate assumptions and abstractions of a PA model (or other less-detailed model).

Bases for Confidence

Given this general perspective on the meaning of "sufficient confidence", several attributes of a model that could contribute to confidence in that model were identified. These bases for confidence in process models include:

1. GOOD SCIENTIFIC BASIS – established and documented scientific principles must be used to develop the model.

2. CORRECT APPLICATION – the scientific principles must be applied appropriately for the system being modelled and the correct scientific principles must be selected.

3. TESTABLE – the model should be somehow testable to determine if its predictions agree with reality. Test results must be traceable and reproducible

4. MULTIPLE LINES OF INDEPENDENT EVIDENCE – demonstrations of the ability of the model to match or predict results consistent with the current conceptual model(s), based on different lines of evidence, improves confidence more than a single line of evidence; e.g. confidence in a transport model that agrees with contaminant migration

tests over relatively short distances will be enhanced if the model also agrees with variations in groundwater geochemistry at the site.

5. CAPTURES PHYSICS AND CHEMISTRY – the model should represent the chemistry and physics of processes occurring at the site, especially those physicochemical processes relevant to radionuclide transport.

6. DEFENDABLE ASSUMPTIONS – the assumptions used to simplify the modelling should be defensible against challenge; they should be internally consistent and have a rational basis.

7. CONSISTENT WITH KNOWN SITE DATA – there will be greater confidence in a model that is consistent with known site data than one that is not; more site data and greater consistency with that data will increase confidence in the model.

8. APPROPRIATE CONCEPTUAL MODEL – the conceptual model upon which the mathematical and numerical models are based should be appropriate to the site and processes modelled. Confidence will increase if alternative conceptual models have been considered and tested at each stage of site characterisation.

9. DEFINED RANGE OF APPLICABILITY – the range of applicability for the model should be defined and that range should be consistent with the application of the model.

10. ROBUSTNESS – the model should be robust in that small changes to parameters, conditions, and structures should not render the model results completely incompatible with the conceptual model and site data.

In any given case a mix of these attributes is likely to be available as a source of confidence. These various attributes may be combined subjectively to approximate an overall level of confidence in a model applied for a particular problem.

The level of confidence needed for a model depends upon a variety of factors, including:

- the application or problem being studied;
- the audience, which may be:
  - technical/scientific
  - regulatory/official
  - public/political
- the stage of repository development;
- the amount of supporting data.

To illustrate these general concepts, the WG examined two examples of how confidence in a particular model evolved in time. These examples of confidence history were (1) A Model of Matrix Diffusion (Finland) and (2) A Model of Diffusion in Clay (France). These are discussed in turn.

Modelling of Matrix Diffusion (Finland)

In the case of Finland, matrix diffusion has been considered as the main retardation mechanism for isolation of waste and is quite important to the safety case. The issue regarding confidence in models of matrix diffusion is whether one can count on them under site conditions. This issue can be divided into several subissues:

Subissue 1: Does it exist?
Subissue 2: How important is it?
Subissue 3: How much available pore space and sorption capacity is there?

53

There were multiple lines of reasoning and various tests that were able to support the use of matrix diffusion in the safety case. For example, laboratory tests demonstrated existence of the phenomenon; testing of the effects of stress fields on matrix diffusion demonstrated a degree of persistence of the phenomenon. Additionally, counter-arguments to its use were successfully responded to. However, challenges still remain to the confidence in this model. Current open issues include: (1), showing, on larger spatial scales, the existence and importance of matrix diffusion for the Finnish concept and (2), determining the ratio of flow-wetted surface to flow rate, which is the essential parameter determining the ability of the phenomenon to retard radionuclide migration to the biosphere. If these issues are resolved in favour of matrix diffusion, the confidence in the use of the model in this context will be enhanced.

Model of Diffusion in Clay (France)

For the clay concept in France, the diffusion coefficient describes the main mechanism for radionuclide transport through clay. The issue regarding confidence in the model is how to obtain defensible values for the diffusion coefficient. This issue can be divided into several subissues:

Subissue 1:   Laboratory measurement may not scale up appropriately to describe field-scale diffusion.

Subissue 2:   Many factors could affect field measurements and invalidate values obtained.

These subissues were examined by experimentally testing, at the Mont-Terri URL, technologies and procedures intended to eliminate measurement difficulties; resolution of the subissues will be sought in the future by carefully designing and performing field experiments, intended to increase confidence, at the Meuse/Haute-Marne URL.

Progress over the last ten years

The WG discussed progress in confidence building during the past ten years, especially in comparison to the time of the INTRAVAL project. Several trends and aspects of progress were noted by the WG:

1. the ambitious goal of absolute model validation has been replaced by the more realistic goal of confidence building;

2. various entities were motivated to compose documents about confidence building; e.g.:

3. (a) NEA;

4. (b) NRC/SKI;

5. since INTRAVAL there has been a greater emphasis on formal peer reviews;

6. since INTRAVAL there has been a greater emphasis on international meetings and co-operation to build confidence in models;

7. the approach to tracer experiments has been refined;

8. the awareness that calibration is not validation has been sharpened; this has led to an avoidance of models with too many fitted parameters;

9. increased awareness that separation of chemical and physical processes in modelling, does not enhance confidence in the resulting suite of models.

Communication and Documentation of Confidence

The WG discussed how to communicate information regarding confidence in models. An overall observation was that how one communicates confidence depends upon the audience addressed; i.e. different approaches would be needed to communicate to such diverse groups as (1) the general public, (2) regulators, and (3) programme managers. Three aspects important in communicating to public were identified as: (1) the credibility of the presenter of the information and the organisation he represents, (2) the precedent for the approach presented, and (3) the quality of the presentation. Additional thoughts on communicating information about confidence, especially to the public, included:

1. communication regarding confidence in models needs to be simple and transparent;

2. scientific explanations must be clear and not obscured by details;

3. disagreement, without resolution, within the scientific community may lead to a loss in confidence;

4. international consensus is useful;

5. the public needs to know that you care, before they care what you know;

6. openness through publication in the scientific and technical literature is important; strive to report all results, not just results that you want the public and scientific community to see.

Another important aspect of confidence building, is the proper documentation of confidence building activities and the resulting confidence in models. Some aspects of appropriate documentation for confidence building include:

1. a clear statement of all significant assumptions, including:

    (a) a clear identification of the conceptual model;
    (b) a clear description of the system;

2. a clear statement of the data used and its source;

3. QA is needed to assure traceability of data; however, this places additional burdens on scientific investigators;

4. formal peer reviews, well documented, may help to accelerate the normal scientific review process;

5. systematic methods to evaluate the completeness of models, if documented, add to confidence.

Identification of Open Issues

A systematic approach to confidence building would be to: (1) identify a desired level of confidence in a model (the desired level would likely be related to the importance of the model in developing and substantiating the safety case), (2) identify the current level of confidence in the model, and (3) then identify open issues. Open issues are those questions, which, if answered, will bring the level of confidence from its current level to the level that is desired.

Spatial and Temporal Extrapolation

A central issue in geosphere-transport is how to extrapolate from confidence building activities on laboratory scales to field scales. Extrapolation to field scales is inherently limited by the possibility that boundary conditions and processes may change with increasing scale. Examples of situations for which increasing scale may be inconsistent with model assumptions include: (1) the use of constant or effective coefficients (e.g. dispersion) in transport models and (2) the use of constant boundary conditions in transport models, which inherently may imply a closed system, a condition which is generally not the case for the systems modelled.

Some of the approaches that may assist in extrapolation of laboratory-scale results to the field-scale domain include:

1. Natural analogues (used in its broadest and most inclusive concept, rather than a strict geochemical analogue)

2. Numerical experiments and sensitivity studies

3. Gradual increase in scales of experiments

4. Knowledge of geological system to provide assurance that fundamental processes do not change over the time period modelled

5. Modelling evolution of a system from its past state to its present conditions.

Conclusions

The conclusions reached by this working group may be summarised as follows:

- Progress has been made in these areas:

    - refining concepts for confidence-building;
    - developing greater confidence in specific process models;
    - understanding how to communicate confidence;

- Progress is needed in these areas:

    - improving understanding of processes and their importance at larger scales;
    - better methods for extrapolation;
    - more effective communication:
        - within the scientific community (peer review);
        - with the public;
        - with better documentation;
    - better representation of heterogeneity.

CONCLUSIONS OF WORKING GROUP D

STRATEGIES FOR INCREASING CONFIDENCE IN TRANSPORT MODELS THROUGH DATA ACQUISITION

Chairman: Richard L. BEAUHEIM (SNL, United States)

Members: *Carmen BAJOS-PARADA (ENRESA, Spain), Jaime GOMEZ-HERNANDEZ (UPV, Spain), Y. IJIRI (JNC, Japan), M. JENSEN (OPG, Canada), P. LALIEUX (NEA), M. MATTHEWS (USDOE/CAO, United States), G. OUZOUNIAN (ANDRA, France), A. POTERI (VTT Energy, Finland), K. SCHELKES (BGR, Germany), J. SOLER (PSI, Switzerland), B. STROMBERG (SKI, Sweden)*

Creating Confidence

The group began by discussing how to secure technical confidence of the disposal community (including the regulators) and within the external scientific community. Four principal ideas were discussed:

1. Confidence is created when multiple lines of evidence (e.g. hydrologic, geologic, chemical, isotopic) support a particular conceptual model. The point here is that any single type of data alone may not be sufficient to distinguish among alternative conceptual models, but that such discrimination among alternatives may be possible when different types of data are combined. Confidence is also increased when no line of evidence is in conflict with the chosen conceptual model.

2. To increase confidence, new data gathering (experiments) should be focused on discriminating between alternative conceptual models. To the extent possible, efforts should be made to distinguish among alternative conceptual models even when the different models have similar consequences for compliance. A position of not knowing which of two (or more) models is correct does not instil confidence.

3. To instil confidence before necessary data have been collected, a step-by-step plan showing how and when the needed data will be collected should be created. No one should expect that a project would have all necessary data from the beginning. In the absence of data, confidence is created by a clear plan showing recognition that the data are needed and a process for obtaining them.

4. Assessing when adequate data have been collected for an acceptable level of confidence is determined through general agreement among project staff, the scientific community, and regulators. Complete confidence (certainty) is never to be attained, but responsible technical people should concur that an adequate level of confidence has been reached, so that the public, in turn, can have a reasonable level of confidence. Important in this regard is recognition of the time-dependence of "adequacy". This time-dependence has two main attributes. First, the level of confidence needed to proceed with site characterisation is less than that needed for licensing of a facility. Thus, the amount of data needed to provide confidence increases as a project progresses. Second, data will

be judged adequate (or not) based on the scientific standards ("state of the art") of the day, not the standards when the data were collected. As scientific methods and theories develop, data and interpretations must keep pace.

General Data Needs

The group discussed general data needs to create confidence in transport models. Five principal needs were noted:

1. More understanding of diffusion and sorption in site-specific media. This need centres around determination of diffusion-accessible porosity, appropriate diffusion rates, and the transferability of lab-derived sorption coefficients to the field.

2. Methods of detecting dominant pathways. This need relates to the difficulty of knowing that the pathways that have been detected and tested are, in fact, the pathways that will dominate the actual behaviour around a repository. New methods, potentially including geophysical and hydraulic techniques, are needed to detect features remotely that have not been intercepted directly.

3. Characterisation of small-scale structure of flow pathways. Even when a dominant pathway, such as a particular fracture, has been identified, questions remain as to what portion of the fracture has the most effect on transport. Thus, studies of channelling within fractures and the effects of gradient changes should continue.

4. Understanding of time-dependence of transport properties arising from both natural and induced changes. Transport properties of a medium may change both through processes that would occur in the absence of a repository and through processes that are either initiated or enhanced by the presence of a repository. For example, natural changes in groundwater chemistry resulting from glaciation may result in dissolution or precipitation of pore and/or fracture fillings. These processes may also be affected by the chemistry of any fluid coming out of a repository.

5. Definition of correlations among different types of data and increased data integration. The oil industry has made great efforts in recent years to define correlations among different types of well logs, core measurements, field tests, and other types of data. These correlations are used to condition geostatistical reservoir models. More efforts along these lines should also be made by the nuclear-waste community.

Ongoing Issues

A number of items were identified as ongoing issues that continue to be of interest:

1. Coupling between process models (e.g. hydrologic, chemical, and mechanical). Coupled hydromechanical modelling is primarily of interest in the stress-affected region around a repository. Coupled hydrochemical modelling is of interest both in terms of understanding the evolution of the groundwater system at a site and for predicting the effects of mixing waters/brines from a repository with native waters.

2. Interest in international forums. International forums such as GEOTRAP and INTRAVAL provide a much-needed opportunity for programmes from different countries to compare ideas, approaches, and experimental concepts. Continuation of international forums is seen as important to building confidence for all countries.

3. Understanding of the past evolution of a system to infer transport processes/parameters on large scales. Studying the past evolution of a system, whether through paleo-hydrologic studies, environmental isotopes, or other methods, is seen as a means of understanding the behaviour of a system on the same time and spatial scales as are of concern to PA. Even though the future evolution of the system may be different from the past, the demonstrated ability to explain and model the past evolution provides confidence in modelling of the future.

4. Derivation of effective parameters for modelling. Parameters that can be measured through experiments, such as transmissivity and hydraulic gradient, are not always the parameters actually desired for modelling, such as permeability and flux or velocity. Thus, confidence in modelling could be enhanced by improving methods of deriving effective parameters from the measured parameters.

Limitations of Further Testing

The working group considered that testing and modelling would always have certain limitations that should be openly acknowledged and discussed when trying to create confidence:

1. Performing field tests with radionuclides of concern to performance-assessment is difficult to impossible, depending on country-specific circumstances. Thus, methods must be developed to apply laboratory-derived data to the field (performance-assessment) scale and/or to extrapolate from the behaviour of other isotopes or elements.

2. Testing will always be limited to temporal and spatial scales smaller than those of concern to performance-assessment, necessitating upscaling. The limitations and assumptions underlying this upscaling should be addressed explicitly.

3. A corollary to the previous point is that upscaling to the performance-assessment scale can never be validated. Therefore, creating confidence in the procedure used for upscaling is critically important.

4. Parameter variability and uncertainty will always exist. Parameters can never be measured exactly or continuously over the domain of interest. Therefore, methods used to treat parameter uncertainty must provide confidence that the variability and uncertainty are being fairly represented in models.

5. Both budgets and time are finite. No repository programme has the time or money to perform experiments or run models until every question is answered. In the end, the technical community must decide how much time and money to spend before it feels confident in licensing a repository.

Feedback from Performance-assessment to Site Characterisation

Performance-assessment gains in credibility as it is used to identify weak areas in our understanding of a system, which are then strengthened through additional experiments. In this way, performance-assessment can be used to rank or prioritise research areas in terms of reducing model or parameter uncertainty. The group noted an increase in the number of such performance-assessment-driven experiments/research around the world, such as:

- YMP unsaturated zone test.
- Development of Posiva flow meter.

- Canadian *in situ* diffusion tests.
- WIPP tracer tests;

The group cautioned, however, that performance-assessment is unable to estimate the value of research on new processes or models not currently included in the performance-assessment framework and should not, therefore, be used to determine the entire scope of site-characterisation activities.

Role of Natural Analogues in Building Confidence

The working group considered the role of natural analogues in building confidence in a repository programme. The group felt that a good natural analogue has the potential to make a strong positive impression on the public, provided that it is explained clearly and not misrepresented as being more of an analogue than it actually is. Analogues involving natural deposits of radioactive materials can be problematic because confidence-building observations of concentration rather than transport of certain radionuclides can be counterbalanced by questions about the fate of unobserved radionuclides that may have been originally present. Natural analogues do not have to involve radionuclides or represent the overall behaviour of the repository in any way, but may just be an analogue for a specific process that is of significance to performance or safety assessment. Their value comes from allowing study of how these processes operate over very long time frames. In some cases, natural analogues may be used to benchmark models that otherwise might not be testable.

Trends

Three trends were noted in the nuclear-waste community:

1. Underground Research Labs (URLs) and experimental programmes are being internationalised as countries are increasingly teaming up to support *in situ* research. This is thought to reflect both the desire to share the high costs of URLs and experimental programmes as well as a desire on the part of countries lacking a URL to be involved in research that will ultimately aid in their own repository programmes.

2. Block-scale tests are becoming increasingly common, especially at crystalline sites. Countries working in fractured crystalline (or tuff) rock seem to be devoting increasing attention to characterising flow and transport behaviour in detail on a scale as large as possible. While the block scale is still smaller than the largest spatial scales considered in performance-assessment modelling, it represents a significant increase over the single-borehole scale and allows processes and properties to be identified that are probably relevant at the largest performance-assessment scales.

3. Some repository programmes are showing more interest in media in which transport is diffusion-dominated, such as clay. This seems to reflect recognition on the part of these programmes of the difficulty of gaining stakeholder confidence when modelling obviously heterogeneous systems.

Conclusions

Four conclusions were discussed by the working group:

1. In the media that have historically been studied the most (saturated, fractured and/or porous media), few fundamentally new ideas or experiments are being developed at the field scale. More new experimental approaches are being developed for unsaturated media and clay. This observation does not represent a criticism of those programmes focusing on saturated, non-clay media, but rather a recognition of the maturity of experimental approaches to those media. Advances are now being made through refinements or modifications of existing experimental equipment and procedures rather than by development of entirely new techniques. An exception is the development of tools, such as the Posiva flow meter, to measure groundwater velocities directly.

2. Maintaining independent, non-performance-assessment-driven research, especially at the process level, is highly important in creating and maintaining confidence. Performance-assessment can be effective in prioritising research efforts in those areas that can be reasonably represented in the performance-assessment models, but it cannot define the value of information on processes not yet included in the models. Independent research, driven by scientific judgement and intuition, is essential in defining the scope of performance-assessment, not the reverse.

3. More consideration should be given to long-term experiments that might be performed over the entire developmental and operational period of a site, such as natural-gradient tracer tests. The primary purpose of such tests would be to confirm predictions made for larger time and distance scales than could be tested for a license application. The tests would take advantage of the decades-long presence of technical capabilities and staff at a repository site. Ideally, the tests would be integrated with whatever monitoring system was set up around a repository and would have modest personnel and equipment costs. Such tests would contribute to increasing confidence in a site for as long as the site (or monitoring programme) was active.

4. Increasing confidence in transport modelling still (always) requires additional data collection. Given the uncertainties that will always exist in any model, additional data that verify or refine model predictions will always have value. This will especially be true as theoretical, measurement, and modelling advances continue to be made. Theoretical advances, in particular, have the potential to motivate entirely new types of data collection to support new models.

PART B:

WORKSHOP PROCEEDINGS

Confidence in the Long-term Safety of Deep Geological Repositories – Approach Developed under the Auspices of the NEA

H. Röthemeyer

Federal Office for Radiation Protection (BfS), Germany

The paper summarizes the results and conclusions of the NEA report NEA/RWM/DOC (99)4 of 26 March 1999 and evaluates their importance by examples linked to radionuclide transport in the geosphere.

A repository system consists of the repository mine, the waste packages and the chosen site. As the geological situation of the site cannot be standardised, the design of the repository mine and the requirements on the waste packages are site specific. They thus depend on an open system that cannot be completely characterised and may be influenced by natural and human-induced factors outside the system boundaries. Whilst the safety of mere engineering systems can be based on natural laws, experience and respective quantitative criteria, the stepwise evaluation of a repository system requires the development and compilation of a safety case that gives adequate confidence or reasonable assurance to support the necessary decisions at any particular stage of a repository development.

These conclusions are the result of a long term human endeavour for a proper understanding of nature. It began with natural philosophy striving at an integral understanding of nature. Since Galilei (1564-1642), however, scientists have isolated natural processes from their environment; the mathematical models were considered to be natural laws and an objective representation of nature (closed systems). Heisenberg (1901-1976) concluded from observations in atomic physics that natural laws are no objective representation of nature but reflect our understanding of (isolated parts) of nature [1]. This understanding is based on models. The understanding and description of the geo-biosphere by means of models is thus in agreement with the general scientific possibilities. Further limitations stem, however, from the openness of the system. This challenge must be met by a confidence building process.

The process must give sufficient confidence, or reasonable assurance, to the decision makers within the relevant organisations that the decision is an appropriate course of action, and consistent with applicable requirements and objectives (e. g. operational safety, flexibility of the disposal concept, post-closure safety, benefit to society). In arriving at the decision, and in determining the level of confidence needed, the decision makers should evaluate the risk and consequences of the decision proving to be incorrect [2]. An example is given in fig. 1.

It is this hitherto unknown dimension of safety-related system evaluation, which may act as a stumbling block in repository development and decision making. It can only be overcome by a confidence building process involving specialists, other scientists, politicians and the general public (fig. 2). Besides the NEA report presented here, the project, Disposition of High-Level Radioactive

Waste Through Geological Isolation: Development, Current Status, and Technical and Policy Challenges under the auspices of the US-National Research Council is another effort striving for a proper understanding of decision making in the face of uncertainty [3].

Figure 1. **Example for risk and consequence evaluation**

Decision:	Acceptance of a backfilling and sealing concept based on further R+D (this proves to be inadequate during operation)
Possible Consequences:	• Acceptance of the risk • New concept based on present day technology • Stop of emplacement and intermediate storage • New site • Loss of public acceptance

Figure 2. **Confidence in decision making for repository development rests on these basic elements**

CONFIDENCE IN DECISION MAKING FOR REPOSITORY DEVELOPMENT

General agreement regarding the ethical, economical and political aspects of the appropriateness of the underground disposal option

Confidence in the practicality and long-term safety of disposal (including safety case and statement of confidence)

Confidence in organisational structures, legal and regulatory framework for repository development; including agreement on development stages

The repository development stages usually start with a conceptual design based on a generic safety strategy for the host rock specific natural and engineered barriers. The next steps are increasingly costly surface and underground site characterisation programmes, and the adaptation of the results in a more and more site specific design of the repository and the waste packages. The final stages are the construction of the repository, the emplacement of waste and the final backfilling and sealing of the repository. These step-wise repository development processes are guided and linked by safety assessments, allowing the incremental development of the safety case. Thus, efforts may need to be made continuously to support the decision-making process. Safety assessment involves (fig. 3):

- the establishment of an assessment basis, i.e. the safety strategy (the strategy for the building of a safety case), the system concept (the selection of a site and design), and the assessment capability (the assembly of relevant information, models and methods to evaluate performance);

- the application of the assessment basis in a performance-assessment, which explores the range of possible evolutions of the repository system and tests compliance of performance with acceptance guidelines;

- the evaluation of confidence in the safety indicated by the assessment.

Figure 3. **Procedure for the development and compilation of a safety case**

Establish an ASSESSMENT BASIS

– safety strategy
– repository
– assessment capability

Carry out a PERFORMANCE-ASSESSMENT

– performance for the assessment cases
– compliance with acceptance guidelines
– sensitivity analyses

Modify, if necessary, the assessment basis

Compile a SAFETY CASE

Interact with decision makers and modify, if necessary, the assessment basis

The "Joint Convention on the Safety of Spent Fuel Management and on the Safety of Radioactive Waste Management" of 1997 assigns the prime responsibility for the safety of spent fuel or radioactive waste to the holder of the relevant license. Therefore the implementer must decide, whether the safety assessment at a given stage has been sufficiently successful to justify the compilation and presentation of the safety case. The safety case then serves as a basis for a further decision, by the implementer, the regulator and/or others, as to whether there is sufficient confidence in safety to justify proceeding to the next repository development stage. A successful safety case should, in general, include:

- a description of the status of development of the assessment basis and the performance-assessment findings and an evaluation of confidence in the safety margins indicated by the findings;

- a description of the approaches adopted to achieve confidence and a formal statement of that confidence;

- feedback to future development stages and, in particular, reasonable assurance that the safety strategy is appropriate to handle remaining, not-fully resolved safety-related issues, during future stages.

The safety case must be presented within a system of documentation. Highlighting, within the documentation, the connections between safety and the role of the various barriers within the multi-barrier concept is a practical and useful approach for improving these aspects and demonstrating that safety can be achieved. It also heightens the role of safety assessment as a support to decision making, and not only an analysis to show compliance with regulatory targets. Such a system of documentation facilitates the evaluation of confidence (e.g. by peer review) and thus promotes acceptance by the scientific community and by stakeholders, including the politicians and the public.

International peer reviews are of increasing importance in a confidence building process. Major countries have and will benefit from the respective services provided by the IAEA and the OECD/NEA. The OECD/NEA report presented here can provide guidance and advice in such a review process. It has done so in the most recent international peer review on behalf of the Japan Nuclear Cycle Development Institute (JNC) [4].

An evaluation of confidence in long-term safety principally entails the evaluation of the robustness of the system concept, the quality of the assessment capability and the reliability of its application in performance-assessment (fig. 4).

Figure 4. **Elements of the evaluation of confidence in long-term safety**

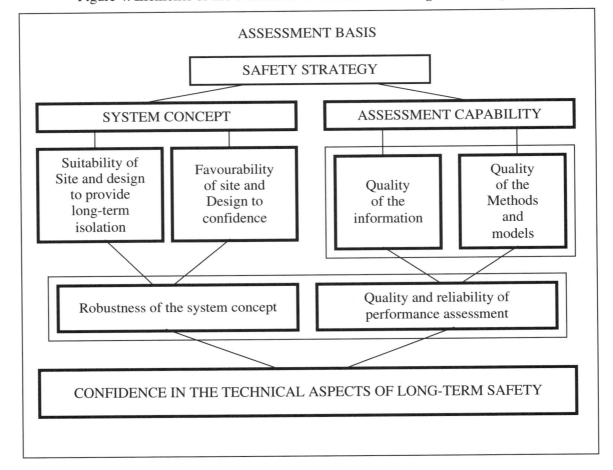

If, either following the safety assessment itself or following the compilation and presentation of a safety case, the evaluated confidence is found to be insufficient, then the assessment basis must be re-evaluated and modified with a view to confidence enhancement and a new assessment carried out. If, following the (repeated) compilation of a safety case, convergence to sufficient confidence is not achieved, then the decision sequence that drives repository development may need to be revised. The iterative process of confidence evaluation and enhancement may be viewed in terms of "confidence cycles". The concept of confidence cycles reflects the current dynamic approach to achieving confidence, especially during the early stages of repository development, when information increases rapidly in quantity and quality.

Measures that enhance the robustness of the system concept can proceed by:

modifying the system concept with a view to increasing safety margins, so that compliance with acceptance guidelines is relatively insensitive to the presence of any unresolved issues and uncertainties, and

selecting a site and design with a view to simplicity, so that uncertainties that could be detrimental to the evaluation and communication of safety are avoided, or forced to very low probability.

Examples for modifying the system concept may be taken from the Yucca mountain and Gorleben site.

The Viability Assessment of a repository at Yucca mountain compares the reference design with that including ceramic coating and backfill. The latter modification delays projected doses to more than 100 000 years and reduces dose rates by an order of magnitude [5]. In evaporites like the Gorleben site an increase in the depth of disposal by about 100 m may increase the isolation potential by 10^6 years [6].

The isolation potential is of special importance, if former raw material mines are used as underground repositories. Here the structural stability and other safety related aspects of the former mine workings may be difficult to assess in the longer term. This may also have an adverse effect on the backfilling and sealing concept. An example is the Morsleben repository. The safety assessment was originally based on a preliminary backfilling and sealing concept; more detailed site specific information and respective safety analyses proved this concept to be not acceptable. Alternative concepts have not yet given sufficient confidence in meeting the safety objectives; this was therefore the main safety related reason for temporarily stopping all emplacement activities in 1998.

Measures that enhance the quality of the assessment capability, and the reliability of its application in performance-assessment, proceed by enhancing current understanding of safety-relevant phenomena, by better characterising or reducing uncertainty and by incorporating features to:

- ensure that assessments take full account of current understanding of phenomena that are relevant to long-term safety, including uncertainty in these phenomena, so that performance is evaluated in a manner that aims at not underestimating consequences, while avoiding excessive simplification;

- ensure that the computational tools for quantitative assessments are free from error.

Nature observation is a reliable base for the evaluation and enhancement of confidence along the lines pointed out in the NEA-report. The reason is that natural systems have evolved over much longer time-scales than the time periods of concern in radioactive waste management. This aspects and respective examples are discussed in greater detail in the working group A report "Role of independent indicators to support transport models" of the Proceedings.

In his "Critic of Pure Reason" the philosopher Kant (1724-1804) states: A notion without a perception is empty and a perception without a notion is blind. The application of this knowledge helps to put the scientific base for confidence in the long-term safety of deep geological repositories in a nutshell: Safety assessment without nature observation is empty, nature observation without safety assessment is blind.

A technology striving at such an approach could be called Geonik (Geologie und Technik) in analogy to the word Bionik (Biologie und Technik, Elektronik) [6].

References

[1]   Heisenberg, W. (1957) Das Naturbild der heutigen Physik. Rowohlt Taschenbuch Verlag GmbH, Hamburg.

[2]   Pescatore, C. (1999) Private Communication.

[3]   National Research Council (1999) Disposition of High-Level Radioactive Waste Through Geological Isolation – Development, Current Status, and Technical and Policy Challengers. National Academy Press, Washington, D.C.

[4]   Nuclear Energy Agency – Organisation for Economic Co-operation and Development (1999) OECD/NEA International Peer Review of the Main Report of JNC's H 12 Project to Establish the Technical Basis for HLW Disposal in Japan. Report of the OECD Nuclear Energy Agency International Review Group, NEA/RWM/Peer(99)2. OECD, Paris.

[5]   Department of Energy (DOE) (1998) Viability Assessment of a Repository at Yucca Mountain; DOE/RW-0508/V3. DOE, Washington, D.C.

[6]   Herrmann, A.G; Röthemeyer, H. (1998) Langfristig sichere Deponien – Situationen, Grundlagen, Realisierung. Springer, Berlin Heidelberg New York.

Evolution of Confidence in Transport Models for Performance-assessment: The WIPP Case

Wendell D. Weart
Sandia National Laboratories, United States

Abstract

Since selection of the WIPP site in 1975 until the present time, understanding of the hydrologic system and the modeling of transport through that system have been under continued development. The earliest studies focused on the potential for ground water dissolution of the salt beds since repository integrity was a prerequisite for site selection. Consequently salt dissolution and transport rather than radionuclide transport was the first concern for site selection. These initial studies relied more on geologic evaluation and interpretation than on hydrology. As more conventional hydrologic studies were pursued, what was initially expected to be a rather simple hydrologic regime turned out to be a complicated system. The earliest hydrologic testing revealed much more variability in the hydrologic properties than anticipated. While the Culebra aquifer was shown to be the only unit of radionuclide transport significance, its transmissivity varied by about six orders of magnitude across the site and fractures within the Culebra were found to be an important aspect of the flow system. Geochemical investigations provided chemistry data for the aquifer brines that were difficult to rationalize with the then existing conceptual hydrologic model of the site and the flow fields for the Culebra. The last dozen years have been devoted to studies that could improve confidence in our ability to model transport of actinide isotopes through the Culebra and predict releases at the site boundary over a period of 10 000 years.

The initial modeling of radionuclide transport performed in the late 1970s for the 1980 Environmental Impact Statement assumed simple porous flow and represented the Culebra and Magenta units as one aquifer. Four regions, each of uniform transmissivity, represented the model area variability over the 3 300 square km model regime. Sorption was uniformly applied using a value for retardation determined by batch Kds measurements. Since that time the WIPP Project has quantified many aspects of the flow and transport. More than 70 boreholes, at 48 locations, have provided hydrologic data to better determine transmissivity over the area modeled. Large scale, long-duration pump tests have provided insight into the regional transmissivity and its asymmetry at the site. Non-sorbing tracer studies have been conducted at several locations to provide additional transport parameter information, especially those bearing on the relative roles of fracture and matrix flow. Retardation, chemical and physical, is now based on extensive laboratory batch and column flow-through tests using the rock, fluid and isotopes appropriate to the WIPP.

Three-dimensional flow modeling, which incorporates the small amount of vertical infiltration, now provides a logical framework to explain the observed variations in Culebra fluid chemistry, previously at variance with the assumption of a confined aquifer. Flow and transport modeling for PA now incorporate the extensive database compiled for the Culebra. Inverse modeling is used to calibrate a spatially variable transmissivity field based on transient head information. Transport is modeled with a double-porosity model that accounts for interactions between advective and diffusive elements of the porosity. On-going work will enable us to incorporate multiple rates of diffusion into the PA calculations.

All these studies benefited from the frequent examination and suggestions obtained from both internal and independent external peer review. Indeed, the required confidence in the ability of the models and codes to adequately predict the role of the natural barrier, represented by the Culebra aquifer, could not have been developed without the participation of these groups. Confidence in transport modeling fluctuated as new issues surfaced and were addressed. This paper will summarize both the technical progress and the role of peer review in achieving the confidence in transport modeling necessary to satisfy technical critics and the regulatory agency.

Confidence in Geosphere Performance-assessment – The Canadian Nuclear Fuel Waste Disposal Programme 1981-1999: A Retrospective

M.R. Jensen
Ontario Power Generation Incorporated, Canada

B.W. Goodwin
GEAC Incorporated, Canada

Abstract

In March 1998 a federal Environmental Assessment Panel completed a public review of the concept proposed for the disposal of Canada's nuclear fuel waste within plutonic crystalline rock of the Canadian Shield. Safety of the geologic disposal concept was evaluated in two Integrated Performance-assessments submitted to the Panel: the Environmental Impact Statement Case Study (EIS) and the Second Case Study (SCS). These assessments examined the behaviour of two hypothetical disposal systems with different engineered designs and geosphere properties. The properties of the geosphere were based primarily on geoscience research and site characterization activities initiated by Atomic Energy Canada Limited in 1981.

An important objective in the Panel's mandate was to review the safety and scientific acceptability of the geological disposal concept. The review process was broad and included submissions by a Panel-appointed Scientific Review Group, federal regulatory agencies, learned geoscience bodies and advocacy groups. While general agreement existed amongst most technical reviewers that the geologic disposal concept was safe and viable, there was also dissatisfaction with the sufficiency and adequacy of the arguments.

This paper examines four geoscience issues that arose during the Panel review, particularly as they affect the application of geosphere Performance Assessment (PA) methodologies used in evaluating concept safety. The issues deal with PA model verification, methods to deal with conceptual flow model uncertainty, the connection between detailed geosphere models and the integrated performance-assessment models, and PA documentation. A goal of this paper is to gain a better understanding of the expectations held by external reviewers of the EIS and SCS, and the elements that most affected confidence in predicted repository performance.

1.   Introduction

The Canadian Nuclear Fuel Waste Management Programme was established in 1978 by the Governments of Canada and Ontario (Joint Statement 1978). The responsibility for research and development in examining the disposal of nuclear fuel waste in the plutonic rock of the Canadian Shield was assigned to Atomic Energy Canada Limited (AECL). In a second joint announcement, the Governments of Canada and Ontario indicated that the siting of a deep geologic repository for used fuel disposal would not be started until the disposal concept had been accepted by government (Joint

Statement 1981). Acceptance would be tested in a public review process which the federal government formally initiated in 1988. As part of this public review process a federal Environmental Assessment Panel (the Panel) was struck in 1988.

Figure 1. **Major milestones in Canadian nuclear fuel waste management programme**

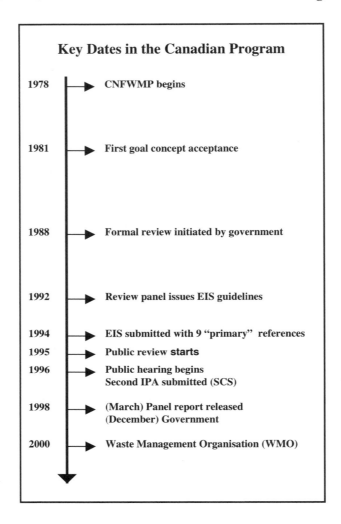

Figure 1 shows some important milestones in the programme. During the period between 1981 and 1994, AECL focused research and development activities on the deep geological disposal of nuclear fuel waste within the crystalline plutonic rock of the Canadian Shield. Many of the geoscience activities associated with the programme were centred at the AECL Whiteshell Laboratories and Whiteshell Research Area (WRA) situated near Lac du Bonnet, in southeastern Manitoba. Detailed characterisation studies and experimental activities were conducted over a 15 year period at the Underground Research Laboratory (URL), coupled with reconnaissance level investigations at research areas in East Bull Lake and Atikokan. These studies provided a basis to understand groundwater flow and mass transport within large domains of fractured plutonic rock. The approaches to site characterisation and mathematical modelling also evolved during this period, and contributed to the development of geosphere Performance Assessment (PA) strategies to predict the fate of radionuclides released from an underground used fuel repository.

In 1994, AECL submitted the Environmental Impact Statement (EIS) (AECL 1994) for the concept to the Panel. The EIS documented an integrated performance-assessment (IPA) for a case study that considered a hypothetical used fuel repository situated 500 m below ground surface within a geologic setting characteristic of the Whiteshell Research Area. AECL also submitted a Second Case Study (SCS) (Wikjord *et al.* 1996) which considered a substantially different combination of engineered and natural barriers. These two studies were meant to illustrate the robustness of the disposal concept. The two assessments also served to illustrate a methodology to evaluate the long-term performance of a used fuel repository and to provide estimates of potential impacts on the environment.

As part of the Panel review process, the EIS and its supporting documents were subject to a formal review in a public forum. Through this process, a wide range of independent review comments were received from the technical community as well as from the public. Technical reviewers included the Scientific Review Group (SRG), an assembly of distinguished independent experts established by the Panel to examine specifically the safety and scientific acceptability of the disposal concept. In addition, Government agencies with oversight responsibilities also provided comment; these agencies included the Atomic Energy Control Board and the Federal Ministries of the Environment and Natural Resources. Other review comments were provided by the Nuclear Energy Agency of the Organisation for Economic Co-operation and Development (OECD/NEA), the Canadian Geoscience Council and the Technical Advisory Committee to the Canadian Nuclear Fuel Waste Management Programme.

These comments provide considerable insight into the degree of confidence held in geosphere PA and the effectiveness of different approaches to convey an understanding of geosphere performance. This paper examines the technical comments submitted to the Panel on geosphere PA. It identifies four commonly stated comments that are judged to most critical and that would require redress in future geosphere PA work. The issues deal with PA model verification, methods to deal with conceptual flow model uncertainty, the connection between geosphere and IPA models, and PA documentation. An intent of examining these issues is to gain insight into the expectations held by external reviewers of the EIS and SCS and to develop improved strategies to enhance confidence.

The public review process ended in 1998 with the submission of the Panel's report (CEAA 1998) to the federal Government and the Government's response which outlined their recommendations and expectations on further progress. The used fuel disposal research and development programme of Ontario Power Generation has now taken the lead role in nuclear fuel waste management studies, pending the creation of a Waste Management Organization in the near future (Figure 1). One of the interim goals is to evaluate issues arising during the public review process, notably issues affecting confidence in geosphere PA.

2.   The EIS and Second Case Study

Within their geoscience programme, AECL focused research and development activities on the development of investigative site characterization methodologies and predictive models necessary to estimate groundwater flow and radionuclide transport within heterogeneous, anisotropic fractured media. This research culminated in 1994 with the submittal of the EIS to the Panel (AECL 1994). The EIS was directly supported by four "primary" references that described the underlying vault model (Johnson *et al.* 1994), geosphere model (Davison *et al.* 1994), biosphere model (Davis *et al.* 1993) and long-term environmental assessment (Goodwin *et al.* 1994).

The geosphere model was based on hydrogeologic flow system studies carried out up to 1985 within the Whiteshell Research Area and, in particular, near the Underground Research Laboratory (Figure 2). These studies indicated that groundwater velocities were very small at the proposed location of a (hypothetical) disposal vault, which was at a depth of 500 m in low-permeability (less than 10^{-19} m^2) sparsely fractured granitic rock. In fact, groundwater velocities were sufficiently small that radionuclide releases to the biosphere were dominated by diffusive transport through the "Waste Exclusion Zone", a region of rock 50 metres or more in thickness immediately surrounding the vault. The results of the long-term assessment showed that this zone of rock was the most effective of all engineered and natural barriers in the disposal system, and played a pivotal role in the safety case.

The apparent strong reliance of safety on the characteristics of the Waste Exclusion Zone prompted AECL to submit a second case study (SCS) (Wikjord *et al.* 1996). The SCS is also supported by references that describe the environmental assessment (Goodwin *et al.* 1996), the vault model (Johnson *et al.* 1996), geosphere model (Stanchell *et al.* 1996) and biosphere model (Zach *et al.* 1996). The SCS assumed a fictitious geosphere setting which had the same flow system geometry as the EIS, but which also had a 100 times more permeable Waste Exclusion Zone (Stanchell *et al.* 1996).

Figure 2. **Illustration of the geosphere near the URL**

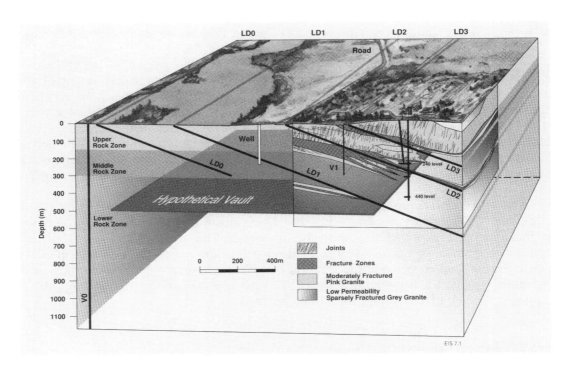

In addition, the SCS used a more robust in-room emplacement design and more durable corrosion-resistant containers (Johnson *et al.* 1996). The results of the long-term environmental assessment indicated that a case for safety could be made for the SCS, in part because the more effective engineered barriers compensated for a less effective Waste Exclusion Zone in the geosphere (Goodwin *et al.* 1996).

A summary comparing the key features of the EIS and SCS integrated performance-assessments is presented in Table 1. The differences having the strongest influences on safety were found to be the container materials and failure mechanisms, the emplacement method and the properties of the rock immediately surrounding the disposal vault (the host rock).

76

Table 1. **Comparison of disposal system features in EIS and second case studies**

Feature	EIS Case Study	Second Case Study
Fuel		
Burnup	685 GJ/kg U	720 GJ/kg U
Mass	1.6×10^8 kg U	8.2×10^7 kg U
Container		
Material	Grade 2 titanium alloy	Copper alloy
Failure mechanisms	Fabrication defects, crevice Corrosion, delayed hydride Cracking	Fabrication defects
Failed after 10^4 years	100%	0.02%
Emplacement method	Boreholes in floor of vault rooms	Within vault rooms
Vault		
Location	500 m deep, located entirely below fracture zone LD1	500 m deep, located above and below LD1
Plan area	3.2 km^2	3.4 km^2
Buffer volume	4.9 m^3 per container	9.4 m^3 per container
Backfill volume	23 m^3 per container	12 m^3 per container
EDZ	Not explicitly modeled	1.4 m thick
Host Rock		
Thickness	200 m	180 m
Permeability	10^{-19} m^2	10^{-17} m^2
Porosity	3×10^{-3}	10^{-5} to 10^{-3}
Typical GW velocity		
In Host rock	6×10^{-6} m/a	0.4 m/a
Fracture zone LD1	1.3 m/a	5.0 m/a
Biosphere		
General properties	Typical of the Canadian shield	Typical of the Canadian shield
Critical group	Rural, self-sufficient community	Rural, self-sufficient community
Well depth	0 to 200 m	0 to 100 m

3.   GEOSPHERE PERFORMANCE-ASSESSMENT METHODOLOGY

The geosphere PA methodology applied in the EIS and SCS case studies is documented in Davison *et al.* (1994) and Stanchell *et al.* (1996). The methodology can be broken down into the following steps, although it should be noted that some activities overlap and that the process is characterised by feedback and iteration.

- The approach started with development of a conceptual model of the flow domain, integrating field and laboratory data to create a quantitative description of sub-surface lithology, structural geology, spatial distribution of physical and chemical hydrogeologic properties and flow system boundary conditions.

- A mathematical approximation of the conceptual model was then derived using the 3-dimensional code MOTIF (Chan and Melnyk 1994). Hydraulic head distributions and groundwater velocities predicted by MOTIF were used to characterise and identify advective flow paths leading from the repository to the biosphere. This process was aided by the particle tracking code, TRACK3D (Nakka and Chan 1994).

- Selected particle trajectories are then combined to construct a 3-dimensional network of interconnected 1-dimensional streamtubes or segments. This network of segments is used by GEONET, the geosphere sub-model used in the long-term IPA. GEONET estimates sub-surface advective-disperse-diffusive mass transport from the vault and discharge to the biosphere (Davison *et al.* 1994).

MOTIF is a finite-element code developed by the Canadian Nuclear Fuel Waste Management Programme as the primary tool to simulate 3-dimensional groundwater flow and solute transport. It is capable of solving coupled problems involving transient and steady state groundwater flow (variable density), solute migration, heat transport and geomechanics in variably saturated fractured porous media. Verification studies, which illustrate the abilities of MOTIF, are documented by Chan *et al.* (1995).

Simulation of flow and mass transport with MOTIF is based on the Equivalent Porous Medium approximation. The code is particularly adept at simulating arbitrary flow system geometries and boundary conditions typical of fractured crystalline rock settings. For example, transmissive planar features of variable strike and dip can be imposed on an otherwise isotropic, sparsely fractured low permeability flow domain. Ophori (1995) provides a recent example of the application of MOTIF in the simulation of the WRA groundwater flow system.

A key aspect in the EIS was the representation of spatial variability characterised by an irregularly fractured crystalline flow system. For the purpose of the EIS case study, the geological and environmental characteristics of the disposal system were assumed to be consistent with observations made at the WRA. Simulation of the groundwater flow system with MOTIF was conducted at the regional (40 km x 50 km) and local repository scale. Regional simulations are used primarily to establish flow system behaviour and local model scale boundary conditions. The local scale model, centered on a hypothetical vault, is used to develop a detailed understanding of groundwater flow directions and magnitudes near the vault. In describing the regional and local flow system geometry, planar transmissive features such as fractures and fracture zones of variable strike and dip and continuity are represented deterministically. Features such as the thin veneer of overburden and lake sediment overlying the bedrock are not represented within the geosphere model.

In each of the groundwater flow simulations performed with MOTIF for input to the PA, steady-state conditions were assumed. For the EIS case study, MOTIF results were calibrated to obtain close agreement with measured WRA hydraulic head distributions. Further validation of the flow model was pursued through simulation of the groundwater perturbation resulting from construction of the Underground Research Laboratory (URL). MOTIF was also used to estimate the affect of geothermal gradients and decay heat within the vault on groundwater flow, and flow system response to water supply well pumping. This latter issue is necessary to predict the effect of well drawdown on solute migration and plume capture in the consequence analyses.

Similar activities apply to the SCS, but with one important difference. The SCS considered the WRA geosphere but with hypothetical, pessimistic physical properties assigned to the zone of rock immediately surrounding the disposal vault. In this circumstance, simulations using MOTIF were performed simply to generate a theoretical groundwater flow field.

The IPA made extensive use of probabilistic analysis and the System Variability Analysis Code (SYVAC) (Goodwin *et al.* 1994, 1996). A probabilistic analysis was selected to deal with the effects of uncertainty in estimating potential impacts over the long time frames of interest. Results from the two IPAs show that it is important to provide a systematic treatment of uncertainty, because the cumulative effect of uncertainty can influence estimates of impact by 10 or more orders of magnitude (Goodwin and Andres, 1998).

It was decided that MOTIF was unsuitable for use with SYVAC. The most important reason is probabilistic analysis typically requires hundreds to thousands of simulations, and it was clear that a large data- and cpu-intensive code like MOTIF was impractical.

GEONET was therefore developed to provide a simplified representation of the geosphere. It uses a network of interconnected flow segments to simulate the transport of radionuclides from the vault to the surface environment. It is connected to MOTIF through the TRACK3D particle tracking code: GEONET segments derive from TRACK3D simulations that are based on steady-state groundwater velocity fields predicted by MOTIF. For the purposes of IPA, GEONET was computationally efficient. It was also argued that GEONET could implicitly include uncertainty and natural variability through the specification of probability density functions (PDFs) instead of single discrete values for parameters affecting sub-surface radionuclide transport.

Figure 3 illustrates the network of segments used by GEONET in the EIS. The network of 1-dimensional segments, connected in 3-dimensional space, are manually selected to:

- represent individual parts of the modeled geosphere that have distinct chemical and physical properties, and

- replicate the local scale groundwater flow fields predicted by MOTIF.

Upon completion, the geosphere network is comprised of a sequence of segments and nodes (points of convergence or divergence) which define a set of sub-surface pathways leading from the disposal vault to discharge areas in the biosphere. The network employed was fixed in time and space for each simulation in the probabilistic analyses, although the EIS and SCS used different networks.

The network geometry is somewhat constrained by segment Peclet numbers that provide an indication of the dominant transport mechanism. For advection dominated flow fields, TRACK3D is used to define pathways (Nakka 1994). In situations where diffusive transport is dominant, the segments are defined to correspond to the maximum concentration gradient and thus had a minimum transport distance. This situation arose in the EIS in the low-permeability sparsely fractured rock surrounding the vault (Davison *et al.* 1994).

Each segment in a GEONET network is assigned a unique set of properties. In the EIS, these properties were based largely on measurements obtained from investigations at the WRA. In the second case study, a fictitious and pessimistic parameter set was used for the rock immediately surrounding the disposal vault. The properties assigned included porosity, tortuosity, dispersivity, permeability, salinity of the groundwater, and the types and amounts of minerals affecting sorption. Segment and node properties are used to calculate the major transport parameters for each segment: the groundwater velocity, dispersion coefficient, path length, diffusive parameters and contaminant retardation factors. Retardation factors represent sorption of contaminants on minerals along the flow path and are equivalent to the ratio of the groundwater velocity to the contaminant transport velocity. Darcy velocities for each segment are estimated from hydraulic heads and temperatures at the inlet and outlet nodes of the segment and the segment permeability.

Figure 3. **Network of segments used by GEONET in the EIS**

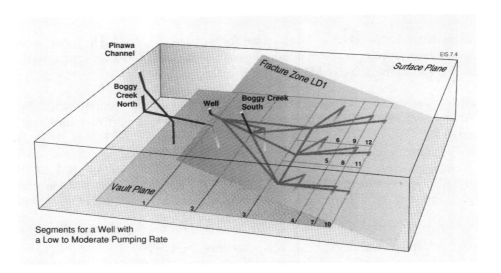

Some of the parameters associated with these properties are uncertain. The source of uncertainty could include spatial and temporal variability as well as inaccurate or poorly postulated measurements. For the most part, an uncertain parameter was characterised using a PDF that would encompass all feasible values of the parameter and assign a likelihood to the realisation of each value. In part, because of the time and spatial invariance of the GEONET network, the uncertainty in groundwater velocity was described using a multiplicative "velocity-scaling factor" in the EIS. In the assessment, this factor allowed a groundwater velocity that could be as little as one tenth and as much as ten times the velocity obtained from the MOTIF simulations. To conserve water mass, this scaling factor was applied uniformly to all of the segments in the network.

4.   Criticisms of the geosphere-performance-assessment

The discussion that follows is based on a review of technical comments made during the Panel hearings. We have identified four issues that appear to be most critical, and that typically were raised by many reviewers. For instance, most of the issues appear (sometimes in different form) in the reports prepared by the AECB (1995), the Canadian Geoscience Council (1995), the NEA (1995), the SRG (1995, 1996) and the SAT (1995). The issues involve:

- verification of the geosphere PA models,

- approaches used to explore and quantify uncertainty in conceptual flow model(s);

- the linkage between detailed models of the geosphere and the corresponding models used in the integrated performance-assessment, and

- documentation of the geosphere PA process and results.

In several respects, it will be evident that overlap between issues is unavoidable.

In the four sections that follow, the discussion examines how each issue has had a negative effect on confidence in the EIS and SCS. A brief description of activities in progress or planned by the UFDP to improve confidence is also provided.

Verification of Geosphere PA Models

Model verification is a fundamental step in developing a sense of model reliability. Verification has been described as a strict comparative exercise in which predictive model output is compared for agreement against well behaved analytical solutions or, for problems of more complicated geometry, against well established and recognised numerical codes. In the context of model development, some researchers differentiate between the stages of model verification by referring to concepts of "process verification" and "complexity verification". In process verification, efforts are made to find evidence that the concepts relied upon in the predictions are correct and adequately encompass the problem being simulated. Complexity verification attempts to demonstrate that model abstractions preserve the essence of the original problem.

Most technical reviewers specifically commented on the role of verification and flow system simplification in terms of demonstrating the adequacy and robustness of the GEONET realisations. It appears that most reviewers were less critical of process verification activities than they were of complexity verification.

The EIS and supplemental documents submitted by AECL to the Panel, described the process verification of the primary predictive tools: MOTIF, TRACK3D and GEONET. Through a systematic approach, each of the model algorithms are shown to replicate known solutions to a variety of test benchmark problems (Davison *et al.* 1994, Chan and Melnyk 1994, and Chan *et al.* 1995). In addition, the suite of EIS codes was applied to international benchmark studies such as INTRACOIN, PSACOIN and HYDROCOIN.

In site-specific applications, however, a comparison of MOTIF and GEONET predictions is required to demonstrate that discretization of the simplified PA model remains faithful to the flow domain geometry and boundary conditions for both flow and transport simulations. Most technical reviewers considered this aspect of model verification (complexity verification) incomplete. That is, most technical reviewers required further effort to show that MOTIF and GEONET yielded equivalent predictions when both are applied to the geosphere described in the EIS and SCS.

The necessity for clarity in the verification of an abstracted geosphere PA model is an important element in developing reviewer confidence. Verification was particularly important in the EIS case study as the geosphere was found to be the single most important barrier to radionuclide migration. An essential issue is the methodology by which the authenticity of an abstracted geosphere model can be reasonably demonstrated to external reviewers. This is a current area of interest, which is to be examined in a forthcoming external review of PA approaches later in 1999 aimed at:

- methods to install confidence in geosphere realizations by developing more explicit links between site characterisation and PA with respect to data collection and synthesis;

- approaches to improve the clarity and technical defense of the processes by which field observations and measurements are used to derive parameters for use in an abstracted PA model;

- application of alternative and complementary predictive PA tools to illustrate and build confidence in geosphere PA outcomes; and

- use of more descriptive (multi-dimensional) codes for the prediction of mass transport in IPA models.

Conceptual Flow Model Uncertainty

Given the time and space over which repository performance must be predicted, the approach used in evaluating conceptual flow model uncertainty is an important element of the PA methodology. Flow system uncertainty arises from sources which include field measurement errors, the inability to completely characterise a complex, heterogeneic flow system, long-term evolution of the flow system and mathematical approximations made necessary by PA requirements to simulate mass transport.

The EIS and SCS treated these uncertainties using a probabilistic analysis that, in fact, took account only of parameter uncertainty. Review comments reflected the concern that implementation of GEONET under the SYVAC framework did not alone provide a basis to adequately capture conceptual flow model uncertainty. In particular, GEONET did not adequately account for uncertainties associated with spatial and temporal flow system variability. In addition, it appeared that neither GEONET nor MOTIF investigated alternative conceptual models of the geosphere that were consistent with the set of observed field and laboratory data. For instance, the validity of the equivalent porous media approximation (which formed the basis for MOTIF and thence GEONET) within fractured crystalline media had not been demonstrated nor explored with alternative modelling techniques.

In attempting to understand this issue, it is important to recognise that the case for safety in the EIS relied almost entirely on the performance of the geosphere (or more accurately on the performance of the geosphere as predicted by GEONET). It became clear that, in the EIS, large domains of low-permeability sparsely fractured rock were required to provide adequate isolation of the disposal vault from regimes where more rapid nuclide transport could occur. As a consequence, many reviewers concluded that the case for safety was inherently dependent on the existence of sufficiently large domains of suitable rock. Many reviewers also concluded that AECL had not provided sufficient evidence that such domains could be found elsewhere in the Canadian Shield.

While the application of probabilistic analyses as implemented by SYVAC offer advantages in exploring the effect of parameter uncertainty, reviewers expressed concern over the lack of transparency in the analyses and the results of the analyses. For instance, criticisms were directed at the derivation of effective transport properties for individual segment pathways in GEONET, coupled with the use of PDFs to capture temporal and spatial flow system variability (NRCan, 1996). Moreover, many reviewers found if difficult to follow the arguments that claimed the IPA models deal with many different scenarios within one set of randomly sampled simulations (SRG 1995, NEA 1995, AECB 1995).

The treatment of uncertainty in conceptual flow modelling is a fundamental concern (Russell *et al.* 1998). In response to technical comments, this issue is being examined as part of the aforementioned independent review of geosphere PA methodologies. The review will examine a variety of model approaches to examine groundwater flow and mass transport uncertainty arising from random and systematic measurement error, parameter definition, flow system boundary conditions and mathematical approximations. Moreover, the used fuel disposal research and development programme is funding the Moderately Fractured Rock and *In situ* Diffusion experiments at the URL. These studies are designed to develop an improved understanding of mass transport by advection and diffusion within fractured crystalline rock.

The Linkage Between Detailed and IPA Models of the Geosphere

One of the more serious issues noted by the AECB (1995) and SRG (1995) is related to the development of the geosphere model used in the long-term safety assessment. The concern is not directly with underlying research codes such as MOTIF, rather, the concern centres on the simplified representation of MOTIF results within GEONET.

GEONET was calibrated with MOTIF studies of the WRA for use in the probabilistic IPA assessment. It used a network of approximately 50 segments to represent the flow regime. Each segment had hydraulic and geochemical properties that characterised different parts of the flow regime. Most of the parameters quantifying these properties were uncertain and parameter values were described by PDFs. The EIS assessment results showed that the natural geosphere barrier was one of the most effective barriers – even for mobile radionuclides such as [129]I. The assessment results also showed that uncertainty had an immense influence as reflected in estimated dose impacts that ranged over more than ten orders of magnitude. A major source of uncertainty could be attributed to uncertainty and variability in the representation of the geosphere.

Reviewers found fault with GEONET that included the following.

- The connection between MOTIF and GEONET was not sufficiently established nor was it transparent. Some details of this issue are discussed under "verification of geosphere PA models'. The issue is further complicated by the use of GEONET in a probabilistic analysis. Reviewers expressed concern over the lack of transparency in the analyses and the results of the analyses. For instance, criticisms were directed at the derivation of effective transport properties for individual segment pathways in GEONET, coupled with the use of PDFs to capture temporal and spatial flow system variability (NRCan 1995). In particular, the groundwater "velocity-scaling factor" was widely criticised as a viable description of flow system spatial and temporal variability. Moreover, many reviewers found if difficult to follow the arguments that claimed the IPA models dealt with many different scenarios within one set of randomly sampled simulations (SRG 1995, NEA 1995, AECB 1995).

- The development of GEONET was adversely affected by its use in a probabilistic analysis. The requirement for thousands of simulations led to over-simplification of important processes in GEONET.

- In the abstraction of GEONET from MOTIF the treatment of uncertainty in the geosphere was regarded as unrealistic and unsatisfactory. In particular, the network of segments was fixed in both space and time. Thus, the safety assessment did not encompass important sources of uncertainty, such as the length of a segment, its linkage to adjacent segments and variability in groundwater velocity within a segment and within adjacent segments.

The review comments reflected a lack of confidence in GEONET that extended to include the results of the IPA. It has become clear that, if the proponent wishes to rely on the efficacy of the geosphere as a transport barrier, there must be a high level of confidence in the models employed.

A re-examination of this issue has led to several recommendations aimed at improving confidence for future safety assessments.

- The importance of a strong connection between detailed and assessment models of the geosphere is well recognized. It is also well recognised that those connections must be thoroughly established, subjected to formal peer review, and be thoroughly documented in a clear and transparent fashion.

- The PA abstraction process needs to follow established methods that are based on objective reasoning. The EIS and SCS Case Studies would have gained wider acceptance had there been a less subjective and more transparent development of the GEONET flow segment network.

- A collection of alternative models must be examined. Several transport models may be needed that offer alternative descriptions of the geosphere, to deal with conceptual model uncertainty, a fundamental source of uncertainty that was not directly examined in the EIS. In addition, it may be necessary to develop more that one safety assessment model that exhibit better fidelity with the underlying research models. For instance, work is underway to determine whether GEONET could implicitly include the groundwater flow algorithms from MOTIF, whether a subset of MOTIF could replace large parts of GEONET or whether some other transport code could replace or augment GEONET.

Acceptance of the geosphere as a major transport barrier requires substantial improvement in the level of confidence attached to the research models, to the simplified IPA models, and to the abstraction procedure(s). This issue continues to be a major challenge.

Documentation of the Geosphere PA Process and Results

This issue affects more than just the technical reviewers and more than just the geosphere. However, focussing solely on the geosphere, it is clear that reviewers were dissatisfied with the extent and quality of documentation. For instance, review criticisms point out that

- the documentation "does not demonstrate any methodology for the translation of realistic field and laboratory data into parameters suitable for .... MOTIF" (SRG 1995, 1996); and
- "insufficient information is provided to permit a detailed review of the three-dimensional ... MOTIF code" (AECB 1995).

That is, key underlying research data was omitted and it was not clear how research data was then used in the geosphere model. Thus, reviewers were unable to replicate analyses or even re-interpret results.

In addition, as indicated in comments from the NEA (1995), there appeared to be inadequate communication between the proponent and external stakeholders in terms of interpretation of the regulatory guidelines and expectations for information requirements.

The expected outcome is that these deficiencies would have a strong negative influence in the degree of confidence the reviewers attached to the geosphere PA approach. One lesson deriving from this issue is the need for technical documentation that is exhaustive and comprehensive. In retrospect, the EIS Case Study may have benefited greatly with the release, prior to the assessment, of documents containing raw research results and documents describing how those results are interpreted for use in a geosphere model. In addition, the Case Study may have benefited if summary reports were submitted for prior technical review and comment by regulators and other decision-makers. A hierarchical set of documents would have improved the clarity and traceability of information.

A recent report by the Integrated Performance Assessment Group (IPAG) of the Nuclear Energy Agency (IPAG2, in preparation) examines in more depth the requirements of clarity, transparency and traceability. It was generally concluded that that traceability was required. In fact, it was suggested that traceability is the most important attribute associated with documentation, based on the expectation that the primary concern is regulatory review. Achieving traceability can be difficult, but one approach now in use appears to be promising: the recent TSPA-VA study is supported by a huge amount of readily accessible information on the Internet.

5.   Summary and Conclusions

A federal Environmental Review Panel in March 1998 completed a public review, which examined the acceptability of a deep geologic disposal concept for Canada's nuclear fuel waste. This concept envisions that a hypothetical used fuel repository would be positioned at a nominal depth of 500 to 1 000m within plutonic rock of the Canadian Shield.

As part of the review process AECL submitted an EIS to the Panel, which was supported by two assessments – the EIS and SCS – that examine the safety and robustness of the used fuel disposal concept. The assessments also serve to illustrate a methodology by means of which the long-term performance of a repository could be evaluated for comparison with regulatory safety criteria and guidelines.

The EIS and SCS received a broad range of critical comments during the public review. Amongst those submitting detailed technical comments to the Panel were the Panel-appointed Scientific Review Group, Canada's nuclear regulator and other affected government agencies, learned geoscience bodies and professional technical organisations. This paper has examined the review comments with the objective to gain insight into issues surrounding the development of confidence in geosphere PA.

The ability to predict radionuclide migration within the sub-surface is a complicated undertaking given the inherent spatial and temporal variability of fractured crystalline groundwater flow systems. Technical comments submitted to the Panel reflect this sentiment and point to the challenges faced by geosphere PA in providing the necessary assurances to external reviewers that flow system uncertainty has been adequately captured in predictions of mass transport.

Within the EIS and SCS, radionuclide migration was predicted with the code GEONET, which simulates transport using an equivalent porous medium approximation through an interconnected network of 1-dimensional geosphere segments. These segments are invariant in time and space. For site specific applications, the geometry of the 1-dimensional segment network is implicitly linked to steady-state 3-dimensional simulations of groundwater flow performed with the finite element code MOTIF and particle-tracking code TRACK3D. Uncertainty in the conceptual flow system model and its effect on radionuclide transport is explored principally through a probabilistic analysis as implemented within the SYVAC framework. The SYVAC framework deals with parameter uncertainty using PDFs to describe the likelihood of feasible values and has proven particularly useful in identifying system parameters that most influence disposal system performance.

It is evident from the technical review comments that GEONET and its implementation within the SYVAC framework did not fully satisfy the expectations of the technical reviewers. Common concerns from many reviewers were on issues related to:

- the verification of the GEONET code, especially for site-specific application;

- the treatment of conceptual model uncertainty;

- the interface between MOTIF and GEONET; and

- the geosphere PA documentation.

As a generic methodology, a method is required to verify or provide assurance that simplified GEONET realisations provided reasonable estimates of mass transport within spatially variable 3-dimensional flow fields typical of a fractured plutonic terrain. Such verification was judged to be incomplete in the EIS and SCS. A second issue was the ability of GEONET to evaluate conceptual flow model uncertainty. Although invoked within a probabilistic analysis framework, GEONET described flow systems in which key properties are invariant in time and space. An associated issue was the justification of the creation of the GEONET network, a manual process based on knowledge of detailed 3-dimensional numerical simulations of advective transport that was not transparent to reviewers. Finally, documentation of the geosphere PA lacked the clarity reviewers expected.

At present an evaluation of geosphere PA strategies based, in part, on comments made during the Panel's public review process is being completed. The evaluation will examine various aspects of geosphere PA including: the role of site characterisation in supporting PA, alternative methods to illustrate geosphere performance, the derivation of effective transport properties in fractured media and methods to communicate PA results to various audiences. It is evident that PA tools should be capable of assessing issues related to long-term flow system evolution. There is also a requirement to demonstrate explicit links with site characterisation, both in regards to data collection and the derivation of conceptual flow system models used by PA.

6. References

AECB (Atomic Energy Control Board). 1995. AECB staff response to the environmental impact statement on the concept for disposal of Canada's nuclear fuel waste. Submitted to the Nuclear Fuel Waste Disposal Concept Review, Canadian Environmental Assessment Agency, July 25, 1995.

AECL (Atomic Energy of Canada Limited). 1994. Environmental impact statement on the concept for disposal of Canada's nuclear fuel waste. Atomic Energy of Canada Report, AECL-10711, COG-93-1. Available in French and English.

CEAA (Canadian Environmental Assessment Agency). 1998. Nuclear fuel waste management and disposal concept. Report of the Nuclear Fuel Waste Management and Disposal Concept Environmental Assessment Panel. Canadian Environmental Assessment Agency. Minister of Public Works and Government Services Canada. Catalogue No. EN-106-30/1-1998E. ISBN: 0-662-26470-3. February 1998, Hull, Québec.

Canadian Geoscience Council. 1995. Review of the environmental impact statement on the concept for disposal of Canada's nuclear fuel waste/ Submitted to the Nuclear Fuel Waste Disposal Concept Review, Canadian Environmental Assessment Agency, August 8, 1999.

Chan, T. and T. Melnyk. 1994. Condensing a detailed groundwater flow contaminant transport model into a geosphere model for environmental and safety assessment. Presented at the 4th International Topical Meeting on Nuclear Thermal Hydraulics, operations and Safety. April 6-8, Taipei, Taiwan.

Chan, T., V. Guvanasen, B.W. Nakka, J.A.K. Reid, N.W. Sheier and F.W. Stanchell. 1995. Verification of the MOTIF code version 3.0. Atomic Energy of Canada Limited Report, AECL-11496, COG-95-561-I. Pinawa. Manitoba.

Davis, P.A., R. Zach, M.E. Stephens, B.D. Amiro, G.A. Bird, J.A.K. Reid, M.I. Sheppard and M. Stephenson. 1993. The disposal of Canada's nuclear fuel waste: The biosphere model, BIOTRAC, for postclosure assessment. Atomic Energy of Canada Limited Report, AECL-10720, COG-93-10.

Davison, C.C., Chan, A. Brown, M. Gascoyne, D.C. Kamineni, G.S. Lodha, T.W. Melnyk, B.W. Nakka, P.A. O'Connor, D.U. Ophori, N.W. Scheier, N.M. Soonawala, F.W. Stanchell, D.R. Stevenson, G.A. Thorne, S.H. Whitaker, T.T. Vandergraaf and P. Vilks. 1994. The disposal of Canada's nuclear fuel waste: The geosphere model for postclosure assessment. Atomic Energy of Canada Limited Report, AECL-10719, COG-93-9.

Goodwin, B.W. and T.H., Andres. 1998. The role of probabilistic systems analysis in the safety assessment of used fuel disposal in Canada. Prepared by Atomic Energy Canada Limited for Ontario Power Generation. Nuclear Waste Management Division Report 06819-REP-012000-0004, Toronto, Ontario.

Goodwin, B.W., T.H. Andres, W.C. Hajas, D.M. LeNeveu, T.W. Melnyk, J.G. Szekely, A.G. Wikjord, D.C. Donahue, S.B. Keeling, C.I. Kitson, S.E. Oliver, K. Witzke and L. Wojciechowski. 1996. The disposal of Canada's nuclear fuel waste: A study of postclosure safety of in-room emplacement of used CANDU fuel in copper containers in permeable plutonic rock. Volume 5: Radiological Assessment. Atomic Energy of Canada Limited Report, AECL-11494-5, COG-95-552-5.

Goodwin, B.W., D.B. McConnell, T.H. Andres, W.C. Hajas, D.M. LeNeveu, T.W. Melnyk, G.R. Sherman, M.E. Stephens, J.G. Szekely, P.C. Bera, C.M. Cosgrove, K.D. Dougan, S.B. Keeling, C.I. Kitson, B.C. Kummen, S.E. Oliver, K. Witzke, L. Wojciechowski and A.G. Wikjord. 1994. The disposal of Canada's nuclear fuel waste: Postclosure assessment of a reference system. Atomic Energy of Canada Limited Report, AECL-10717, COG-93-7.

Johnson, L.H., D.M. LeNeveu, D.W. Shoesmith, D.W. Oscarson, M.N. Gray, R.J. Lemire and N.C. Garisto. 1994. The disposal of Canada's nuclear fuel waste: The vault model for postclosure assessment. Atomic Energy of Canada Limited Report, AECL-10714, COG - 93 - 4.

Johnson, L.H., D.M. LeNeveu, D.W. Shoesmith, F. King, M. Kolar, D.W. Oscarson, S. Sunder, C. Onofrei and J.L. Crosthwaite. 1996. The disposal of Canada's nuclear fuel waste: A study of postclosure safety of in-room emplacement of used CANDU fuel in copper containers in permeable plutonic rock. Volume 2: Vault Model. Atomic Energy of Canada Limited Report, AECL-11494-2, COG-95-552-2.

Joint Statement. 1978. Joint statement by the Minister of Energy, Mines and Resources Canada and the Ontario Minister of Energy, June 5, 1978. Printing and Publishing, Supply and Services Canada, Ottawa, Canada K1A 0S9.

Joint Statement. 1981. Joint Statement by the Minister of Energy, Mines and Resources Canada and the Ontario Minister of Energy, August 4, 1981. Printing and Publishing, Supply and Services Canada, Ottawa, Canada K1A 0S9.

NRCan (Natural Resources Canada). 1996. Natural Resources Canada's submission to the environmental assessment panel: nuclear fuel waste management and disposal concept – phase II – technical session. June 5, 1996.

Nakka, B.W. and T. Chan. 1994. A Particle-tracking code (TRACK3D) for convective solute transport modelling in the geosphere: Description and user's manual. Atomic Energy of Canada Limited Report, AECL-10881, COG-93-216. Pinawa, Manitoba.

NEA (Nuclear Energy Agency of the Organisation for Economic Co-operation and Development). 1995. The disposal of Canada's nuclear fuel waste. Report submitted to Natural Resources Canada. April 27, 1995.

Ophori, D.U., D.R. Stevenson, M. Gascoyne, A. Brown, C.C. Davison, T. Chan and F.W. Stanchell. 1995. Revised model of regional groundwater flow of the Whiteshell Research Area: Summary. Atomic Energy Canada Limited Report AECL-11286.

Russell, S.B., P.J. Gierszewski, M.R. Jensen and J.E. Villagran. 1998. Safety assessment plan for the Used Fuel Disposal Programme: 1998. Ontario Hydro Report No. 06819-REP-01200-0072 - R00.

SRG (Scientific Review Group). 1995. An evaluation of the environmental impact statement on Atomic Energy of Canada Limited's concept for the disposal of Canada's nuclear fuel waste. Report of the Scientific Review Group, advisory to the Nuclear Fuel Waste Management and Disposal Concept Environmental Assessment Panel. Available from the Minister of Supply and Services Canada, Cat. No. En 106-30/1995E.

SRG (Scientific Review Group). 1996. An evaluation of the environmental impact statement on the concept for disposal of Canada's nuclear fuel waste. An addendum to the report of the Scientific Review Group. Canadian Environmental Assessment Agency. Hull. Quebec.

SAT (Subsurface Advisory Team). 1995. SAT review of AECL's environmental impact statement on the concept for disposal of Canada's nuclear fuel waste. Submitted to Environment Canada, April 26, 1995.

Stanchell, F.W., C.C. Davison, T.W. Melnyk, N.W. Scheier and T. Chan. 1996. The disposal of Canada's nuclear fuel waste: A study of postclosure safety of in-room emplacement of used CANDU fuel in copper containers in permeable plutonic rock. Volume 3: Geosphere Model. Atomic Energy of Canada Limited Report, AECL-11494-3, COG-95-552-3.

Wikjord, A.G., P. Baumgartner, L.H. Johnson, F.W. Stanchell, R. Zach and B.W. Goodwin. 1996. The Disposal of Canada's Nuclear Fuel Waste: A study of postclosure safety of in-room emplacement of used CANDU fuel in copper containers in permeable plutonic rock. Volume 1: Summary. Atomic Energy of Canada Limited Report, AECL-11494-1, COG-95-552-1.

Zach, R., B.D. Amiro, G.A. Bird, C.R. Macdonald, M.I. Sheppard, S.C. Sheppard and J.G. Szekely. 1996. The disposal of Canada's nuclear fuel waste: A study of postclosure safety of in-room emplacement of used CANDU fuel in copper containers in permeable plutonic rock. Volume 4: Biosphere Model. Atomic Energy of Canada Limited Report, AECL-11494-4, COG-95-552-4.

Radionuclide Transport-related Aspects of the Radiological Evaluation of the Mururoa and Fangataufa Atolls

Ghislain de Marsily
University Paris VI (France)

Jörg Hadermann
PSI, (Switzerland)

Abstract

The evaluation of the potential environmental consequences of the French nuclear weapon tests in the Pacific atolls of Mururoa and Fangataufa was requested by Jacques Chirac, the President of France, at the end of the testing period in 1996. He commissioned the IAEA to independently evaluate the potential radiological impact of the tests, and also asked Professor C. Fairhurst to form a Committee to evaluate the mechanical stability and hydrological situation of the atolls. Both groups worked together to determine the flow pattern into the atolls, subsequently used to evaluate the radionuclide transport in the underground, the release in the environment and the resulting doses to man and fauna (1,2).

This presentation will focus on the use of models to first understand the present hydrological situation of the atolls, its predicted evolution with time, and the related transport of radionuclides out of the rock and into the environment. Attempts at validating these predictions based on radionuclide concentration measurements will also be described. The existence and role of data will be emphasized.

Geology

The Mururoa and Fangataufa atolls, situated in the Tuamotu Archipelago in the Pacific, have a similar structure: a volcanic base, made of basaltic lava that was deposited either underwater or above sea level, and a carbonate cover resulting from coral development and destruction on the rim of the volcano. Due to the initial erosion and also subsidence of the volcano when it became extinct, by the cooling of the oceanic plate, the volcanic rocks lie at about 350 to 500 m below sea level; the carbonate cover is highly heterogeneous, with dolomite, limestone, chalky limestone, karstic features and fractures, making it highly pervious. The underlying volcanic structure is of very low permeability, but a set of natural fractures exist which make the flow of water possible. The fracturing is certainly enhanced, in an unknown fashion, by the mechanical effects of the tests. The site is thus believed to be representative of a fractured igneous rock, covered by a permeable sedimentary unit and a sea water surface. Although numerous boreholes were drilled in the atolls, only a few geological or geophysical logs and cores were available, and it was not possible to determine any fracture pattern to use in a discrete fracture modeling approach. All nuclear tests were in principle performed in the basalt, however some tests did not have a thick enough or strong enough volcanic cover above the "chimney" resulting from the explosion, thus making a permeable path for radionuclides to "leak" from the volcanics to the carbonates; some so-called "safety tests", where a nuclear device is submitted to a chemical explosive to test if the device will go critical (which it should not), were performed in the carbonates, a fraction of which did go critical. In both cases however, Pu was dispersed in the carbonate rocks.

Pre Test Hydrology

Although some permeability measurements on cores samples were available (giving values on the order of 10^{-10} m/s or less), surprisingly there were no in situ testing providing estimates of natural pre-tests large scale permeability (no pumping nor injection tests). The estimation of the in situ large scale flow properties of the medium were all based on the use of a natural tracer: temperature. It so happens that the temperature profile in two boreholes was available, one on the rim and one on the lagoon, and that they showed unusual shapes, with initially decreasing temperature from the surface to a depth of several hundred meters, and then increasing again due to the natural geothermal gradient. This type of profile is a sign of a particular type of groundwater flow in an atoll, called "endo-upwelling". The cold deep seawater is "sucked" into the flanks of the atoll, mostly in the carbonate cover but also in the volcanic rocks, then flows quasi-horizontally towards the center of the atoll, while being heated by the rock thanks to the natural geothermal gradient in the rock, and then flows upwards due to buoyancy. A density dependent flow model was thus fitted to the temperature measurements, and provided indirect estimates of the large scale permeability distribution. An upper limit of 10^{-7} m/s was calibrated for the volcanic part, and values ranging from 10^{-3} to 10^{-4} m/s for the carbonate part. Sensitivity studies were done to estimate the uncertainty range of these values, and several "acceptable" models of permeability distributions were thus obtained. The estimated Darcy velocity in the system ranged from mm/y in the volcanics to m/y in the carbonates.

Effects of the tests

A nuclear test first creates a spherical cavity, in which the device and the surrounding rock are vaporized, and later assemble in a molten lava which cools at the bottom of the cavity. Because the surrounding rock is fractured by the mechanical effect of the test, the roof of the spherical cavity collapses between minutes and hours after the test, and the falling rubbles form a vertical "chimney" which extends upwards on a height about 5 times the radius of the cavity. The chimney becomes stable when the rubbles fill the void space and support the roof. As the heat of the explosion slowly dissipates, the hot gases inside the cavity and chimney cool and the pressure falls below atmospheric. Water is then "sucked" into the chimney and eventually fills it up. The rate of filling of the chimney is an indication of the permeability value in the area of the rock surrounding the chimney and affected by the test. Some measurements of this rate of filling are available through boreholes which are systematically drilled after the explosion to core the lava at the bottom of the cavity, as part of the analysis of the test. The filling of a cavity was modeled, and an estimate of the increase of the permeability of the volcanic rock compared to the intact value was thus obtained. On the lava cores, leaching experiments were conducted from which radionuclide dissolution rates could be derived.

Post Test Hydrology

The rate of release of radionuclides from the volcanic rocks was assessed based on the modeling of the hydrologic situation after the tests. The dominant mechanism after the test is the buoyancy effect of the heat generated by the test, which creates first a strong mixing loop inside the chimney, then a vertical flux of the water towards the volcanic-carbonate interface. The flow in the volcanics is therefore dominated by the effect of the test, for periods of up to 500 years; however, in the carbonates, the dominant mechanism is still the natural endo-upwelling flow, the heat and water released by the volcanics has in general a negligible impact. An additional source of uncertainty in the flow and transport in the carbonates is the effect of the tide, which can be felt even in boreholes at the center of the atolls, probably because of the presence of karstic layers in the carbonates. This alternating motion may spread the radionuclide plume transported to the carbonates, both in the horizontal and vertical directions. Although the general flow direction in the carbonates is from the flanks of the atoll towards the center and up to the lagoon, it was felt possible that the tidal effects could generate radionuclide releases directly to the ocean, mostly for the tests performed on the rim of the atolls. Modeling precisely the release of nuclides would require the knowledge of the position of each test, the yield, the inventory of generated nuclides, and data on the integrity or properties of the volcanic cover separating the top of the chimney from the base of the carbonates. Non of this was available, being classified information. The study had to make up for this missing information as follows. The yield of the tests was estimated by international groups based on the seismic measurements. The inventory was estimated by the same group. For the position, three scenarios were considered : (i) "good" tests where a thickness of 25 to 250 m of undamaged volcanic cover existed between the top of the chimney and the base of the carbonates; 121 tests were reported by the French authorities in this category; (ii) "poor" tests where this volcanic cover was damaged and a high permeability fractured zone existed between the top of chimney and the carbonates; 4 such tests were reported; (iii) tests where the top of the chimney reached the carbonates, with 12 cases reported.

The modeling of the flow was done using either a 2-D radially symmetric cross-section, or in some cases a 3-D model of the volcanic rock. The permeability of this rock was taken either as that identified through the interpretation of the filling rate of chimneys, or as higher values in a worst case approach. It was found that the yield of the test did not produce very different figures for the resulting vertical velocity above the chimney, as the size of the cavity also scaled with the yield, and thus the initial increase in temperature of the water in the chimney (which governed the buoyancy effect) was not yield-dependent. The calculated velocity decreased with time, but was taken as an average value for the radionuclide transport calculations. Depending on the case, these Darcy velocities ranged from 0.1 m/y to 100 m/y.

Source term and radionuclide transport modeling

The radionuclides to be considered come from four sources: the lava, dissolved in chimney water, sorbed on chimney rubble and, for safety tests, from the nuclear device. The lava contains refractory elements (e.g. most of the actinides, in particular Pu, and fission as well as activation products to a varying degree). It dissolves slowly and congruent release of radionuclides into the chimney was assumed. Within the chimney, radionuclides are distributed between water and solid phase according to their sorption distribution ratio (e.g. tritium is present quantitatively in the form of HTO). Based on considerations of inventory, half-life, sorption properties and toxicity, 35 radionuclides were considered with emphasis on 239-Pu, 137-Cs, 90-Sr and 3-H because these isotopes were of special concern and/or measurements of concentration were available. Little data on sorption distribution data was known for the sites; a range of values was assumed based on literature and rock composition. It was further assumed that the chimney is a well mixed compartment, and radionuclides are transported advectively from the chimney top into the surrounding rock. The "good" tests were grouped into four yield classes (5, 25, 60, 100 kt yield) at five nominal depths (25 m, 75 m, 100 m, 150 m and 250 m of good volcanic cover above the top of the chimney); the four "poor" tests were analyzed individually assuming direct release into the carbonates. For the safety tests, Pu is directly released from the device at the solubility limit.

Transport in the volcanic rock was described by a double porosity model accounting for vertical 1D advection with constant Darcy velocity (0.1 m/y, 1 m/y or 10 m/y) as determined by the hydraulics calculations, matrix diffusion (data mainly from general literature), sorption (same Kd as for the source term), radioactive decay and build-up. The reason for this choice of model was, firstly, the fractured nature of the chimney near-field and of the volcanic rock and, secondly, the conservativity of this approach compared to a single porosity medium. It must, however, be recognized that a proper characterization of flow paths does not exist, and assumptions on fracture frequency and width had to be made (10 per m and 1 mm, respectively). It was assumed that all Darcy flux is in the fractures; this overestimates consequences. All short-lived sorbing nuclides, including 90-Sr and 137-Cs, decay to insignificance during transport in the volcanic rocks. For the further fate of radionuclides, two scenarios were considered: either direct release, through the karsts, to the ocean, or vertical transport through a nominal 300 m thick carbonate cover into the lagoon. Transport through the carbonates was described by an advection-dispersion model with the exception of tritium (see below). The main results are that doses are dominated by releases from "poor" tests and (for Pu) from safety trials. Tritium and the other non-sorbing nuclides form presently an underground source for the lagoons; their release will decrease in the future. Sorbing nuclides will breakthrough to the lagoons in the future at low level if their half-life permits. Plutonium and other strongly sorbing nuclides will be confined in the carbonates for very long times. Calculated doses to individuals at the atolls are indeed very low, of the order of 6 μSv/y presently, and 4 μSv/y in some thousand years (mainly from Pu). Doses are much lower in the ocean scenario.

Attempts at validating the calculations

Radionuclide concentration measurements were available in the atoll lagoons, and in some boreholes drilled in the carbonates. Tritium, 90-Sr and 137-Cs measurements were available in 5 chimney waters, 241-Am, Pu, 36-Cl, and 14-C in two chimney waters. The calculated concentrations for tritium, Sr, Cs, I, and Cl do agree fairly well. C, Pu and Am are strongly over-predicted. The reasons may lie in a strong underestimate of sorption and, for the actinides, in a too conservative assumption on the initial distribution between lava and rubble. At 10 m distance from a safety trial in the carbonates, Pu has been measured. Given the sorption distribution ratio, the measured concentration yields a lower limit for water velocities consistent with the hydraulics calculations. The water of the lagoons is flushed daily by the tide, and the average residence time in them is known. It is thus possible to estimate the flux of tritium released from each atoll to the lagoons, from all tests. Several transport models through the carbonates were tested to convert the flux of tritiated water released by the volcanics into a flux of tritium into the lagoons. A piston flow model, a convection-dispersion model, and a mixing tank model were tested to represent this transfer through the carbonates. It was found that the mixing tank model was the best model to account for the few observations of tritium concentration in the lagoons, and in the available boreholes. With this model, it was found that the calculated velocities of tritium release from the volcanics were in reasonable agreement with the observed concentrations. Although not a perfect validation, this result was felt reassuring. For Cs the transport model predicts negligible contribution of underground sources to the lagoon concentrations. This is consistent with measurements which show a decrease in time (Mururoa) or essentially constant (Fangataufa) concentration in the lagoons stemming from global fallout and leaching of lagoon sediments. Release of Pu into the lagoons is predicted to arise far in the future. Again, this is consistent with measurements. For Sr, the model predicts a maximum release rate of 30 GBq/y at 150 years after the tests. Whereas the French measurements show a decreasing concentration in the lagoons, consistent with the model calculations, a new IAEA measurement points to an increase in lagoon concentration. The reasons for this discrepancy are unknown.

Glaciation scenario

A series of alternative scenarios was also considered. Among them, the glaciation scenario in 10 000 or 20 000 years proved to be the most significant. Because of a 100 to 150 m sea level drop, the atolls would be converted into carbonate islands, in which a freshwater lens could develop in the future. Assuming a recharge of the aquifer of 0.5 m/y, the position of the freshwater lens could be such that the location of the safety trials (that were unfortunately shot in the carbonates at shallow depth) could be within the freshwater lens. The only nuclide of interest at this date is Pu. Whether the safety trials went critical or not in fact did not matter, since in both cases the Pu would still be left in the rock. Rough calculations showed that this Pu would not have migrated away from the location of the tests because of groundwater flow more than some tens of meters, because of the assumed high Kd of Pu in the carbonates. If a well was drilled in the Pu plume, then the concentration of Pu in the water would be several thousands time the limit giving an acceptable dose through the drinking water pathway. This scenario was however considered to be of low probability, given that there were only 7 such tests in the atolls. Another potential consequence was that the Pu released from the volcanics to the carbonates during the pre-glaciation years would in fact not be dispersed in the ocean, but sorbed in the carbonates. The concentration of this Pu in the wells was not easy to estimate, but was thought to be below the acceptable limit.

Lessons learned

The amount of "waste" generated by a nuclear explosion is really trivial compared to the amount of waste generated by nuclear power. The very small doses resulting form the analysis of the Mururoa and Fangataufa atolls are therefore not comparable to those of a waste disposal site. In addition, the sites have the great advantage of the strong dilution provided by a release in a coastal environment and of a big distance to the next inhabited islands. The modeling was made with highly insufficient data, and pessimistic assumptions were thus required, which could be used because of the small radiological impact. The calibration of the models was made with environmental tracers (in this case temperature) and a small number of in situ tests (rate of cavity filling). The other parameters (fracture density and aperture, porosity and diffusion coefficient in matrix, sorption, lava leaching rate...) were based on general literature with a few exceptions (fracture density, matrix porosity). The attempt at validating the global transport models at a large scale was possible here because of the existing flux of tritium, Cs and Sr, which would not be available in a repository. The abnormal scenario (glaciation) proved to be more significant than the normal operation, but was dismissed due to the low probability of having water wells drilled in the future at the exact location of safety tests. Confidence in the results was mostly based on the possibility to observe the present behavior, the large experience from performance assessment of civilian waste repositories and the absence of any significant impact.

Finally, it is worth mentioning that the major environmental impact does not stem from the 121 "good" tests but from the few others, and there is a need for a better scientific understanding of transport in karsts under the influence of tides.

References

(1)   IAEA, The Radiological Situation at the Atolls of Mururoa and Fangataufa, Main Report and Technical Vols 1-6, Vienna, 1998.

(2)   IGC (C. Fairhurst, E.T. Brown, E. Detournay, G. de Marsily, V. Nikolaevskiy, J.R.A. Pearson, L.Townley), Stability and Hydrology Issues Related to Underground Nuclear Testing at French Polynesia, 2 Vols, Paris, 1998.

Regulatory Views and Experiences Regarding Confidence Building in Geosphere-transport Models

B. Strömberg and B. Dverstorp
Swedish Nuclear Power Inspectorate, Sweden

J. Geier
Oregon State University, USA

Abstract

During the last two decades, the Swedish Nuclear Power Inspectorate (SKI) has developed regulatory views regarding transport models and their use in performance assessment, through international collaboration and independent performance assessment projects. These views should be effectively communicated such that the implementor and other affected parties have a clear view of what the regulators will expect after each phase of repository development. Topics addressed in this paper relate to the structure of utilised models, their scientific basis and the associated uncertainties. Of particular relevance is how confidence can be judged during each stage of site investigations and repository construction, since these activities may commence in Sweden in the not-too-distant future. The diversity of data from site measurements and the development site-scale models for hydrogeology and transport are discussed briefly. Finally, several applications of radionuclide transport modelling in the overall performance assessment are mentioned.

1. Introduction

In this paper, we give an overview of regulatory aspects and experiences related to the role of radionuclide retardation in performance assessment (PA), the evaluation of geosphere-transport models and how transport models may be improved during the different stages of repository development. Important parts of this include methodology, documentation, confidence enhancement and integration in the overall PA. The Swedish regulatory context is such that very detailed requirements on an individual component of a PA, such as radionuclide transport models, would be inappropriate. This is because the implementor should have the sole responsibility for the development of the safety case. Swedish law states that the producers of the nuclear waste, through their jointly owned company, the Swedish Nuclear Fuel and Waste Management Co. (SKB), should have full responsibility for providing a safe solution for the final disposal of spent nuclear fuel and nuclear wastes. Nevertheless, a number of indirect requirements follow from demands and suggestions related to, for instance, comprehensiveness in the choice of scenarios and in the description of the scientific basis for models, quantitative modelling capability and a systematic and thorough treatment of uncertainties.

The regulatory views expressed in this paper were initially developed through experiences from international collaboration on modelling of groundwater flow and radionuclide transport during the last two decades [1-3]. In order to develop a more comprehensive understanding of PA methodology, the SKI conducted two independent performance-assessment projects. The goal of the first PA-project, Project-90 [4], was to develop and implement PA models and methods using mainly generic and hypothetical data. This work was followed up by the SITE-94 [5] project, in which the emphasis was placed on the treatment of site-specific data, which came from SKB's site investigation at the Äspö Hard Rock Laboratory.

2. Transparent structure of models

A complete performance-assessment for a nuclear waste repository represents a complex system of data, models, assumptions and arguments of different types. To gain insight in such a system, it is essential that the utilised models, assumptions, arguments and flow of information have been clearly documented. A quality assurance programme should therefore be implemented such that the vast number of decisions that are taken before the PA is completed can be traced, e.g. by the regulator.

Figure 1 illustrates essential connections between data acquisition and different types of models utilised in PA. The top-level models should provide a representation of the total system performance, such that compliance with regulatory criteria can be evaluated. Such a model is by necessity a gross simplification that must be motivated by more detailed underlying submodels that only cover a part of the process system. Since the top-level models are likely to be evaluated from many different perspectives, they should if possible be kept reasonably simple, robust and transparent. A higher degree of complexity can be expected for detailed process models that are used to test the validity of different simplifying assumptions used in the higher level models. In addition, effective parameter values utilised for the top-level models cannot usually be directly measured but are obtained by abstraction from lower level models. Simplifying assumptions can also be motivated by laboratory experiments or field-measurements.

Figure 1. **Structure of different types of models in PA with examples of how models and data may be used**

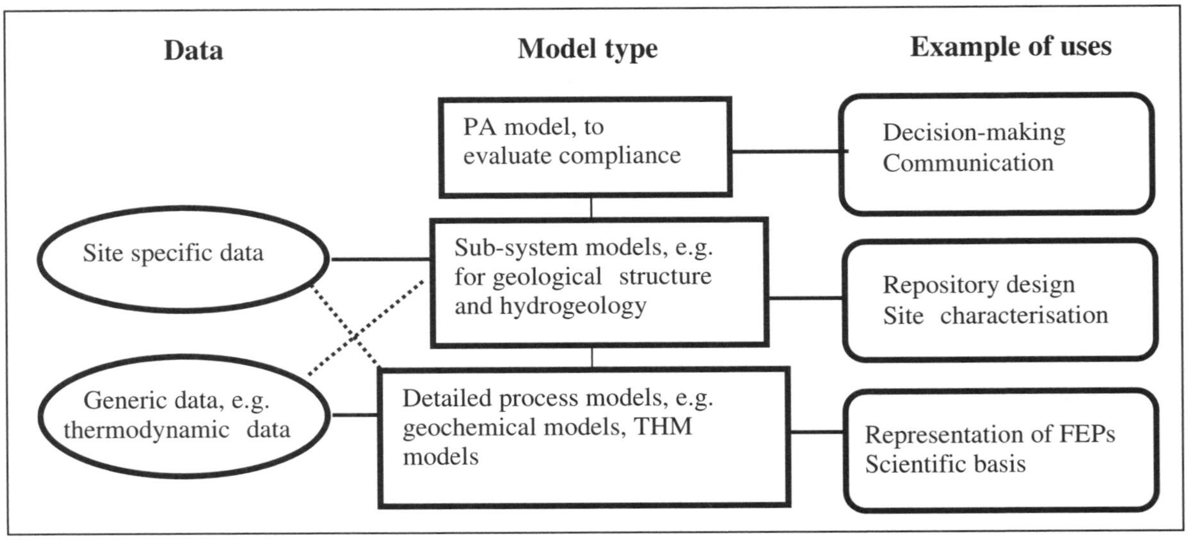

Confidence will be improved if a structured approach with a hierarchy of model complexity is used [6], such that the PA is accessible to all the affected parties (Figure 1). At the top level, it is essential that decision makers and the general public have a chance to understand the fundamental approaches and results. Top-level models should refer to subsystem models for those who wish to obtain a more detailed description. Subsystem models will also be of use by persons involved in the repository design or site characterisation, while the detailed models should represent the scientific basis of individual processes that are to be scrutinised by specialists.

3. Confidence from scientific basis and understanding of uncertainties

To obtain confidence for the utilised models, parameters and conditions, the scientific basis must be well established. This needs to be demonstrated, e.g. through a comprehensive inventory and description of the associated features, events and processes (FEPs). Because it will never be possible, or necessary, to represent all processes with the same level of detail, the development and selection of models to represent these FEPs, by necessity, involves subjective judgements. Firstly, different FEPs may be judged to be of lesser or greater importance to radionuclide transport problems in PA. Secondly, the knowledge base and availability of quantitative models to represent different transport processes varies significantly. To achieve confidence in PA, all simplifications and assumptions associated with the representation of different FEPs should be carefully documented. In addition to this, one must also ensure that the accuracy of a model is sufficient for the specific context in the PA. If a particular model, boundary condition or parameter value has a large impact on the end result, this would of course suggest that a more thorough evaluation is required. Since the importance of parts of the PA may not be apparent, calculation cases to illustrate this can be helpful (e.g. by including sensitivity analysis and what-if calculation cases). The requirements stated above are related to the previously often discussed issue of "model validation" [7]. A strict formulaic interpretation of the word suggests that no radionuclide transport model can be validated. Such a conclusion is not particularly useful, since it is apparent that models can be more or less appropriate for a particular context.

Uncertainties will always prevail in spite of continuous research, ambitious site characterisation efforts, and a gradual refinement of the repository concept. In order to understand the importance of these, it is important that all types of uncertainties are comprehensively evaluated and discussed. This could be simplified by dividing the different uncertainties into different categories, e.g. parameter uncertainty, conceptual model uncertainty, and system and scenario uncertainty. Parameter uncertainty may be dealt with by using available site-specific information to establish reasonable ranges for parameter values. Further data acquisition may then be directed to reducing the parameter ranges that have the largest impact on the consequence calculations. Remaining parameter uncertainty can be handled by assigning conservative values or by using stochastic models to represent the uncertainty or variability. Conceptual model uncertainty occurs when multiple modelling approaches can explain a given data set equally well. An example of conceptual model uncertainty was found in the SITE-94 project, where the available data did not allow discrimination between two strongly contrasting models for pore structure in the major hydrogeologic features. System uncertainty is related to the completeness of the features, events and processes (FEPs) that should describe the process system, while scenario uncertainty is related to the comprehensiveness in the selection of scenarios (see section 7).

4. Role of geosphere radionuclide retardation in PA

The long-term safety of a spent fuel repository in crystalline rock is likely to rely more heavily on the performance of the engineered barriers than on the retardation of the radionuclides in the geosphere [8]. This does not mean that a comprehensive evaluation of the geosphere performance is unnecessary. The PA should be based on multiple barrier functions and should illustrate the level of confidence that has been established for each barrier and barrier functions. It needs to be demonstrated that deficiencies in a single barrier function will not jeopardise the safety of the entire system. For instance, it is not sufficient to assume that all the canisters in the repository would remain intact for a very long time, although it may be argued reasonably that this is the most probable case.

Confidence in PA results will be enhanced if redundancy between the different barrier functions can be demonstrated. This could show that unavoidable uncertainties and possible deficiencies in the barriers may be of limited importance for the whole system. A PA that contains a systematic evaluation of the effects of uncertainties and the possible deficiencies in barriers and/or barrier-functions, will be transparent in this respect.

The required level of confidence for a geosphere-transport model depends on (1) the relative importance of the modelled process(es) in the proponent's safety case, and (2) the stage of repository development (see section 4.). The proponent will have the freedom to put more or less weight on the different barriers (e.g. spent fuel, buffer, near-field rock, far-field rock) and/or barrier functions (e.g. sorption, matrix diffusion, solubility constraints) that are included in the PA. For instance, if the uncertainty in the extent of radionuclide retardation in the geosphere is very difficult to limit, this could be balanced by limiting the uncertainty of safety enhancing features of the engineered barriers. Experiences from the SITE-94 project illustrate the difficulty to limit uncertainty of radionuclide retardation in the geosphere [5]. This was mainly due to the fact that the available field measurements provided little data in direct support of radionuclide transport models.

5. Stepwise confidence building during repository development

One would naturally expect a gradual improvement in the confidence in models and characterisation of site-specific features, during the different stages of a site investigation programme (Figure 2). At an early stage, when few data are available, there should at least be confidence in the strategy and the methods that will be utilised to characterise the site. The implementor needs a sufficiently good understanding of the site to judge the likely effectiveness of alternative methods of characterisation. Basic information that is required includes a good conceptual understanding of the types of heterogeneity that are present on a local measurement scale and of the basic boundary conditions required for the geosphere-transport models. Experience from SITE-94 suggests that characterisation of the regional scale hydrology and hydrochemistry should be prioritised. Two crucial questions to be answered are: 1) whether the site lies in a regional discharge area or a regional recharge area, and 2) how the regional hydrology has evolved over a time scale comparable to that considered in PA.

In the early stages of surface-based site investigations, it is essential that a broad suite of conceptual models and measurement techniques be used to understand the geology and the hydrogeology of the site. The models should be flexible enough to incorporate a variety of data from the site characterisation. At the most basic level, site models should take into account information from regional hydrologic modelling and detailed hydrologic tests in boreholes. However, these types of information are not likely to sufficiently constrain the site models. Hence data collection should be diversified and developers of site models should seek to incorporate other types of data, e.g. from tracer tests, groundwater monitoring, and natural geochemical tracers. In the SITE-94 project, data from hydrologic tests were indeed found to be insufficient, since transport parameters could only be derived from such tests if one could define a conceptual model for the pore structure of water conducting features. Multiple conceptual models for pore structure had to be used since no unique model could be defined. The consequence was a large uncertainty in the performance of the geological barrier, since it was not possible to discriminate among the alternative conceptual models.

Confidence in geosphere-transport models is enhanced if it can be demonstrated that the results from all models used in the PA are compatible with all the site data. However, beyond this it needs to be demonstrated that the site data are sufficiently comprehensive to allow discrimination among alternative conceptual models that would yield significantly different results in PA. In order to achieve this, an analysis of what type of additional information that would best discriminate among alternative conceptual models should be made at an early stage of the site characterisation. Feedback from the SITE-94 project suggested, for example, that significant conceptual uncertainties could be addressed in future site investigations by characterisation of the rock mass in between major fracture zones, characterisation of fracture structure and fracture fillings, and the use of multiple tracer tests (with non-sorbing and weakly sorbing tracers).

During the latter stages of repository development, a large amount of data from underground measurements will be available. These data may be used both for choosing waste emplacement locations within the repository and for evaluation and refinement of the geologic and hydrogeologic models, especially with regard to the near-field rock. However, uncertainty regarding large-scale variation of hydrologic properties, e.g. within fracture zones, will probably be difficult to resolve by underground measurements, since the (repository) tunnels and shaft(s) will be located such that these features are avoided so far as possible. Possibilities to address such uncertainties should nevertheless be explored and may be possible by conducting underground multi-tracer tests. If conceptual model uncertainty prevails despite gathering all of the data that are reasonably obtainable, then it will be necessary to apply alternative models in the consequence calculations in order to illustrate the importance of remaining conceptual uncertainty.

Figure 2. Example of measures to develop confidence during different phases (preparation*, surface-based investigations' detailed underground characterisation and construction°) of a site investigation programme

Stages of confidence building in site characterisation……
1. Strategy and methods*
  2. Broad picture of heterogeneity types'
    3. Regional scale hydrology and hydrochemistry'
      4. Multiple site-scale models for hydrogeology and transport'
        5. Detailed hydrologic tests in boreholes'
          6. Tracer tests and/or groundwater monitoring in boreholes'
            7. Discrimination of alternative models'
              8. Underground measurements°
                9. Confirmation of site models°
                  10. Near-field rock characterisation°
                    11. Waste emplacement locations°

6. Iterative approach of model development and data acquisition

The development of models, planning of field measurements and derivation of data from laboratory experiments can most effectively be done using an iterative approach. The derivation of e.g. preliminary hydrogeologic models may provide valuable insight in the planning of a site characterisation programme. Early interpretation of site data may help to focus modelling efforts, which may in turn help to direct additional measurements, etc. For instance, during the SITE-94 project it was found that additional boreholes in the Äspö area could probably have provided the most valuable data if they were placed in a more regular pattern. This is because the existing boreholes that were placed to investigate major fracture zones did not provide sufficient information on the low and moderately fractured proportion of the rock. Moreover, additional measurements to better characterise the regional groundwater system would have improved the confidence in the boundary conditions utilised for the site scale hydrogeologic models [9].

Confidence in parameter values used in transport codes can be improved by *in situ* measurements or by laboratory measurements on site-specific materials. For a preliminary performance-assessment, it would probably in many cases be acceptable to use an appropriate "generic" database. As regulatory authorisation is approached, however, confidence can be built by obtaining supporting data through experiments on site-specific material. A good example of such parameter values are K_d-values measured by batch sorption experiments, which may vary by orders of magnitudes depending on either the water chemistry, the characteristics of the solid phase or both [10]. Since the impact of some radionuclides may be small despite such variation, sensitivity studies should play a role in deciding the level of effort to be devoted to each radioelement. In defining the water chemistry relevant for a particular site, one must account for future changes in the groundwater chemistry, e.g. due to climate evolution.

7. Applications of radionuclide transport modelling in PA

A complete PA will be required after different phases of repository development. The required level of confidence in the PA should progressively increase with increasing amount of data and understanding from site characterisation and repository development. A deterministic prediction of the repository evolution over a time scale of several thousand years or more is nevertheless an unreasonable target. The possible repository evolution should rather be investigated by evaluating a number of scenarios quantitatively. These evaluations have a predominantly heuristic value and should be considered as illustrations of possible consequences of the repository evolution.

The choice of scenarios should give a broad picture of the possible detrimental consequences. In order to achieve this, they need to cover all essential aspects of repository safety. An important aid in the scenario development is the compilation of features, events and processes (FEPs) that may be relevant to the performance of the repository system. These FEPs may then be assigned to the different scenarios that are considered. Since the treatment of the individual FEPs does by necessity involve judgements, expert elicitation may be used to resolve different opinions and to ensure that key decisions are documented. The evaluation of scenarios is not only of scientific value but also a tool to communicate the risk associated with a repository. In order to ensure that commonly occurring concerns are appropriately addressed, the selection should be based on dialogue between the implementor, the regulator and other affected parties.

Radionuclide transport modelling should if possible and reasonable be used in the evaluation of a scenario, such that compliance with dose or risk criteria [11] can be addressed. However, since the analysis of some scenarios may be used solely to gain an improved understanding of the repository system and its surroundings, compliance with such criteria is not always relevant. Since the significance of different scenarios may vary, it is convenient to divide them into different categories, e.g. a main scenario, less likely scenarios and scenarios of specialised interest. The main scenario should be based on what is currently perceived as the most probable repository evolution. It should include events that can not be shown to be of low probability, such as glaciation and canister failure. The second category of scenarios should also include less likely events such as reduced barrier function(s) caused by human actions. The third category may cover events that are studied regardless of their likelihood, for instance the redundancy of barrier functions or human intrusion.

For Swedish conditions, the most important single external event that may affect the repository safety is probably the onset of the next glaciation. The next glaciation cycle is expected to last about 100 000 years, which can therefore be regarded as a minimum period to cover for a scenario that includes glaciation. Doses or risk that is estimated for such long periods should be considered as hypothetical and not predictions of health impacts. For even longer time periods, the ability of the different barriers to retain the radionuclides may be a preferred indicator of the repository performance. Qualitative reasoning may complement the calculations, in order to describe the expected performance of a repository. Alternative safety indicators such as comparisons with fluxes and concentrations of natural radionuclides in the environment have been discussed and may be used as a complement, but their usefulness needs to be further evaluated.

The time scale during which there is a need include calculations of radionuclide transport from its source to the biosphere is a matter of judgement. According to recently published PA studies, a reasonable time scale to cover is between 10 000 years and 1 000 000 years [12]. A condition that may be imposed if the shorter range is chosen is that there should be no indication of increased dose or risk beyond the time period that is covered.

8. Stable conditions in geosphere

The geosphere has several other functions in PA apart from its ability to retard radionuclides. First of all, it needs to provide a stable and favourable environment, such that the engineered barriers will function as intended. Secondly, it is a barrier that provides a certain degree of protection from human intrusion and effects of human actions.

Regardless of the level of confidence obtained for the geosphere as a barrier for radionuclide transport, the role of the geosphere in facilitating the integrity of the engineered barriers is likely to be equally or more important. The regulator needs to carefully consider whether the proponent has convincingly demonstrated that stable and favourable conditions are expected to be maintained such that the barriers will function as intended. The bedrock should provide an environment with a limited groundwater flux, a chemical environment that is compatible with assumptions made in the analysis of the engineered barriers, and stable geological conditions to prevent mechanical failure of the barriers. It may be difficult to judge whether a sufficient knowledge of the possible long-term alterations of the site hydrology, geochemistry and geology has been obtained. Nevertheless, the understanding of the previous evolution of the site over a time scale similar to that required in PA is likely to be the best indicator. Investigations of the site paleohydrology as a means of achieving this goal should therefore be given priority in future site characterisation work.

9. Discussion

The regulatory views that are discussed here, must be effectively communicated such that the implementor, and other affected parties have a clear view about what will be required for different stages of repository development. This goal can best be achieved by combining different activities, with a good example being international collaboration through e.g. the NEA. The outcome of independent PA projects conducted by a regulator, such as SITE-94, may also have some usefulness in this respect by providing an example of what components of the PA that will be asked for. In Sweden, the research, development and demonstration programme that is presented by SKB every third year provides one effective means of communication between the implementor and regulator [13]. This programme is required by Swedish law and is reviewed by SKI with input from the Swedish Radiation Protection Institute (SSI) and other organisations with an interest in the issue. Finally, the most basic requirements can probably be most effectively communicated through regulatory criteria and guidelines. These should ensure that the basic requirements on the implementor should be made clear for all parties involved. SSI has recently published regulations concerning the protection of human health and the environment from the final disposal of nuclear wastes [11]. SKI has produced draft regulations concerning the long-term safety of a nuclear waste repository [14]. The current aim is that they should be published and come into force during the year 2000.

10.   References

[1]   INTRAVAL, The international INTRAVAL project, Phase 2, Summary Report, Swedish Nuclear Power Inspectorate, OECD/NEA, Paris, France, 1997.

[2]   INTRACOIN, International Nuclide Transport Code Intercomparison Study, Final report levels 2 and 3, Swedish Nuclear Power Inspectorate, Stockholm, Sweden, 1986.

[3]   HYDROCOIN, The International Hydrocoin Project, Groundwater Hydrology Modelling Strategies for Performance Assessment of Nuclear Waste Disposal, Summary Report, OECD/NEA, Paris, France, 1987.

[4]   SKI Project-90, SKI Technical Report 91:23, The Swedish Nuclear Power Inspectorate, 1991.

[5]   SKI SITE-94, Deep Repository Performance Assessment Project, SKI Report 96:36, The Swedish Nuclear Power Inspectorate, 1991.

[6]   Eisenberg N.A., Lee M.P, Federline M.V., Wingefors S., Andersson J., Norrby S., Sagar B., Wittmeyer G.W., Regulatory perspectives on model validation in high-level radioactive waste management programmes: A joint NRC/SKI White paper, US Nuclear Regulatory Commission, NUREG-1636, 1998.

[7]   INTRAVAL, The International Intraval Project, Developing Groundwater Flow and Transport Models for Radioactive Waste Disposal, Final Results, OECD/NEA, Paris, France, 1996.

[8]   Vieno T., and Nordman H., Safety assessment of spent fuel disposal in Hästholmen, Kivetty, Olkiluoto and Romuvaara TILA-99, POSIVA 99-07, 1999.

[9]   Dverstorp B. and Geier J., Feedback from Performance Assessment to Site Characterisation: The SITE-94 Example. Water-conducting Features in Radionuclide Migration, Workshop Proceedings, Barcelona, Spain 10-12 June 1998, GEOTRAP Project, OECD/NEA, Paris, France, 1998.

[10]   Krupka K.M., Kaplan D.I., Whelan G., Serne R.J., and Mattigod S.V., Understanding Variation in Partition Coefficient, K_d, Values. United States Environmental Protection Agency EPA 402-R-99-004A, 1999.

[11]   SSI FS 1998:1, Statens strålskyddsinstituts föreskrifter om skydd av människors hälsa och miljön vid slutligt omhändertagande av använt kärnbränsle och kärnavfall (in Swedish), 1998.

[12]   McCombie, C., Nuclear waste management world-wide, *Physics Today*, Vol. 50, No. 6, pp. 56-62, 1997.

[13]   SKB, RD&D-Programme 98, Treatment and final disposal of nuclear waste, Programme for research, development and demonstration of encapsulation and geological disposal, SKB, September, 1998.

[14]   Statens kärnkraftinspektions föreskrifter om säkerhet för slutförvaring i berg av kärnavfall efter förslutning av slutförvarsanläggning, Preliminär version av föreskrifter (in Swedish), Swedish Nuclear Power Inspectorate, 1999.

Confidence Building in the Nagra Approach to Geosphere-transport Modelling in Performance Assessment

P.A. Smith
Safety Assessment Management Ltd., United Kingdom

M. Mazurek
University of Bern, Switzerland

A. Gautschi, J.W. Schneider and P. Zuidema
NAGRA, Switzerland

1. Introduction

It is inherent in science that the absolute correctness of a theory can never be proven. The general consensus is that, for a theory to be accepted, it should explain a large number of observations and experiments, and should have the capability to make predictions (or at least bounding estimates) that can be tested. The larger the base of empirical evidence, and the more thorough and wide ranging the tests, the higher the degree of confidence in the theory. These principles also apply to performance-assessment models for radioactive waste disposal. A number of examples will be presented illustrating how the degree of confidence in geosphere-transport modelling, and its conceptual basis, has been raised with time in the Nagra programme. The examples cover:

- confidence in the understanding of the geological setting and its evolution;
- confidence in the understanding of transport-relevant processes; and
- confidence in model abstraction for performance assessment.

Conclusions at the end of this paper are placed in the context of wider aspects of confidence building.

2. Confidence in the understanding of the geological setting and its evolution

Proper understanding of the geological environment and its evolution over time is a necessary pre-requisite for a credible safety case. One of the major challenges in this respect is the ability to bound the system evolution far into the future (e.g. 1 Ma) by a set of scenarios. Such an extrapolation into the future can be constrained by characterising the past system evolution, i.e. by studying the geological history of the system, including paleo-hydrogeology. The advantage of using paleo-hydrogeological arguments lies in the fact that the time scales of observation are comparable or longer than the time scales of the extrapolation (e.g. 1 Ma for a typical performance assessment for high-level waste such as Kristallin-I [1]; see Figure 1).

Figure 1. **The use of paleo-hydrogeology to predict future evolution**

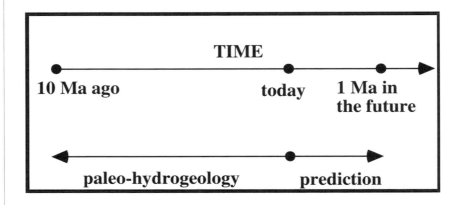

PALEO-HYDROGEOLOGY

Investigation techniques

- Characterisation of the present-day hydraulic regime (spatial distribution of permeability, pressure, hydraulic gradients) and its past evolution

- Study of the chemical and isotopic composition of groundwaters (incl. environmental tracers)

- Study of water/rock interactions recorded by the rock (e.g. veins)

Different methods have been applied in order to study the paleo-hydrogeology of the proposed host formations for possible repositories in Switzerland. Table 1 shows the paleo-hydrogeological implications of geological, hydrogeological and hydrochemical investigations at Wellenberg in Central Switzerland, which has been proposed as a repository site for low- and intermediate-level waste.

Some of the relevant findings include:

- Metamorphic, 20 Ma old fluids trapped in fluid inclusions are chemically and isotopically very similar to the fluids sampled in the central parts of the host-rock body today [2], [3].

- The central parts of the host-rock body are underpressured [4], [5], and the same feature is also found in 20 Ma old fluid inclusions [2].

- The host-rock body is heavily faulted (ca. 5-15 faults per 100 m along vertical hole). Only a small fraction of the faults are, however, hydraulically active in the boreholes [3], and zones with high fault densities do not clearly correlate with higher hydraulic conductivities. It follows that the larger-scale connectivity of the fault system is limited.

- The salinity of porewaters in the central parts of the host-rock body is approximately half that of seawater and is explained by *in situ* mixing of connate water with water derived from metamorphic dehydration reactions [6], [7]. Salinities of pore- and groundwaters in units over- and underlying this zone, as well as salinities in the laterally adjacent limestones, are much lower. It follows that fluid exchange with the surrounding units has been very limited over millions of years. One single pore volume exchange in the host-rock body would have resulted in a substantial decrease of salinity [6].

The overall conclusion from the field-based investigations at Wellenberg is that three different lines of evidence relating to paleo-hydrogeology exist (geology, hydraulics, hydrochemistry) that indicate that permeabilities and fluid fluxes have been very limited over the last 10-20 Ma. The basis to extrapolate these findings into the future along the lines of Figure 1 is the principle of actualism, i.e. the assumption that the same processes that governed fluid evolution in the past will also do so in future. This principle is widely used in Earth sciences, but its validity is not universal. In the specific example of Wellenberg, it needs to be mentioned that continued uplift and decompaction will eventually lead to the exhumation of the repository zone within a time scale shorter than that over which fluid evolution in the past was observed. By the time the repository zone is exhumed, however, the radiological hazard associated with the waste will have declined to insignificant levels.

Table 1. **Use of geological, hydraulic and hydrochemical evidence to constrain the paleo-hydrogeology of the Wellenberg site**

discipline	*observations, conclusions*	*paleo-hydrogeological implications*	*references*
geology	• vein mineralisations indicate very small water/rock ratios throughout geological evolution • vein fluids were buffered locally • almost no evidence of post-Alpine water/rock interaction	closed-system behaviour; large-scale connectivity of existing water-conducting features is limited	[7], [8]
hydraulics	• low transmissivities in central parts of host-rock body • sub-hydrostatic pressures • high hydraulic gradients	small hydraulic conductivity of the host formation over long periods of time	[2], [3], [4], [5], [8], [9], [10]
hydro-chemistry	• central part of host-rock body contains very old waters (several Ma) • no evidence for fluid exchange with adjacent formations or the surface	the central part of the formation is unaffected by waters from the surface, from the base, and from the laterally adjacent limestones	[2], [6], [7], [8], [11], [12]

3. Confidence in the understanding of transport-relevant processes

Transport-relevant processes in the geosphere-transport models employed in Nagra performance assessments include advection, hydrodynamic dispersion, matrix diffusion, sorption and radioactive decay. For some potential host rocks (e.g. the Opalinus Clay), "coupled processes" are also relevant. Confidence is required in the adequacy of the catalogue of relevant processes, in the understanding and modelling of individual processes and in the modelling of combined processes in a transport model. The following example focuses on confidence in the understanding and modelling of the matrix-diffusion process in fractured media.

In fractured media, the modelling of the retardation mechanism of matrix diffusion presupposes the existence of an interconnected porosity in the rock matrix adjacent to the fractures. Many laboratory techniques are available to characterise the geometry and value of matrix porosity. All these methods have one common uncertainty related to the unquantified effects of stress release, drilling and sample preparation. In order to constrain this uncertainty, the "Connected Porosity" (CP) experiment was run in the granitic rocks of the Grimsel Test Site (Switzerland) [13].

The principle of the experiment is shown in Figure 2. A thin borehole was drilled beyond the excavation-disturbed zone of the tunnel. After packing-off, the borehole was filled with an acrylic resin (with viscosity and wetting properties similar to water). By applying a pressure of 10 bar for ca. 1 month, the resin was injected into the rock matrix. After impregnation, a heater was installed in the borehole, in order to polymerise the resin. After several weeks of heating, the injected rock volume was overcored and studied in the laboratory. The main findings include:

- Impregnation of the granitic matrix under *in situ* stresses was successful, and the resin penetrated several cm into the rock, well beyond the zone affected by borehole drilling.

- The architecture of the microfracture network was studied under the microscope and was found to be very similar to that observed in samples that were impregnated in the laboratory. It appears that stress release, and the preparation of samples in the laboratory, may increase pore space by the widening of existing microfractures, but do not create new types of microfractures.

- Microfracture apertures cannot be measured with confidence under the microscope. Qualitatively, however, apertures are generally smaller in samples impregnated *in situ*, when compared to laboratory impregnations. Thus, stress release and sample preparation may result in the widening of existing fractures.

- In order to quantify the water-filled porosity, gravimetric water-loss measurements were performed in the laboratory, using both non-impregnated and *in situ* impregnated samples (all of them water-saturated initially). Non-impregnated samples typically yield water-loss porosities of around 1 vol%, which is consistent with data derived from other methods, e.g. Hg injection [14]. Water loss of *in situ* impregnated samples translates into a porosity of around 0.6 vol%. Because connected *in situ* porosity was filled with resin in these samples, this measurement quantifies the pore space either created or connected by stress release and sample preparation. It appears that more than half of the porosity measured by standard laboratory techniques is an artefact of stress release and sample preparation, whereas 0.4 vol% represents the connected *in situ* porosity. The proportion of artificially-created porosity is probably much lower in higher-porosity crystalline rocks, such as the crystalline basement of Northern Switzerland, where porosity is predominantly in the form of micro-pores caused by hydrothermal alteration of clay minerals.

In line with a large number of natural-analogue studies, the Connected Porosity experiment substantiates the existence of interconnected matrix porosity in crystalline rocks and can also be used to assess the effects of stress release on laboratory measurements.

Figure 2. **The connected porosity (CP) experiment**

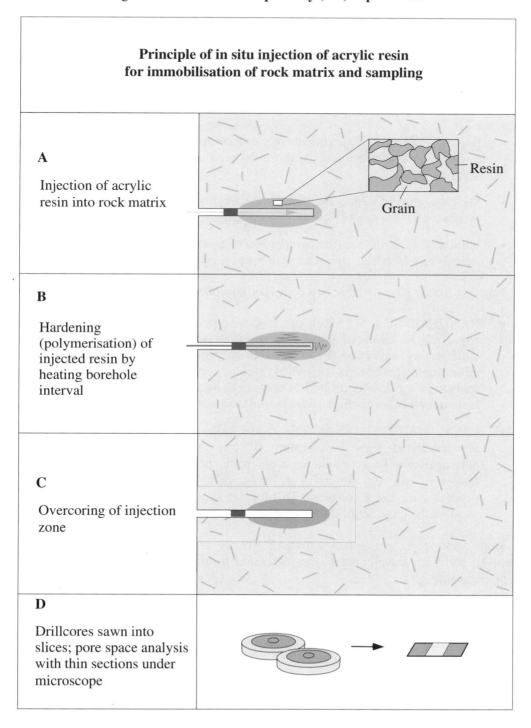

Principle of in situ injection of acrylic resin for immobilisation of rock matrix and sampling

A Injection of acrylic resin into rock matrix

Resin

Grain

B Hardening (polymerisation) of injected resin by heating borehole interval

C Overcoring of injection zone

D Drillcores sawn into slices; pore space analysis with thin sections under microscope

4. Confidence in model abstraction for performance assessment

Geosphere-transport models for Nagra performance assessments are designed to represent the key features and processes of the geosphere-transport barrier in such a way as to take advantage of information in which there is confidence, e.g. from site characterisation and from laboratory studies. These studies can, however, never fully characterise all of the transport-relevant features and processes and, at any stage of site characterisation, there exists uncertainty. The representation of these features and processes in performance assessment models thus generally involves simplification, or model abstraction. The model abstraction process must be shown either not to affect the prediction of geosphere releases, or to yield releases that err on the side of conservatism. It is also desirable to understand the degree to which releases are over-estimated by conservative model abstraction (e.g. in order to gain an indication of the level of conservatism and to focus future R&D and site characterisation). The methods to examine the effects of the abstraction process include:

- field and laboratory experiments (e.g. the comparison of break-though curves from field tracer transport experiments with model predictions, based, for example, on laboratory sorption data);

- logical argument (e.g. it may, in some cases, be argued that omitting certain features of processes from a model *must* be conservative);

- numerical experiments (the comparison of models that include a feature or process with models in which the same feature or process is omitted).

Nagra currently has available two codes for geosphere-transport modelling:

- RANCHMDNL [15], which can model sorption of fracture- and matrix-pore-surfaces expressed in terms of non-linear sorption isotherms, but is relatively simple in its treatment of advection, hydrodynamic dispersion and matrix diffusion. In particular, advection/dispersion is modelled along a single, representative transport path (a transport "leg"), with properties that are constant along its length, and matrix diffusion is modelled as a one-dimensional process in a homogeneous domain.

- ICNIC [16], [17], which can model advection/dispersion through a network of interconnected legs. Furthermore, it can model diffusion into a matrix that is heterogeneous in a two-dimensional plane normal to the flow direction. PICNIC is, however, relatively simple in its treatment of sorption, which is modelled as a linear process using the K_d concept. This limitation is imposed by the numerical method used to solve the large set of transport equations for a network efficiently.

Both codes represent a move towards greater realism with respect to their predecessor (RANCHMD), and their development was motivated by the desire to take advantage of available information regarding:

- the distribution of transmissivities of advective transport paths in potential host rocks;

- the small-scale, heterogeneous structure of the rock adjacent to these paths into which matrix diffusion may occur;

- sorption behaviour, in particular that of caesium, which, in the case of the crystalline basement of Northern Switzerland, is well described by a non-linear, Freundlich sorption isotherm [18].

The use of numerical experiments to evaluate model abstraction effects associated with these aspects is described below.

Distribution of transport path transmissivity

In a fractured host rock, where flow is typically channelled within the fractures, there will be a large number of paths along which radionuclides released from a repository can migrate; much larger, for example, than the number of "legs" that can, in practice, be modelled by a code such as PICNIC. Thus, each leg should be considered, not as a single transport path, but rather as a collection of paths with similar transport properties, the most important nuclide-independent property being the transmissivity. Starting from a distribution of transmissivities measured in the field, the assignment of transmissivities to PICNIC legs (or to a single, RANCHMDNL leg) is discussed in [19]. Key considerations are that:

- the transmissivity of the leg should be chosen conservatively – i.e. it should be biased towards the upper end of the range of transmissivities of the constituent paths;

- the flow through a leg should equal the combined flow through the paths that the leg represents;

- the relatively small number of "fast paths" that dominate the performance of the geosphere-transport barrier should be well represented (using most of the legs) – conversely, only a few legs are required to represent the large number of slow paths, within which most radionuclides decay to insignificant levels.

A numerical experiment can be performed to investigate how many legs, with suitably chosen parameters, are required in order to closely (and conservatively) approximate the behaviour of a much larger number of paths. It is assumed, hypothetically, that:

- transport paths through the geosphere are identical in all aspects other than their transmissivity;

- the transmissivity distribution is log-normal; and

- the release from the near-field is slowly varying, so that radionuclide concentrations in the geosphere approximate to a steady state.

An analytical solution can be obtained for the nuclide-dependent "relative geosphere release". This is defined as the steady-state geosphere release of a nuclide, divided by the steady-state near-field release, and thus gives an indication of the effectiveness of the geosphere in attenuating the near-field release by radioactive decay during transport. Relevant nuclide-dependent parameters are the half life and the K_d. The analytical solution for the relative geosphere release (see Appendix 8 of [1]) can be expressed in terms of the half-life to K_d ratio, and is plotted as the thick solid line in Figure 3. The figure is based on the transmissivity distribution measured in the crystalline basement of Northern Switzerland [20].

Figure 3. Relative geosphere release as function of half-life to K_d ratio and as a function of the number of transport legs considered (see text for further explanation)

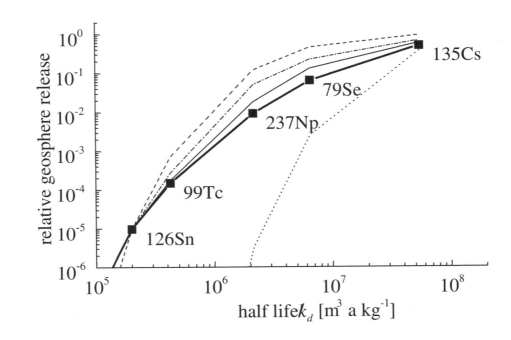

The analytical solution can be viewed as corresponding to a hypothetical, "infinite" number of transport paths, each conveying an infinitesimal amount of water.

In this numerical experiment, the same distribution is modelled by 1 (dashed line), 5 (dot-dash line) and 10 (thin solid line) PICNIC legs, with transmissivities chosen such that the relative geosphere release is conservative (overestimated) over the safety-relevant range of the half-life to K_d ratio. The figure also shows the case where the arithmetic mean transmissivity is assigned to a single leg (dotted line).

The results of this numerical experiment indicate that:

- a single leg, if assigned the arithmetic mean transmissivity, leads to non-conservatively evaluated releases, that can underestimate the solution for the full distribution by several orders of magnitude for shorter-lived and/or higher sorbing nuclides;

but that, if a suitable methodology is adopted for assigning parameters:

- a single leg, if assigned a conservatively chosen transmissivity, can overestimate releases of some nuclides with intermediate half-life to K_d ratios, by about an order of magnitude in this example;

- the solution for the full distribution can be closely and conservatively approximated (to within a factor of about 2) by as few as 10 PICNIC legs.

Small-scale heterogeneity

The channels within a fractured host rock may be embedded in a surrounding, diffusion-accessible matrix that is heterogeneous in its transport properties (e.g. porosity, sorption parameters). For example, as illustrated schematically in Figure 4 (which is based on the conceptual models for water-conducting features, developed from observations of borecores taken from the crystalline basement of Northern Switzerland [1]), fractures may be partly infilled and wallrock immediately adjacent to the fractures may have undergone alteration.

If the matrix is represented using RANCHMDNL, then only the shaded region in the right-hand sketch in Figure 4 is considered to be diffusion accessible (i.e. the altered wallrock above and below the channel, Case A0). The model assumption that the remainder of the matrix is inaccessible, and thus does not contribute to retardation due to matrix diffusion and sorption, is clearly conservative. At a given location along the channel, diffusion into this homogeneous region is a one-dimension process, and is thus amenable to modelling using this code. RANCHMDNL also allows sorption to be represented as a non-linear isotherm, rather than as a conservatively selected K_d. PICNIC, while limited to a K_d representation of sorption, allows credit to be taken for matrix diffusion and sorption throughout the entire matrix, and thus allows more realistic modelling in this respect, and better advantage to be taken of information on small-scale heterogeneity (Cases A1 and B1).

A numerical experiment can be performed to investigate the degree to which the calculated performance of the geosphere barrier is affected by the choice of using either PICNIC or RANCHMDNL. PICNIC calculations use either one or ten non-intersecting legs to model the distribution of path transmissivity, with diffusion into a 2-D, inhomogeneous matrix (Cases A1 and B1, respectively). The RANCHMDNL calculation uses a single leg, with conservatively chosen transmissivity, and with diffusion into a 1-D, homogeneous matrix. The RANCHMDNL calculations are calculated both with, and without, non-linear sorption. Although the full range of safety-relevant nuclides are calculated, the effects of non-linear sorption are considered only for the element caesium, for which adequate sorption data are available to support the use of a Freundlich isotherm.

Results are shown in Figure 5, where both codes are applied to a set of geosphere data based on the characterisation of the crystalline basement of Northern Switzerland. Near-field release and biosphere models and data are from the Kristallin-I safety assessment for a repository for reprocessed, high-level waste [1]. The results indicate a small reduction in the doses calculated by RANCHMDNL, summed over all nuclides, if the non-linear sorption of caesium is modelled (Case A0 linear and non-linear). A more significant reduction in the calculated dose is obtained if PICNIC is used to model 2D matrix diffusion (Case A1), with the maximum reduced by more than an order of magnitude. A further, but smaller, reduction in the calculated dose is obtained if ten PICNIC legs are used (Case B1) in place of one.

Figure 4. **Simplification of the conceptual model of small-scale matrix heterogeneity in order to model the matrix-diffusion process using the codes RANCHMDNL and PICNIC**

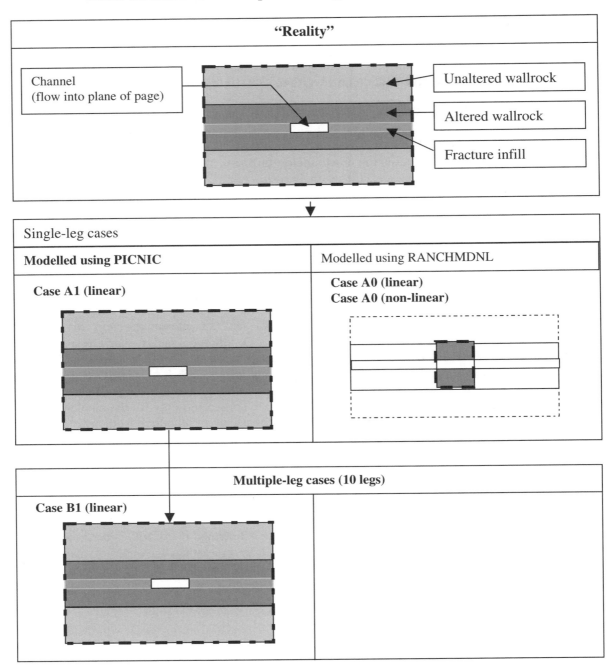

Figure 5. Dose vs. time curves for the four model representations of the geosphere illustrated in Figure 4

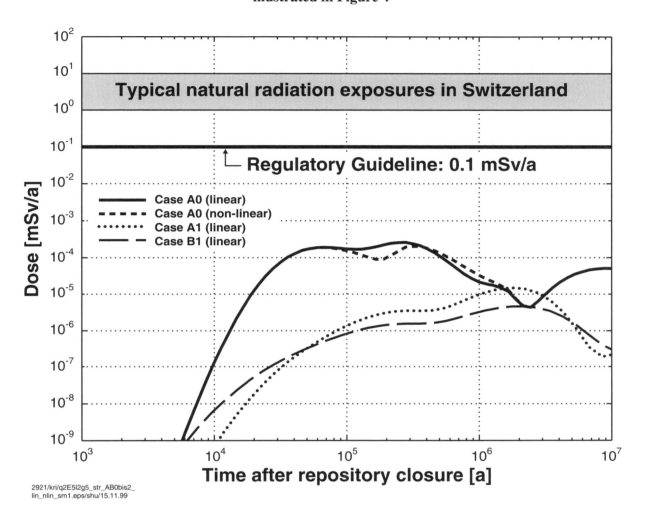

2921/kri/q2E5l2g5_str_AB0bis2_
lin_nlin_sm1.eps/shu/15.11.99

5.   Discussion and conclusions

Performance assessment models of the geosphere (as well as of other aspects of a disposal concept) are based on the available understanding of safety-relevant features and processes. In order to achieve confidence in these models, therefore, confidence is required in the underlying understanding of the geological setting and its evolution, and in the understanding of transport-relevant processes, as well as in model abstraction for performance assessment.

It is now generally accepted that a high degree of system understanding, well beyond the specific needs of direct input to performance-assessment models, is required for a proposed disposal site. Not only is such understanding beneficial to confidence building in the scientific community, but it can also be used to constrain the spectrum of model calculations. At the Wellenberg site, for example, the age structure of groundwaters provided an independent check of calculated travel times. It was shown that initial transport model calculations used substantially underestimated travel times.

The consistency of relevant paleo-hydrogeological conclusions based on geological, hydraulic and hydrochemical evidence provides a tool to assess future evolution that is largely independent of groundwater flow model predictions of travel times.

The identification of all processes that are relevant for transport of contaminants through the geosphere is a pre-requisite for the design of appropriate conceptual models that underlie performance-assessment calculations. In line with other evidence, the "Connected Porosity" experiment at the Grimsel Test Site demonstrated that an interconnected pore space exists in unperturbed granitic rock matrix under *in situ* conditions, and thus confirmed that matrix diffusion is a process that needs to be considered in contaminant transport calculations. The experiment also provided information relevant to the extrapolation of laboratory measurements to *in situ* conditions for unaltered wallrock.

Confidence in model abstraction for performance assessment can be based on field and laboratory experiments, but also on logical argument and numerical experimentation. The latter has a role, not only in demonstrating that the treatment of a feature or process is conservative, but also the degree to which a conservative treatment affects model results. If there is sufficient information to support a more realistic treatment of a feature or process, and if it is also shown (e.g. by numerical experiment) that a more realistic treatment will significantly improve the calculated performance, this can provide a strong motivation for further model development.

The arguments illustrated by these examples do not necessarily provide appropriate means for building confidence outside the technical radioactive waste management community. In order to build confidence in a broader sense, additional arguments can be employed, which may include simplified or less quantitative arguments, such as the "age" and origin of groundwaters (e.g. old sea water, glacial water), or dry tunnel sections in underground excavations.

6.   References

[1]   NAGRA (1994): Kristallin-I Safety Assessment Report; Nagra Technical Report Series, NTB 93-22, Nagra, Wettingen, Switzerland.

[2]   DIAMOND L.W. (1998): Underpressured Paleofluids and Future Fluid Flow in a Planned Radwaste Host Rock; Proc. 9th Int. Symp. Water-Rock Interaction, pp. 769-772.

[3]   MAZUREK M., LANYON G.W., VOMVORIS S. & GAUTSCHI A. (1998): Derivation and Application of a Geologic Dataset for Flow Modelling by Discrete Fracture Networks in Low-permeability Argillaceous Rocks; J. Contam. Hydrol., 35, pp. 1-17.

[4]   VINARD P., BLÜMLING P., MCCORD J.P. & ARISTONENAS G. (1993) Evaluation of Hydraulic Underpressures at Wellenberg, Switzerland; Internat. J. Rock Mech. Mining Sci., 30, pp. 1143-1150.

[5]   VINARD P., VOMVORIS S. & MARSCHALL P. (1998) A Comprehensive Hydrodynamic Modelling Approach for the Hydrogeological Site Characterization and for Deriving Input to Performance Assessment; Mat. Res. Soc. Symp. Proc. 506, pp. 857-864.

[6]   PEARSON F.J., WABER H.N., & SCHOLTIS A. (1998): Modelling the Chemical Evolution of Porewater in the Palfris Marl, Wellenberg, Central Switzerland; Mat. Res. Soc. Symp. 506, pp. 789-796.

[7]   MAZUREK, M. (1999): Evolution of Gas and Aqueous Fluid in Low-permeability Shales during Uplift and Exhumation of the Central Swiss Alps; Applied Geochemistry (in press).

[8]   NAGRA (1997): Geosynthese Wellenberg 1996 - Ergebnisse der Untersuchungsphasen I und II; Nagra Technical Report Series, NTB 96-01, Nagra, Wettingen, Switzerland.

[9]   LÖW S., GUYONNET D., LAVANCHY J.M., VOBORNY O. & VINARD P. (1994): From Field Measurements to Effective Hydraulic Properties of a Fractured Aquitard – a Case Study; in Transport and Reactive Processes in Aquifers (eds. T. H. Dracos and F. Stauffer), pp. 191-196, Balkema.

[10]   MARSCHALL P., VOMVORIS S., JAQUET O., LANYON G.W. & VINARD P. (1998): The Wellenberg K-model: A Geostatistical Description of Hydraulic Conductivity Distribution in the Host Rock for Site Characterization and Performance Assessment Purposes; Mat. Res. Soc. Symp. 506, pp. 749-756.

[11]   PEARSON F. J. & SCHOLTIS A. (1995): Controls on the Chemistry of Porewater in a Marl of very low Permeability; Proc. 8th Int. Symp. Water-Rock Interaction, pp. 35-38.

[12]   SCHOLTIS A., PEARSON F.J., LOOSLI H.H., EICHINGER L., WABER H.N. & LEHMANN B.E. (1996): Integration of Environmental Isotopes, Hydrochemical and Mineralogical Data to Characterize Groundwaters from a Potential Repository Site in Central Switzerland; in Isotopes in Water Resources Management (Int. Atomic Energy Agency), pp. 263-280.

[13]   FRIEG, B., ALEXANDER, W.R., DOLLINGER, H., BÜHLER, C., HAAG, P., MÖRI, A. & OTA, K. (1998): In Situ Resin Impregnation for Investigating Radionuclide Retardation in Fractured Repository Host Rocks; J. Contam. Hydrol. 35, pp. 115-130.

[14]   BOSSART, P. & MAZUREK, M. (1991): Grimsel Test Site: Structural Geology and Water Flow-paths in the Migration Shear-zone; Nagra Technical Report Series, NTB 91-12, Nagra, Wettingen, Switzerland.

[15]   JAKOB, A., HADERMANN, J. & ROESEL, F. (1989): Radionuclide Chain Transport with Matrix Diffusion and Non-linear Sorption; PSI Report No. 54, PSI, Würenlingen and Villigen, Switzerland and Nagra Technical Report Series, NTB 90-13, Nagra, Wettingen, Switzerland.

[16]   BARTEN, W. & ROBINSON, P.C. (1996): PICNIC: a Code to Model Migration of Radionuclides in Fracture Network Systems with Surrounding Rock Matrix; in Hydroinformatics '96, A. Müller (ed.), pp. 541-548, A. A. Balkema, Rotterdam.

[17]   BARTEN, W., ROBINSON, P.C. & SCHNEIDER, J.W. (1998): PICNIC-II – A Code to Simulate Contaminant Transport in Fracture Networks with Heterogeneous Rock Matrices; in Advances in Hydro-science and engineering, Vol. III, proceedings of the 3rd International Conference on Hydroscience and Engineering, K.P. Holz *et al.* (eds.), p. 152, Cottbus, Berlin, Germany.

[18]   STENHOUSE, M. (1995): Sorption databases for Crystalline, Marl and Bentonite for Performance Assessment; Nagra Technical Report Series, NTB 93-06, Nagra, Wettingen, Switzerland.

[19]   SCHNEIDER, J.W., ZUIDEMA, P., SMITH, P.A., GRIBI, P., HUGI, M. & NIEMEYER, M. (1998): Novel and Practicable Approach to Modelling Radionuclide Transport through Heterogeneous Geological Media; Scientific Basis for Nuclear Waste Management XXI, Davos, Materials Research Society, Mat. Res. Proc. Vol. 506, 821-828.

[20]   THURY, M., GAUTSCHI, A., MAZUREK, M., MÜLLER, W.H., NAEF, H., PEARSON, F.J., VOMVORIS, S. & WILSON, W. (1994): Geology and Hydrogeology of the Crystalline Basement of Northern Switzerland – Synthesis of Regional Investigations 1981-1993 within the Nagra Radioactive Waste Disposal Programme; Nagra Technical Report Series, NTB 93-01, Nagra, Wettingen, Switzerland.

The Posiva/VTT Approach to Simplification of Geosphere-transport Models, and the Role and Assessment of Conservatism

T. Vieno, H. Nordman, A. Poteri
VTT Energy, Finland

A. Hautojärvi, J. Vira
Posiva Oy, Finland

Abstract

Posiva's geosphere-transport model is similiar to those used in most recent safety assessments on crystalline rock. The model simulates transport of radionuclides along fractures of a streamtube and matrix diffusion is the only phenomenon causing dispersion and retardation. The main parameters are the ratio of the flow wetted surface and flow rate (WL/Q) along the migration route and diffusivities and K_d values in the rock matrix. The results show that the geosphere is an efficient transport barrier for strongly and moderately sorbing nuclides. For non-sorbing and weakly sorbing nuclides the geosphere appears, however, to be a poor transport barrier. The short transit times used in PA calculations seem to be in some contradiction with the interpreted geochemical ages of deep groundwaters. The dilemma is related to that the PA approach and the "physical" transport models, where radionuclides are modelled as particles "swimming" in the groundwater stream and interacting with the surrounding medium by a means of a simple K_d reaction, do not aim to simulate realistically mass transport in the geosphere. In reality the transport of trace elements is retarded by geochemical reactions which we are not able to incorporate in a defensible and realistic way in PA transport models. The aims and tools of PA lead to that the modelling of flow and transport is heavily biased towards overestimating mass transport from the repository into the biosphere.

1. Modelling of geosphere-transport: advection in fractures + matrix diffusion

The conceptual model of geosphere-transport of radionuclides applied in the TVO-92, TILA-96 and TILA-99 safety assessments (Vieno *et al.* 1992, Vieno & Nordman 1996, 1999) on spent fuel disposal at the Finnish candidate sites in the crystalline bedrock is simple: The model considers advection along fractures and matrix diffusion is the only phenomenon assumed to cause dispersion and retardation. Diffusion and sorption in the rock matrix are modelled according to the porous medium and linear equilibrium sorption approaches. The essential characteristics of a transport route can thus be presented by means of a single parameter (u) taking into account the distribution of groundwater flow in the fracture system and the effects of matrix diffusion. For the case where a constant water phase concentration C_o of a stable species is prevailing at the inlet of the fracture beginning from t = 0, the water phase concentration at the distance of L in the fracture is:

$$C_f(L,t) = C_o \, erfc\left[u\,t^{-1/2}\right] \qquad (1)$$

where u is a parameter describing the transport properties of the migration route for the given species

$$u = \left[\varepsilon_p D_e R_p\right]^{1/2} \cdot \frac{W\,L}{Q} \qquad (2)$$

where

ε_p	is the porosity of the rock matrix (-),
D_e	is the effective diffusion coefficient from the fracture into the rock matrix (m²/s),
R_p	is the retardation factor of the species in the rock matrix (-),
W	is the width of the flow channel (m),
L	is the transport distance (m),
Q	is the flow rate in the channel (m³/yr),
t	is the time (yr).

The transport of the dissolved species is retarded and dispersed when u is increased (Figure 1). In the case of a delta pulse input of a stable species, the peak output (m_{max}) and its occurrence time (t_{max}) are (Poteri & Laitinen 1999):

$$t_{max} = t_w + \frac{2\,u^2}{3} \qquad\qquad m_{max} \approx \frac{0.23}{u^2} \qquad (3)$$

where t_w is the groundwater transit time (yr). The effective value of the u-parameter of a transport route is the integral of the local values along the route.

The first factor of the u-parameter is related to diffusion and sorption in the rock matrix. For a non-sorbing species $R_p = 1$, and for a moderately or strongly sorbing species $R_p \approx K_d \rho_s / \varepsilon_p$. The u-parameter is thus reduced to:

$$u_{non\text{-}sorbing} = \left[\varepsilon_p D_e\right]^{1/2} \cdot \frac{W\,L}{Q} \qquad (4)$$

$$u_{sorbing} \approx \left[D_e K_d \rho_s\right]^{1/2} \cdot \frac{W\,L}{Q} \qquad (5)$$

where K_d is the volume-based distribution coefficient (m³/kg) and ρ_s is the density of the rock (kg/m³).

Figure 1.   **The output flux from a delta pulse input of a stable species with different values of the u-parameter (year$^{1/2}$). To incorporate the groundwater transit time t$_w$, the pulses should be shifted to the right with t$_w$. (Vieno *et al.* 1992)**

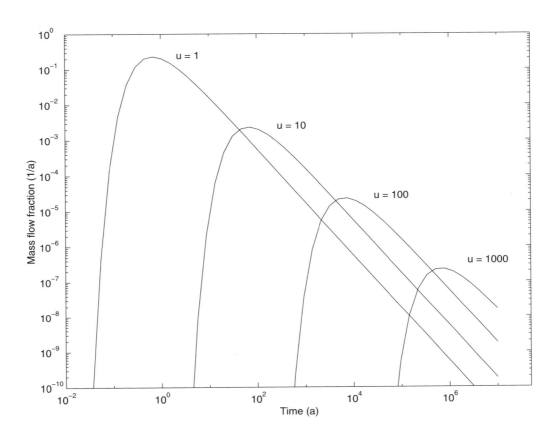

The second factor (WL/Q) of the u-parameter can be expressed also in terms of the groundwater transit time and the volume aperture of the flow channel:

$$\frac{W\,L}{Q} = \frac{W\,L}{v\,W\,2b_V} = \frac{t_w}{2\,b_V} \qquad (6)$$

where v is the advection velocity of the groundwater in the channel (m/yr) and $2b_V$ is the volume aperture (m) of the channel. WL represents the "flow wetted surface" (Elert 1997), i.e. the rock surface area in direct contact with the groundwater flowing in the fractures. In SITE-94 (SKI 1996) the transport characteristics are represented by the so-called F-ratio, $F = a_r/q = 2\,WL/Q$ where q is the Darcy velocity (m/yr) and a_r is the wetted fracture surface area per volume of rock (m^2/m^3).

Sorption on fracture fillings and diffusion into stagnant pools in the fractures are conservatively omitted. Furthermore, no route dispersion term is used in the reference scenarios. All our PA calculations are deterministic and we think that it would not be appropriate to "dilute" the selected WL/Q values by adding a more or less arbitrary dispersion coefficient to represent variability. The effects of a dispersion coefficient have, however, been scrutinised in the sensitivity analyses.

All geosphere-transport analyses of radionuclides are performed for a single canister and employ deterministic, time-invariant parameter values for a single-leg migration route. The numerical analyses are performed with a dual-porosity model (finite element code FTRANS), which can take into account decay chains and the heterogeneity of the rock matrix adjacent to the water-conducting fracture. The code employs the conventional parameters for advection along the fracture: length of the migration path, velocity of the water flowing in the fracture and aperture of the fracture. These input parameters are fixed in such a way that we get (according to Equation 6) the chosen value for our primary input parameter WL/Q.

2. Problems with data: WL/Q (flow wetted surface/flow rate) and matrix diffusion parameters

The characteristics of the radionuclide transport routes for PA calculations are selected on the basis of the regional-to-site scale groundwater flow analyses at the four Finnish candidate sites (Kattilakoski & Koskinen 1999, Kattilakoski & Mészáros 1999, Löfman 1999a-b) and of the site-to-canister scale fracture network modelling based flow and transport analyses (Poteri & Laitinen 1999). The approaches of flow and transport analyses are illustrated in Figures 2 to 4.

Figure 2. **Analyses of groundwater flow and solute transport in the regional-to-site and site-to-canister scales (Poteri & Laitinen 1999)**

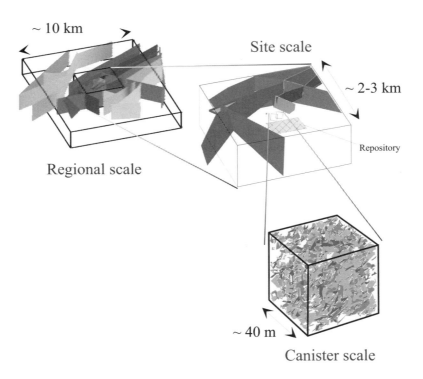

Figure 3. **An example of a realisation of a transport route in the near-field (Poteri & Laitinen 1999)**

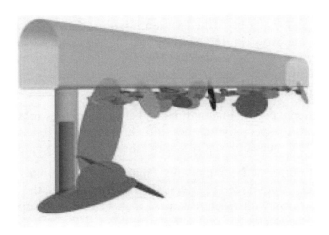

Figure 4.   **Estimation of the ratio of the flow wetted fracture surface area (WL) and the flow rate (Q) along a flowpath in the fracture network (Poteri & Laitinen 1999)**

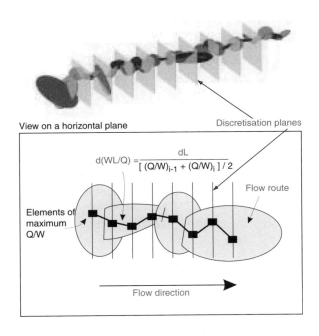

The problems for deriving WL/Q values for PA transport analysis are related to:

- Identifying of relevant transport routes, especially in the vicinity of the repository. The transport route from a canister deposition hole in a KBS-3 type repository may in principle consist of four types of legs: i) sparsely-fractured rock between fracture zones, ii) fracture zones, iii) excavation damaged zone (EDZ) around repository excavations, and iv) tunnels and shafts backfilled with a mixture of crushed rock and bentonite. The distribution of the flow in the vicinity of the repository depends on i) hydraulic conductivities of rock, EDZ and tunnel backfill, ii) orientation of tunnels relative to the hydraulic gradient, iii) fracture zones intersecting tunnels, iv) long-term properties of backfill and sealings.

- There are significant conceptual and parameter uncertainties related to the modelling of flow and transport especially in the EDZ and fracture zones. In highly-conductive fracture zones flow rates are usually so high that their WL/Q values are very low.

- In the analyses performed for the four Finnish candidate sites the differences in the estimated WL/Q values are more related to the details of the preliminary repository layouts (intersections with fracture zones) than to the local topography and site scale properties of the bedrock.

Taking the above uncertainties into consideration, WL/Q values ranging from $5 \cdot 10^3$ to $5 \cdot 10^4$ yr/m were selected for the reference scenarios of TILA-99. Taking into account that WL/Q = $t_w/2b_V$ (Equation 6), the WL/Q of $5 \cdot 10^4$ yr/m can be illustrated as follows: With a volume aperture of $2b_V = 5 \cdot 10^{-4}$ m, it corresponds to a groundwater transit time $t_w = 25$ years. Secondly, assuming a flow channel with a width of W = 0.1 metres and a length of L = 600 metres, the flow rate Q is 1.2 litres/yr. WL/Q of $5 \cdot 10^3$ yr/m corresponds, respectively, to a transit time of 2.5 years and a flow rate of 12 litres/yr in the channel.

A noteworthy feature is that according to the groundwater flow simulations there are no large differences in the flow rates and WL/Q values between the islands of Hästholmen and Olkiluoto, where groundwater is brackish or saline at the disposal depth, and the inland sites Kivetty and Romuvaara, where groundwater is non-saline. At the coastal sites flow is brought about by the displacement of saline groundwater by fresh water as a consequence of the ongoing postglacial land uplift and at Hästholmen also by topographical gradients from the mainland and a neighbouring island (Löfman 1999a-b).

A comparison of the WL/Q values in reference scenarios (or equivalents) of recent safety assessments on crystalline rock (Table 1) shows that the typical values are within the range covered by the reference scenarios of TILA-99. H3 (PNC 1992) is an interesting outlier in this comparison as the WL/Q value is clearly higher than in other assessments although the length of the migration path is no more than 10 metres. This is caused by the high fracture surface area (the frequency of open, parallel plate fractures is 10 m^{-1}) used in this particular case of H3. Therefore, the 10 metres of rock is a very efficient transport barrier, which can been seen also from the respective results of H3.

Table 1. WL/Q (= ½ x **flow wetted surface / flow rate) of the migration routes in the reference scenarios of safety assessments on crystalline rock**

Assessment	Source of data	Length of migration path (m)	WL/Q (yr/m)
Project 90 (SKI 1991)	Scenario VG0	200	10 000
SITE-94 (SKI 1996)	Zero Variant	500	33 000
SKB 91 (SKB 1992)	1)	1 000	50 000
SR 95 (SKB 1995)	1)	1 000	10 000
Kristallin-I (NAGRA 1994)	Table 5.3.4	200	20 000
H3 (PNC 1992)	Page 4-144	10	600 000
GRS/SPA (Baudoin *et al.* 1999)	Ref. scenario	200	60 000
TVO-92	Ref. scenario	400	10 000
TILA-96	Ref. scenario	600	20 000
TILA-99	Ref. scenarios	600	5 000 – 50 000

1) In SKB 91 and SR 95, groundwater flow modelling is stochastic. The mean value of WL/Q has been estimated on the basis of following data: specific surface area 0.1 m^2/m^3, flow porosity 10^{-4}, mean groundwater travel time 100 years in SKB 91 and 20 years in SR 95, length of flow path about 1 000 m.

Defining of representative matrix diffusion and sorption parameters for a transport route with a typical length of a few hundred metres is difficult, too. There are naturally large variations in the values measured in small-scale rock samples, but no significant, systematic differences in the porosities and diffusivities have been observed between the rocks at the four candidate sites. As concerns sorption, there are differences, for example, in the cation exchange capacity and contents of iron oxides between the various rock types found at the different sites. However, for most elements, the water chemistry is of much greater importance for sorption than the rock composition. Therefore, the sorption and matrix diffusion parameters are related to the geochemical conditions (reducing vs. oxidising, saline vs. non-saline) independent of the specific type of the crystalline rock.

In the rock matrix the diffusivities of radionuclides are lowered by a factor of ten in the unaltered rock beyond one centimetre from the fracture. In the numerical model, the maximum penetration depth of matrix diffusion is limited to 10 cm due to computational reasons. Anionic exclusion is taken into account by reducing diffusivities and effective porosities for anions. The phenomenon is more pronounced in non-saline water than in saline water. Accordingly, the effective porosities and diffusion coefficients in the rock matrix depend on 1) the distance from the water-conducting fracture (0-1 cm vs. 1-10 cm), 2) speciation (non-anions vs. anions) and 3) salinity of groundwater (non-saline vs. saline).

3. Problems with results: too short transit times?

The results show that the geosphere is an efficient transport barrier for strongly and moderately sorbing nuclides ($K_d \geq 0.1$ m^3/kg). For non-sorbing and weakly sorbing nuclides (anions in all conditions, cations in the saline water) the geosphere appears, however, to be a poor transport barrier. As the groundwater transit time is a few decades at most, these nuclides migrate rapidly through the geosphere in the PA model. It should, however, be noted that the scenarios assuming high flow of saline groundwater present also some artefacts of the steady state modelling because in reality the saline groundwater would gradually be displaced by fresh water with the more favourable retardation conditions.

The high flow rates and short transit times used in the PA transport analysis seem to be in some contradiction with the interpreted geochemical ages of deep groundwaters, which are typically some thousands of years or even much higher in the case of saline waters (Pitkänen et al. 1996, 1998a-b, Luukkonen et al. 1999). However it should be noted that in PA the focus of interest is on the small fraction of the groundwater which may move rapidly in the fastest flow channels in the fractured bedrock. Furthermore, geochemical studies indicate that deep groundwaters are usually mixtures containing several components with different origins and ages.

4.   "Overly simplified and too conservative" says the regulator

In their review of TILA-96, the safety authority STUK noted that overly simplified and conservative modelling undermines the role of the geosphere as a transport barrier. They further suggested that use of time-dependent parameters should be studied to take into consideration larger uncertainties in the long term. Scientists and consultants have also proposed to increase realism of PA modelling by means of:

- time-dependent modelling to take into account the effects of the operation and resaturation phases, heat generation, land uplift and evolving flow conditions at the coastal sites, and glaciation;
- 3-D geosphere-transport modelling covering the whole repository and site to study variability;
- 2-D or 3-D modelling of the transport within individual flow channels
- space/time-dependent parameter values;
- coupled modelling of groundwater flow and transport, geochemical evolution and radionuclide transport.

In our opinion, one should be very careful when adding more sophisticated features in PA models. There is a danger that a too complex model turns into a black box which generates artefacts and obscurity instead of realism. PA assessors should always be aware of the underlying key phenomena and parameters and the related uncertainties of the model. The uncertainties can best be reduced by means of reliable measurements of these key quantities. In Posiva's programme, an instrument (flowmeter) has been developed and used to measure in situ flow rates in fractures under the natural gradient.

There are some phenomena and effects, for example related to the flow field inside and/or around the repository and long-term changes, which cannot be conclusively resolved by means of measurements or modelling exercises. PA needs to employ conservative assumptions to cover the uncertainties. In the steady state modelling approach, conservative assumptions and parameter values are used to cover long-term uncertainties as well as uncertainties related to the early phase phenomena which are not explicitly modelled.

We should also recognise that the PA approach and the "physical" transport models, where radionuclides are modelled as particles "swimming" in the groundwater stream and interacting with the surrounding medium by a means of a simple K_d reaction, do not aim to simulate realistically mass transport in the geosphere. In reality the behaviour of trace elements is affected by geochemical reactions, such as redox and microbial reactions, coprecipitation, flocculation, mineralisation, and interactions with fracture fillings and colloids which we are not able to incorporate in a defensible and realistic way in PA transport models. The aims and tools of PA lead to that the modelling of flow and transport is heavily biased towards overestimating mass transport from the repository into the biosphere.

5. References

Baudoin, P., Gay, D., Certes, C., Serres, C., Alonso, J., Lührmann, L., Martens, K.-H., Dodd, D., Marivoet, J. & Vieno, T. 1999. Spent fuel disposal performance assessment (SPA project) – Final Report. Luxembourg, European Commission. (To be published in the EUR series).

Elert, M. 1997. Retention mechanisms and the flow wetted surface – implications for safety analysis. SKB Technical Report 97-01.

Kattilakoski, E. & Koskinen, L. 1999. Regional-to-site scale groundwater flow in Romuvaara. POSIVA 99-14.

Kattilakoski, E. & Mészáros, F. 1999. Regional-to-site scale groundwater flow in Kivetty. POSIVA 99-13.

Löfman, J. 1999a. Site scale groundwater flow in Olkiluoto. POSIVA 99-03.

Löfman, J. 1999b. Site scale groundwater flow in Hästholmen. POSIVA 99-12.

Luukkonen, A., Pitkänen, P., Ruotsalainen, P., Leino-Forsman, H. & Snellman, M., 1999. Hydrogeochemical conditions at the Hästholmen site. POSIVA 99-26.

NAGRA 1994. Kristallin-I safety assessment report. NAGRA Technical Report 93-22.

Pitkänen, P., Snellman, M., Vuorinen, U. & Leino-Forsman, H. 1996. Geochemical modelling study on the age and evolution of the groundwater at the Romuvaara site. POSIVA 96-06.

Pitkänen, P., Luukkonen, A., Ruotsalainen, P., Leino-Forsman, H. & Vuorinen, U. 1998a. Geochemical modelling of groundwater evolution and residence time at the Kivetty site. POSIVA 98-07.

Pitkänen, P., Luukkonen, A., Ruotsalainen, P., Leino-Forsman, H. & Vuorinen, U. 1998b. Geochemical modelling of groundwater evolution and residence time at the Olkiluoto site. POSIVA 98-10.

PNC 1992. Research and development on geological disposal of high-level radioactive waste – First progress report. PNC Report TN1410 93-059.

Poteri, A. & Laitinen, M. 1999. Site-to-canister scale flow and transport at the Hästholmen, Kivetty, Olkiluoto and Romuvaara sites. POSIVA 99-15.

SKB 1992. SKB 91 – Final disposal of spent nuclear fuel – Importance of the bedrock for safety. SKB Technical Report 92-20.

SKB 1995. SR 95 – Template for safety reports with descriptive example. SKB Technical Report 96-05.

SKI 1991. SKI Project-90. SKI Technical Report 91:23.

SKI 1996. SKI SITE-94 Deep repository performance assessment project. SKI Technical Report 96:36.

Vieno, T., Hautojärvi, A., Koskinen, L. & Nordman, H. 1992. TVO-92 safety analysis of spent fuel disposal. Nuclear Waste Commission of Finnish Power Companies, Report YJT-92-33E.

Vieno, T. & Nordman, H. 1996. Interim report on safety assessment of spent fuel disposal TILA-96. POSIVA 96-17.

Vieno, T. & Nordman, H. 1999. Safety assessment of spent fuel disposal in Hästholmen, Kivetty, Olkiluoto and Romuvaara TILA-99. POSIVA 99-07.

The Transport Model for the Safety Case of the Konrad Repository and Supporting Investigations

E. Fein, R. Storck
Gesellschaft für Anlagen- und Reaktorsicherheit (GRS) mbH, Germany

H. Klinge, K. Schelkes
Federal Institute for Geosciences and Natural Resources (BGR), Germany

J. Wollrath
Federal Office for Radiation Protection (BfS), Germany

Abstract

Licensing of a nuclear waste repository requires the demonstration of the post-closure safety of the facility. This involves the extensive use of models to assess the potential evolution of the system with respect to behaviour of the waste, release of radionuclides, and their migration through the geosphere and uptake by man.

The Konrad mine, a disused iron ore mine, which is located in Northern Germany, has been selected to serve as a repository for radioactive waste with negligible heat generation. The depths of the mine workings are 800-1 300 m below the surface level. The host rock is an oolithic iron ore deposit which is covered by low-permeability Lower Cretaceous claystones with thicknesses up to 500 m. It has a south to north extension of nearly 60 km and the width varies between 4 km to 15 km. A transport of radionuclides from the repository into the biosphere will mainly be possible by moving groundwater.

It will be shown how compliance with the objectives of the Radiological Protection Ordinance was demonstrated. Special emphasis will be laid on the discussion of supporting investigations as measures for the increase of confidence in the modelling of the radionuclide migration.

1. Introduction

The planned Konrad repository will be constructed in a disused iron ore mine which is located in the northern part of Germany (Figure 1). The iron ore deposit was already discovered in 1933 during oil prospection activities. From 1957 till 1962 the two shafts of the Konrad mine were sunk to a depth of 1 232 m and 999 m, respectively. In 1961 the extraction of iron ore started. On 30 September 1976 the production was discontinued for economic reasons. Since then no exploitation activities have been carried out [6], [8].

Figure 1. **Location of the Konrad site**

For the research and development work necessary to prove the suitability of the Konrad mine to serve as a repository for radioactive waste the mine was kept open on behalf of the German Federal Government who is responsible for the final disposal of radioactive waste in Germany. Until 1990 site characterisation and performance assessment were carried out. On 31 August 1982 an application for the initiation of a plan-approval procedure was submitted by the PTB (Federal Institute for Physics and Metrology) which is the legal predecessor of the BfS (Federal Office for Radiation Protection) to the licensing authority NMU (Ministry for the Environment of Lower Saxony). In 1990 the licensing documents, the so-called "Plan" Konrad was submitted. As a part of the licensing procedure public involvement is required. For that reason in 1991 the public display of the "Plan" and in 1992-1993 the public hearing, which lasted 75 days, was performed. Finally in September 1997 the licensing authority submitted a draft for a license.

In Germany the disposal of radioactive waste in an underground repository is, in particular, governed by the following regulations [3]:

- Atomic Energy Act (Atomgesetz – AtG).

- Radiological Protection Ordinance (Strahlenschutzverordnung – StrlSchV).

- Safety Criteria for the Final Disposal of Radioactive Waste in a Mine.
  (Sicherheitskriterien für die Endlagerung radioaktiver Abfälle in einem Bergwerk).

- Federal Mining Act (Bundesberggesetz – BBergG).

The fundamental objective of radioactive waste disposal in repositories is to ensure that waste is disposed of in such a way that human health and the environment are protected now and in future without imposing undue burdens to future generations. That means that radioactive waste must be managed in such a way that the predicted impact on future generations will not be greater than the relevant levels of impact that are accepted today.

The following issue of the long-term safety criteria is considered to be the most important one: In the post-closure phase, the radionuclides which might reach the biosphere via the groundwater as a result of transport processes must not lead to individual annual doses which exceed the limit of 0.3 mSv/y specified in paragraph 45 of the Radiological Protection Ordinance. For the assessment of long-term safety the safety criteria have to be met taking into account all relevant long-term safety indicators.

It is the aim of this paper to present the transport model for the safety case of the planned Konrad repository and the supporting investigations which were performed with the intention to build confidence in the used models. The paper is organised as follows. In section 0, a short site characterisation is given. The hydrogeology and the flow modelling and its results is described in section 0 as well as additional measures to enhance confidence in the flow model. In section 0, the radionuclide transport and its results together with supporting investigations are presented. Finally, in section 0 the results are summarised.

2.   Site Characterisation

The Konrad study area is part of the North German Basin containing thick accumulations of Mesozoic marine sediments and of about 1 000 m of Permian and Triassic evaporites deposited during periods of restricted marine circulations. The evaporite units have formed diapirs that have migrated towards the land surface. The Triassic and Jurassic strata have a total thickness of 2 000 m. They are built up to the major part of marl and claystone aquitards and of sandstone and carbonate aquifers. The base of the system is formed by nearly impermeable Triassic and Permian salt deposits. During the Upper Jurassic the sedimentary iron ores of the Konrad mine have been accumulated in rim synclines of salt diapirs. The iron ore is a medium to low permeable fissured rock with a generally low matrix permeability. The Jurassic iron ores are covered by low permeable Lower Cretaceous claystones up to 500 m thick. Where salt diapirs have reached the land surface, they have sometimes displaced overlying sedimentary strata and generate topographic heights where meteoric fluids can infiltrate into displaced Mesozoic strata.

Figure 2. **Salt concentration of water samples as a function of depth at the Konrad mine**

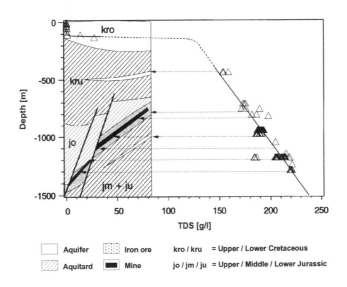

As common in northern Germany, fresh water is underlain by salt water at the Konrad site. There is an upper regime of low mineralised freshwater present down to a depth of 150-250 m. It is controlled by convection and recharge of meteoric water. Below this upper regime there is, as investigations in the Konrad mine have shown, a linear concentration increase from 160 g/l TDS (Total Dissolved Solids) at a depth of 450 m to 220 g/l at depth of 1 300 m (Figure 2). The trend of the salt concentration regression line indicates that saturated brines can be expected at a depth of approximately 2 300 m below the surface at the top of the first bedded evaporite layer. The distribution of the salt concentration generally indicates a horizontal salinity stratification [5].

3. Flow Modelling

3.1 Reference Model

Due to its importance for the safety of the repository main efforts were laid on the assessment of the groundwater flow field. In the case of the planned Konrad repository two- and three-dimensional groundwater flow modelling was performed to achieve information on the flow field. A three-dimensional model with extensions of about 50 km in SN-, about 15 km in EW-direction, and about 2 km in depth was set up. The model boundaries were adopted to the geological and hydrogeological situation: in the South the water divide of the Salzgitter ridge, in the North the flat of the Aller river, and in the East and in the West salt structures running from South to North. The groundwater level close to the surface was used as the upper boundary. The very low permeable clay stones and salt deposits of the Middle Muschelkalk serve as lower boundary. They are overlain by a 2 000 m-thick sequence of sediments ranging from the permeable layer of the Upper Muschelkalk (mo) to the Upper Cretaceous (Figure 3). The sediments consist of very low permeable marls and claystones (e.g. (k) Upper Triassic, (jm) Middle Triassic) with intercalations of permeable sandstone and limestone layers (e.g. Rhät sandstone (ko), Cornbrash sandstone (c). Oolithic iron ores are present in the Oxfordian (ox) of the low permeable Upper Jurassic (jo). These sediments are covered by very low permeable Lower Cretaceous claystones (kru) up to 500 m thick. On top of these claystones locally high permeable sediments of the Upper Cretaceous (kro) or the Quaternary (q) and the low permeable Emscherian marl (krcc-sa) can be found.

Hydraulic conductivities and effective porosities were known from core and *in situ* measurements or could be deduced by analogy from literature data. Since the derived data were not unique band widths were assigned to each formation by expert judgement. The decision which values of conductivity or porosity were used in the flow model was made by taking into account the principle of conservativity. If sensitivity analyses indicated a strong dependency of the travel time on the considered parameter the value resulting in the shortest travel time was used. The hydraulic conductivities of the various hydrogeological units of the so-called "Layer"-Model are given in Table 1.

It was assumed that the aquifer system is confined and that the porous medium approach is valid, i.e. joint aquifers were approximated as equivalent-porous media. Furthermore the water movement was regarded as to be stationary, and Darcy's law was assumed to be applicable. Because of limited computer resources and unavailibity of appropriate numerical tools the three-dimensional groundwater model calculations were performed under freshwater conditions although the investigations at the Konrad site have shown that the deep aquifers contain highly saline connate waters. Hence the influences of salinity on the water density and the associated ground water field were neglected.

Figure 3. **Hydrogeological SW-NE profile of the Konrad site**

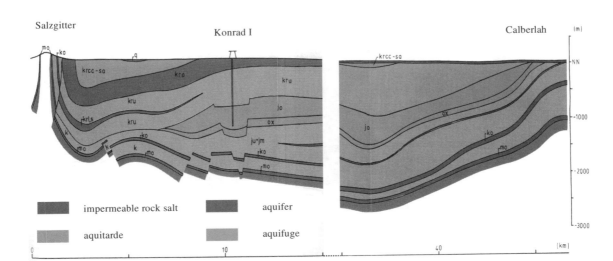

The groundwater flow and the particle tracking simulations were performed with the finite-difference code SWIFT [6]. Due to limitations in computer resources the complex hydrogeological structure of the Konrad site was only coarsely modelled. A rectangular finite difference grid was used with 30 blocks in SN-, 15 blocks in EW-, and 23 blocks in vertical direction. The block lengths in horizontal directions varied between 0.75 km and 2.5 km. In the vertical direction the length was 100 m.

To determine the travel pathways from the repository to the surface, particles were started from each grid block which represented the repository [1]. The analysis of the large number of streamlines allowed a classification of all pathways into three representative flow pathways which differed significantly with respect to path lengths and travel times (Figure 4). If a hydraulic conductivity of 10^{-10} m/s is used for the Lower Cretaceous claystones the first identified pathway led the direct way upwards to the surface passing the Lower Cretaceous formation. The travel time was about 430 000 y. However, if a lower hydraulic conductivity of 10^{-12} m/s is used other identified pathways led through the Oxfordian or through the Oxfordian and the Cornbrash sandstone. The travel times were about 300 000 y for the migration through the Oxfordian and about 1 100 000 y to pass the Oxfordian and Cornbrash sandstone [14].

Figure 4. **Representative travel pathways**

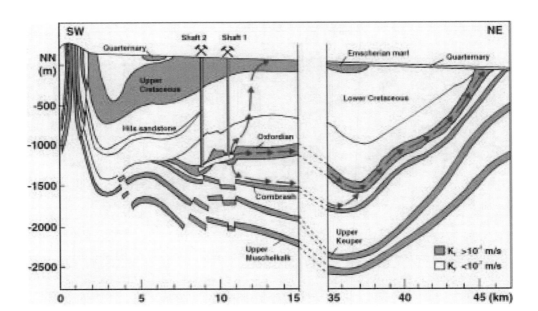

3.2 Confidence Building Measures

3.2.1 Alternative Models and Uncertainty Analyses

Several additional measures were performed to achieve confidence into the results of the models. Due to uncertainties in the measured or deduced hydrogeological parameters like conductivities etc, band widths were assigned to each parameter. These band widths were based on the variances of measurements and on expert judgements. The parameter values which were actually used in simulations were fixed by uncertainty and sensitivty analyses. Always that value that causes the shortest travel time was used. In general, in any case of doubt the worst-case values were considered.

To include not only the parameter uncertainty but model uncertainty as well, an alternative hydrogeological model, the so-called "Fracture"-Model was developed. In the "Layer"-Model one constant parameter set, which was conservatively determined, was assumed to be valid for a whole hydrogeological unit. In contrast to that more realistic values for the hydraulic conductivities, which could be locally changed for fracture zones, were taken into account in the "Fracture"-Model. In Figure 5 the considered fracture zones and in 3 the corresponding factors to modify the conductivities are shown.

Figure 5. **The fracture zones considered in the "Fracture"-Model**

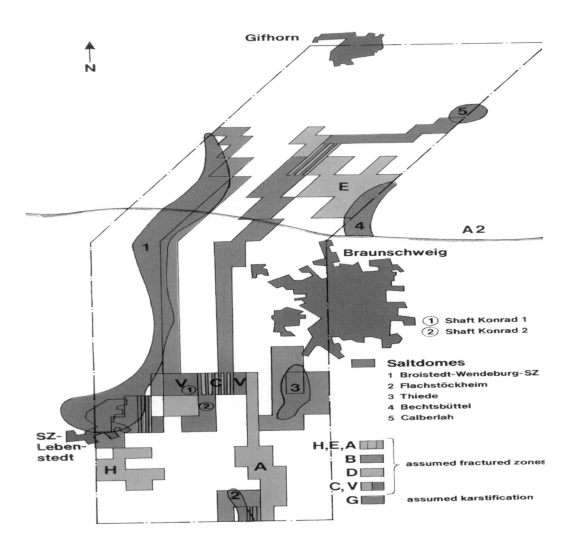

Furthermore various numerical codes were used by the applicant as well as by the reviewing experts. In addition to the above mentioned computer programme SWIFT on the applicant's side the finite element code FEM301 was used [7]. The applicant's safety assessments were not only seriously scrutinised but also recalculated with the applicant's codes by the experts as well as calculated with a different code. In the case of the Konrad licensing procedure the expert's code was NAMMU [3]. Although two different hydrogeological models and various numerical models were used by different modelling groups the results showed that in any case the shortest water travel time from the repository to the surface is longer than 300 000 y.

The shortest travel times of all models considered were found with the SWIFT simulator for the case of the "Layer"-Model.

Table 1. **Hydraulic conductivities and porosities of the "Layer"- and the "Fracture"-Model**

Layer	"Layer"-Model k_f [m/s]	n_{eff} [%]	"Fracture"-Model[*] k_f [m/s]
Quarternary, Tertiary	10^{-5}	25	10^{-5}
Emscherian marl	10^{-8}	20	10^{-8}
Planer limestone Upper Cretaceous	10^{-7}	5	S 10^{-7} N 10^{-6}
Albian			10^{-12}
Lower Cretaceous	$10^{-10}/10^{-12}$	10	10^{-11}
Hils sandstone	10^{-5}	25	10^{-5}
Kimmeridgian, Portlandian, Wealden	10^{-8}	10	$5 \cdot 10^{-9}$
Oxfordian	10^{-7}	2	10^{-8}
Cornbrash sandstone	10^{-6}	5	S 10^{-7} N 10^{-6}
Rhät sandstone	10^{-6}	20	10^{-7}
Upper Trissic	10^{-10}	10	H 10^{-10} V 10^{-12}
Upper Muschelkalk	10^{-6}	2	10^{-7}

To demonstrate the influence of the parameter band widths on the results uncertainty analyses were carried out by using the software system SUSA [3]. The comparison of the distribution of shortest travel times and their mean value with the shortest travel time of the deterministic calculation reflected the conservative character of the deterministic results.

Table 2: **"Fracture"-Model – changing of conductivities**

Fractured Zones	Factors for Changing the Conductivities				Band Widths
	reference case	variation 1	variation 2	variation 3	
A	50	50	50	50	0.1 to 75
B	25	25	25	25	0.1 to 50
C	10	10	20	.5	0.5 to 20
D	10	10	10	10	0.5 to 15
E	50	50	50	50	0.1 to 75
G	10	10	10	10	1 to 50
H	25	25	25	25	0.5 to 25
V	60	500	500	500	0.5 to 500

[*]S: South, N: North, H: horizontal, V: vertical

3.2.2. *Hydrochemical Data and Density-dependent Models*

Hydrochemical and isotopic data helped in building a transient conceptual model of the evolution of the flow system. In general, models can be derived from hydrochemical data, which explain the composition and the different sources of the constituents in the groundwater, allowing comparison with the results of different flow models and permit allowing conclusions to be drawn on the important processes involved. This procedure was very successful for the Konrad area, where the deep layered aquifer system is not very much influenced by near surface time-dependent processes [11].

In the Konrad area, all brines are Na-Ca-Cl type waters. Generally, all constituents positively correlate with depth. There is a linear vertical increase of the Ca/Cl, Mg/Cl and Br/Cl ratios and a corresponding decrease of the Na/Cl ratio. As an example, Figure 6 shows the Na/Cl and Br/Cl ratios versus TDI (Total Dissolved Ions).

The linear vertical increase of TDS and of relevant ion ratios indicates that the mine waters have been formed from a two component mixture of a saturated residual brine expelled from deep lying bedded evaporites and a more dilute Na/Cl brine. The isotope signature defines the mine waters as formation waters which have undergone isotope exchange between water and rock minerals. They originally may have been old meteoric waters, which infiltrated during a warmer climatic period, or a mixture of meteoric water and sea water, and/or of a residual brine [5].

Figure 6. **Chemical composition of Konrad mine waters**

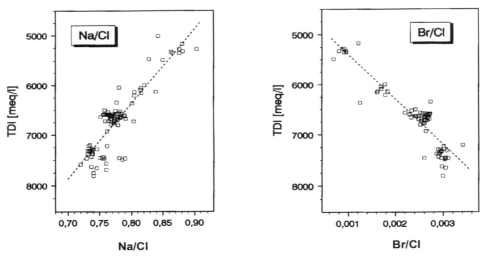

From the hydrochemical findings it can be concluded that the observed situation is in general a result of predominantly diffusive processes, which does not correspond to the results of conventional flow models assuming constant freshwater density. In these models advection is the dominant process, which would not explain the results of the hydrochemistry. The groundwater models have to take into account the salinity-dependent density of the groundwater.

To identify the effects of the variable groundwater density on the flow field and to prove the conservativity of using constant density models time-dependent two-dimensional model studies with variable density were carried out. These salt-/freshwater model calculations were only possible on simplified two dimensional cross sections especially due to limitations in computer resources. In a first step, only essential elements of the hydraulics of the deep groundwater were taken into account (size and shape of the study area, location of the deep aquifers). The results provided a basis for a second step in which calculations were undertaken along a more realistic cross-section. This cross section (46.5 km long and 2 000 m deep) is shown in. It was assumed that the water at the base of the model is in contact with salt formations. All hydraulical parameters, as well as a freshwater boundary condition at the top and the salt boundary condition at the bottom of the models, were derived from the hydrogeological situation. The calculations started from physically reasonable initial conditions. They were stopped at a situation where a dynamic equilibrium was nearly reached.

Figure 7. **Cross sectional model of the Konrad area with permeability distribtion and isolines of the solute mass fraction after about four million years of model time**

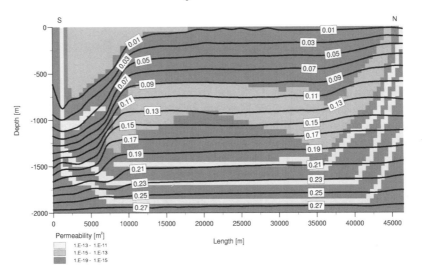

A comparison of the results with the corresponding freshwater system reveals a major difference in the deep groundwater hydraulics. The freshwater system is dominated by groundwater advection. In the state of dynamic equilibrium, the saltwater system, apart from a fresh water lens below the recharge area, exhibits a linear increase in density and salinity with depth (7). It is characterised by quite different groundwater flow patterns and a reduction of the interstitial velocity by one to two orders of magnitude relative to the constant density model. A linear vertical increase in density indicates that, in the saltwater system, diffusion is the dominant mechanism of solute transport from the bedded salt to the surface. The calculated salinity gradient and density stratification are in reasonable agreement with the empirical data (Figure 2) [133], [12].

From the hydrochemical interpretation and the variable density groundwater models it could be concluded that the observed linear increase of porewater salinity and the mixing of fluids is in general a result of predominantly diffusive processes and only to a minor extend a result of vertical saline water convection [5]. Moreover, an important conclusion for the licensing procedure was that the results of constant density models which were used in the safety analyses for the Konrad mine were shown to be conservative with regard to groundwater flow path ways and travel times [11].

Additional benefit from the use of the hydrochemical information was the determination of flow and transport parameters, often directly combined with the model calculations. The comparison between measured and calculated density data for example showed the importance and correctness of the implementation of hydrogeological structures and parameters in the flow and transport models. It also revealed the limits for hydrologic parameters like dispersivity and diffusivity. A combination of hydrochemical data interpretation and density dependent flow modelling allowed conclusions to be drawn for the important transport processes at the Konrad site.

4. Transport Modelling

4.1. Modelling of Transport Processes

Since the shortest travel times of all models considered were determined with the SWIFT simulator for the case of the "Layer"-Model also the transport modelling was based on these pathways using the computer code SWIFT. The travel times were about 430 000 y and 300 000 y for the pathways through the Lower Cretaceous and through Oxfordian, respectively (Figure 4). The pathway through Oxfordian and Cornbrash with a traveltime of about 1 100 000 y was no longer taken into consideration.

Using the above defined shortest pathways and the lengths of the covered way through the various formations one-dimensional transport modelling was performed taking into account advection, diffusion and dispersion as well as radioactive decay chains and sorption. The error which results from the one-dimensional approach was partially compensated by introducing dilution factors. The activity release from the repository was determined for a proposed activity inventory and acted as source term for the transport modelling. This release was modelled to be dependent on the fill-up of the repository with water after the closure of the repository, the mobilisation of the radionuclides, and the water flow through the repository.

The advective part of the radionuclide transport was already fixed by the flow velocity of the water on its considered pathway. This velocity depended on the formation which was passed. The dispersive transport was modelled using the common Scheidegger approach. Since site specific dispersion lengths were not available they were chosen to be proportional to the length of the pathway in the most important formation. But for conservativity reasons even smaller values were used. They were assumed to be constant for each representative pathway, i.e. dispersion length were not site specific. The dispersion length were selected to be 200 m and 30 m for the pathway through the Oxfordian and the Lower Cretaceous, respectively. In order to determine molecular diffusion laboratory experiments with rock samples of the Konrad site were performed. But since the available data base was insufficient the diffusion was fixed by expert judgement to be 10^{-11} m^2/s for each rock type and each radionuclide.

Table 3. **Transport-related and sorption-related porosities, respectively**

Formation	$n_{transport}$ [%]	$n_{sorption}$ [%]
Upper Cretaceous	5	15
Lower Cretaceous	10	20
Kimmeridgian	10	10
Oxfordian	2	20
Middle Jurassic	10	20
Cornbrash sandstone	5	10

The retention of radionuclides was considered in the transport model by taking into account precipitation and sorption. Since the concentration of radionuclides in the geosphere was expected to be small solubility limits were not considered in the geosphere. In contrast to that in the model of the transport through the repository workings solubility limits were used. Sorption was modelled by use of the K_d-concept. Using rock samples of the most important aquifers and water samples from the Quaternary, Upper Cretaceous, Hils, and Oxfordian laboratory experiments were conducted to evaluate the K_d-value. These experiments generally showed that desorption is slower than adsorption.

Hence the K_d-values were fixed on the basis of the measured adsorption coefficients. With that, element specific K_d-values could be used for the various hydrogeological formations. In cases where no experimental data were available K_d-values were fixed by analogy to chemical homologues. The pore volumes which are available for the movement of water and for the movement of radionuclides are different. This is especially valid for joint aquifers where the porosities were assumed to be smaller in order to achieve larger flow velocities. To be conservative in the sorption modelling larger porosities were used (Table 3), i.e. the retardation factor became smaller. Additionally the influence of com-plexing agents, like EDTA, and of the radionuclide concentration was assessed by experts judgement.

Figure 8. **Determination of dilution factors**

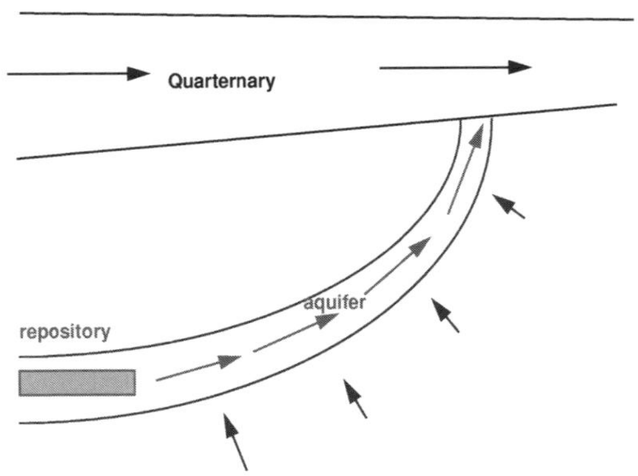

In Figure 8 the situation is shown where dilution is important. In cases where the fluid flow field indicates an essential inflow into an aquifer without altering the absolute value of the velocity, i.e. after entering the aquifer approximately the same amount of water flows out, the radionuclides are diluted. By calculating the water balance in the flow model dilution factors were evaluated.

The above described processes were included in the transport model and the simulations were performed using the computer code SWIFT. The results were temporal functions of the surfacial concentrations of the radionuclides considered. These concentrations were taken as input for the biosphere model where the annual radiation exposure to man was determined (Figure 9). In no case the annual radiation exposure to man did exceed the limiting value of 0.3 mSv/y [2].

Figure 9. **Annual radiation exposure to man**

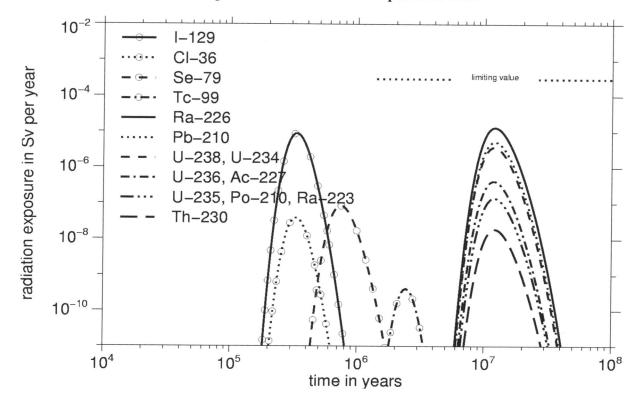

4.2   *Confidence Building Measures*

Besides the conceptual transport model with its conservative parameters described above the same procedures as in the flow modelling were used to gain confidence that the results of the transport modelling overestimate the radionuclide concentration occurring in the biosphere. The applicant's safety assessments were seriously scrutinised and recalculated with the applicant's code SWIFT and the reviewing expert's code MARNIE by the experts [3], too. So the radionuclide transport through the geosphere was modelled by different modelling groups using different computer codes. Since the results showed fairly good agreement the confidence in the modelling was enhanced.

Site specific safety indicators based on nature observation have been extensively used for confidence building in performance assessment for the planned Konrad repository [9], [10]. Table 4 gives some major examples of such safety indicators and site specific phenomena used at the Konrad site [4]. With the Konrad repository they were of utmost value in the licensing procedure to give reasonable assurance that the performance assessment based on nature observation cannot lead to erroneous predictions. The time frame for performance assessment are about 1% of those considered in the geological past of the site.

Table 4. Safety indicators and site specific phenomena used for performance assessment at the Konrad site

Safety Indicator	Phenomenon	Contribution to Performance Assessment
Time scale of prcesses, flux through barriers	Age and salt concentration of natural deep groundwater	Age of groundwater at least 10^7 years, possibly $1.5 \cdot 10^8$ years; the latter is the age of the geological formation. This indicates groundwater movement of less than about 1 cm in 10^3 years or even stagnating groundwater. Proof of a conservative ground-water model (groundwater travel time and dilution).
Flux through barriers	Self sealing effects of clay barriers (drillings, fracture zones)	Proof of adequate modelling of radionuclide transport
Radiotoxicity	Natural radiotoxicity of host rock	Radiotoxicity of the waste decreases to natural levels after about $3 \cdot 10^5$ years. Proof that longer consequences reflect natural risks.

5. Summary

The safety assessment which was performed for the planned Konrad repository was shortly explained. Emphasis was laid on the modelling of the radionuclide transport through the geosphere. For the proposed nuclear inventory the long-term safety of the repository was demonstrated showing the compliance with the objectives of the Radiological Protection Ordinance.

Furthermore, it was shown how confidence in the long-term safety was built by additional measures which were often the outcome of several serious scrutinies of the models and of peer reviews by the applicant, the licensing authority and its experts. The process of confidence building was supported and even stimulated by the intensive interaction between the applicant and the licensing authority and its experts. The main contribution for the gains in confidence were beneath the interpretation of hydrochemical and other site specific data the use of several differing hydrogeological and numerical constant- and variable-density models which were used by the different modelling groups. The performing of supporting investigations and the use of worst-case parameters in cases of doubt, i.e. to assess long-term safety in a conservative manner, substantially contributed to the enhancement of confidence.

The safety assessment procedure and its results were confirmed by safety indicators.

6. Acknowledgments

This work was carried out in the site investigation programme, which is being conducted by the Federal Office for Radiation Protection (BfS) and scientificly directed by the Federal Institute of Geosciences and Natural Resources (BGR), as well as in research projects. The long term safety assessment studies were performed on behalf of Federal Office for Radiation Protection by the German Centre for Environmental and Health Research (GSF). Many scientists have worked in these investigations. The authors like to thank the official organisations for their support in this research and our colleagues in the different organisations for the possibility to use their data and results for this publication.

7. References

[1]   Arens, G., Fein, E.: A Comparison of Results of Groundwater Flow Modelling for Two Conceptual Hydrogeological Models for the Konrad Site. Proceedings GEOVAL 90, 1991.

[2]   Arens, G.: Regulating Long-Term Safety. The Konrad Safety Case. Proceedings of an NEA international workshop "Regulating the Long-Term Safety of Radioactive Waste Disposal", Córdoba, Spain, 20-23 January 1997.

[3]   Baltes, B., Bogorinski, P., Larue, J.: Review the long-term safety of the Konrad nuclear waste repository. Proceedings of an NEA international workshop "Regulating the Long-Term Safety of Radioactive Waste Disposal", Córdoba, Spain, 20-23 January 1997.

[4]   Confidence in the Long-Term safery of Deep Geological Repositories – Its Development and Communication. OECD/NEA, Paris, 1999.

[5]   Klinge, H., Vogel, P., Schelkes, K.: Chemical Composition and Origin of Saline Formation Waters from the Konrad Mine, Germany. In: Water-Rock Interaction, Kharaka & Maest (eds), Balkema, Rotterdam, 1992.

[6]   Plan. Endlager für radioaktive Abfälle. Kurzfassung. Stand: 9/86 in der Fassung 4/90. Schachtanlage Konrad, Salzgitter. Bundesamt für Strahlenschutz, 1990.

[7]   Rivera, A., Johns, R.T., Schindler, M., Löw, S., Resele, G.: Modelling of Strongly Coupled Groundwater Brine Flow and Transport at the Konrad Radioactive Waste Site in Germany. In "Calibration and Reliabity in Groundwater Modelling", Volume of Poster Papers, ModelCare 96, Golden, Colerado, USA, 1990.

[8]   Röthemeyer, H.: Endlagerung radioaktiver Abfälle. VCH Verlagsgesellschaft mbH, Weinheim, 1991.

[9]   Röthemeyer, H.: Langzeitsicherheit von Endlagern. Sicherheitsindikatoren und natürliche Analoga. Atomwirtschaft, Februar 1994.

[10]   Safety indicators in different time frames for the safety assessment of underground radioactive waste repositories. IAEA-TECDOC-767, International Atomic Energy Agency, Vienna, Austria, 1994.

[11]   Schelkes, K., Klinge, H., Vogel, P., Wollrath, J.: Aspects of the Use and Importance of Hydrochemical Data for Groundwater Flow Modelling at Radioactive Waste Disposal Sites in Germany. In: "Use of Hydrogeochemical Information in Testing Groundwater Flow Models", OECD/NEA Workshop, Borgholm, Sweden 1-3 September 1997, Paris, 1999.

[12]   Schelkes, K., Vogel, P., Klinge, H., Knoop, R.-M.: Modelling of Variable-Density Groundwater Flow with Respect to Planned Radioactive Waste Disposal Sites in West Germany – Validation Activities and First Results. In: "Validation of Geosphere-transport Models (Geoval-1990)", NEA/SKI Symposium, Stockholm. May 1990, Paris, 1991.

[13]   Vogel, P., Schelkes, K., Klinge, H., Geissler, N.: Analysis of Density-Dependent Deep Groundwater Movement in Northern Germany Influenced by High Salinity. In "Calibration and Reliability in Groundwater Modelling", Volume of Poster Papers, ModelCare 90, The Hague, The Netherlands, 1990.

[14]   Wollrath, J.: Prediction of Contaminant Migration around Radioactive Waste Repositories in Salt Formations in Germany. In proceedings of 5th International Conference on Radioactive Waste Management and Environmental Remediation (ICEM '95), Berlin, September 3-8, 1995.

Regulatory Experience with Confidence-building in Geosphere-transport Models

N.A. Eisenberg[1], R.B. Codell[1], L.S. Hamdan[1], T.J. Nicholson[1], B.Sagar[2]
[1]U.S. Nuclear Regulatory Commission, United States

[2]Center for Nuclear Waste Regulatory Analyses, United States

1.   Regulatory Perspective

The regulatory imperative is to make decisions. Regulatory agencies, such as the United States Nuclear Regulatory Commission (NRC) are routinely asked to decide such questions as whether a proposed radioactive waste disposal facility should be licensed, or whether a regulated site, appropriately remediated, should be released for other uses. The licensee or applicant has the burden, not only to provide information to support regulatory decisions, but also to propose the means of dealing with residual uncertainties in the information provided. In some cases, the information provided is sufficiently definitive that a positive or negative response to the proposed action clearly emerges. More frequently the information provided has significant uncertainties. In that case NRC needs to decide whether the uncertainties are sufficiently constrained, i.e. that the confidence is high enough, such that a decision can be made. For the NRC, the criterion for sufficient confidence is "reasonable assurance" that the proposed action is safe. The regulator usually chooses one of the following: (1) the proposed action is unacceptable, based on current information; (2) the proposed action is acceptable based on the information provided; and (3) the proposed action may be acceptable, but insufficient information has been provided to provide confidence in the decision and more information is needed.

The goal of the regulator is protection of public health, public safety, and the environment. This is achieved by promulgation of protective regulations and standards and by evaluation of specific actions against those standards.[1] These evaluations routinely involve the implicit and explicit consideration of uncertainty; but such considerations tend to be focused and limited. The ultimate goal of scientific investigation is to pursue all uncertainties and seek to answer virtually all questions; the regulatory process, on the other hand, is a decision-making process with significant time and budget constraints that adopts an applied science approach, focusing on uncertainties central to the decision to be made. Frequently methods that can be demonstrated to be bounding or conservative yield results suitable for decision making, thereby bypassing the need for more precise methods and detailed scientific analyses.

---

1. In the United States regulations may be promulgated by one Federal regulatory agency (e.g. the U.S. Environmental Protection Agency [EPA]), and implemented by another, (e.g. the NRC).

To illustrate how a limited evaluation of uncertainty in geosphere-transport is feasible, consider the case where the regulatory measure of interest is individual dose and the dose pathway is groundwater uptake through well pumping. For these conditions only the concentration of a contaminant at the wellhead is of interest; hence details of the contaminant migration in the aquifer (fast flow paths, lenses of adsorptive minerals) will only be important to the extent that they affect this concentration. Another aspect of regulatory decision-making is that uncertainty frequently needs only to be bounded in one direction. For example, with a dose limit as the regulatory objective, one need only show that the anticipated dose will be below the limit; one need not refine the estimate of dose to some absolute accuracy. These regulatory expectations can have a profound effect on the choice and application of methods for treating uncertainty.

2.   Types of Uncertainty

For purposes of regulatory analysis, geosphere-transport models are generally incorporated into a performance assessment, which is a type of systematic safety analysis used to quantify the impacts of a waste facility; both the magnitude and likelihood of these impacts are considered in the performance assessment. Performance assessments must address a variety of uncertainties that may include: (1) parameter, (2) model (including conceptual model uncertainty) (3) future states, (4) exposure scenario, and (5) programmatic factors (e.g. QA). Generally these categories are useful for describing the origin or manifestation of the uncertainty. However, the classification of uncertainties is not always simple and these categories are not necessarily mutually exclusive. For example, hydrologic parameters such as hydraulic conductivity, porosity, and distribution coefficient are derived from analyses of field and laboratory data and then used for input into geosphere-transport models. However, these parameter values are estimated by interpreting data according to a model of the physicochemical processes occurring in the characterized material. Thus, parameter uncertainty may also contain a large element of conceptual model uncertainty.

3.   Methods for Describing, and Propagating, Evaluating Uncertainties

The science of hydrology has made substantial strides in understanding and evaluating uncertainties related to groundwater flow and contaminant transport. Many methods have been developed to treat uncertainties in hydrologic and transport parameter values that arise from the use of limited data sets to describe complex processes occurring over large distances, e.g. Kriging and conditional simulations. Recent work has also explored the definition and evaluation of uncertainties in conceptual models relating to the nature of the processes and hydrologic features occurring at a particular site. Another uncertainty to receive attention recently is the uncertainty introduced when moving from a detailed hydrologic model to an abstracted model of flow and transport for use in a performance assessment. Recent research has made some progress in defining and evaluating these uncertainties. A wide variety of methods are available for propagating uncertainties, mostly for those related to parameters in models. Methods for propagating uncertainties related to conceptual models are less advanced, even though these uncertainties may have a more pronounced effect on the predicted performance than those related to parameters.

Specifically, methods for describing spatial variability and uncertainty in hydrologic parameters include: (1) simple statistics, (2) Kriging, (3) Bayesian updating, and (4) inverse methods. To evaluate uncertainties, the modeling needs to reflect the uncertainties present using an appropriate method. For example, Monte Carlo analysis may be performed to investigate the effect of uncertainties in model inputs on the output of the model. Conditional simulations, where flow and transport solutions are constrained by data at measured locations, may be used to explore the impact on model output of uncertainties at locations not measured. The effects of alternative conceptual models may be explored by simulating each alternative conceptual model and noting the effect on the simulated output. If alternative conceptual models can be parameterized, then this type of uncertainty may be propagated using those techniques developed for parameter uncertainty. Various methods are used to evaluate the significance of uncertainties. Typically the result of propagating uncertainties is a distribution of performance, e.g. a distribution of dose or a distribution of concentration. One may compare the simulated performance of the system, i.e. the distribution of performance, to the performance limit. Under circumstances where all results, which reflect the significant uncertainties, fall within the limit of acceptable performance, the proposed action is likely to be acceptable, even though the simulated performance may be highly uncertain. If the performance distributions, based on propagating input parameter uncertainties, are narrow, then input parameter uncertainties are not significant. Another means of evaluating uncertainties is to compare the distribution of results to field measurements. However, in those cases where the comparison of the model output, e.g. dose, is not readily compared to field measurements, intermediate outputs of the model, e.g. concentration, may be compared instead, in order to evaluate uncertainties. Finally formal quantitative methods for sensitivity, uncertainty, and importance analysis may be used to help evaluate the significance of various types of uncertainty.

4. Regulatory experiences

The NRC staff: (1) reviews license termination requests that involve analyses to show compliance with the NRC rule for license termination; (2) reviews remediation of mill tailing sites; (3) evaluates prelicensing analyses associated with the proposed Yucca Mountain repository for HLW; and (4) provides assistance to the Agreement States regulating low-level waste facilities. Groundwater flow and transport may be significant issues under certain conditions: (1) radioactive contaminants are widely dispersed in the soil and, in some cases, the groundwater at the site, or (2) radioactive waste has been buried or is proposed for burial on site. In some cases the hazard potential from such sites is small; hence, for some cases, it is possible to make regulatory decisions based on limited data. However prudence dictates consideration of uncertainties in hydrologic parameters and consideration of alternative conceptual models, in some cases. As examples of regulatory experiences in building confidence in geosphere-transport models, the following discussions of regulatory experiences in the uranium recovery programme and the HLW programme are provided.

4.1. Regulatory Experiences in the Uranium Recovery Programme

In the uranium recovery (UR) programme, the NRC staff reviews applications: for new facilities, for amendment of existing licenses, and for site reclamation and termination of licensed operations. The reviews are conducted for three types of facilities: inactive and active conventional processing and byproduct materials (tailings) disposal sites, and *in situ* leach (ISL) facilities. Inactive sites are licensed to the U.S. Department of Energy (DOE); active sites and ISL facilities are licensed to private companies.

4.1.1. Performance Measures for UR Facilities

Long-term performance measures for licensed UR facilities involve compliance with Federal and state water protection standards, particularly the environmental groundwater and drinking water standards set by the U.S. Environmental Protection Agency (EPA) and NRC requirements. Licensees must comply with these concentration limits for individual radioactive and non-radioactive constituents in the groundwater, at a point of compliance specified in these standards. Where applicable, effluent discharge into surface water must also meet concentration limits for individual constituents. The concentration limits are provided in the applicable regulations for facility type; for some facilities the concentration limits may be set equal to natural background concentrations.

4.1.2. Role of Models in Regulatory Compliance

The use of performance models that are based on dose assessment at UR sites is limited, because the performance measures for these operations are generally represented by constituent concentration limits in the groundwater or in the effluent discharge, which are monitored to ascertain that they are within the established limits. However, licensees and applicants still rely on conceptual hydrologic site models, as well as flow and transport modeling, to determine the locations of the point of compliance and, where applicable, the point of exposure, and to demonstrate that the concentration limit standards will be met.

The conceptual hydrologic site model plays an important role in nearly every decision made regarding site decommissioning and safe long-term disposal. The conceptual models are used to characterize the groundwater flow regime, to delineate the groundwater flow and transport paths, and to identify features and processes that may influence flow and contaminant transport offsite. From a regulatory perspective, the conceptual models are mainly used to (1) determine the locations of the point of compliance and the point of exposure at conventional UR facilities and (2) establish monitoring locations of interest at ISL facilities. The conceptual models are developed based on regional and site data. At sites with existing ground-water contamination, particularly inactive mill sites, the conceptual model must be sufficiently detailed to provide a technical basis for selection and implementation of the appropriate remedial action. The need to develop conceptual hydrologic site models, including specific acceptance criteria for formulating such models, are provided in a staff guidance for implementation of the UR standards at inactive mills. The conceptual models are expected to include groundwater and surface water conditions, processes, and interactions.

Flow and transport models are sometimes used to support remedial and other proposed actions by the licensees. Licensees use these models to demonstrate compliance, as permitted under the regulations. These uses include: (1) the use of alternate concentration limits at conventional mill disposal sites; (2) the use of an alternative point of exposure, i.e. at or near the property boundary instead of near the disposal facility; (3) the support of applications which rely on natural flushing to meet the standards over an extended period after a facility closure, (conventional inactive mill and disposal sites); (4) the determination of the locations and spacing of wells needed to monitor offsite concentrations (new ISL licenses); and (5) the demonstration that the proposed remedial action is likely to succeed and that the proposed financial sureties are adequate to cover the remediation costs.

4.1.3. Confidence Building in Hydrologic Data and Models for Licensed UR Facilities

The most important uncertainties, commonly encountered in making regulatory decisions at UR sites, pertain to data uncertainty. Modeling uncertainties are generally not as important, because the regulatory requirement that concentration limits be met is demonstrated by direct measurement. For conventional mills, DOE implements long-term surveillance, including groundwater monitoring. Conceptual hydrologic models play an important role in locating the point of compliance and selecting the locations of the monitoring wells; however, the development of a conceptual hydrologic model or viable alternative models for UR sites, in most cases, is usually not very difficult from a regulatory standpoint. The development, validation, and calibration of representative flow and transport models for UR sites can be as demanding as any other site. But the application of flow and transport models at regulated UR sites to date has generally been limited.

Examples of the data uncertainties commonly encountered at UR reviews include: (1) insufficiency of data for site characterization, including data needed to identify the hydrostratigraphic units and their hydraulic properties, and/or structural controls on flow and transport; (2) uncertainty about the quality and usefulness of the existing data bases, particularly background groundwater quality, that may have been developed before the UR regulations went into effect; (3) uncertainty related to acquisition of additional data, including (i) sampling intervals, (ii) laboratory analysis methods for radioactive content and hydraulic properties, (iii) interpretation of the tests results (e.g. well and aquifer pumping tests), and (iv) quality control measures. NRC staff response to hydrologic data uncertainty in the UR programme generally includes requiring licensees to: (1) provide additional data, including additional on-site testing and measurement; (2) validate the existing database; and/or (3) demonstrate that the regulations can be met with the existing database and that no additional data are required to comply with the regulations (e.g. parameter bounding, sensitivity analysis).

Examples of specific techniques used by the NRC staff to overcome data uncertainty and to build confidence in the hydrologic include requiring a licensee: (1) to establish a point of compliance at a number of locations around a disposal facility, because the available data could not establish the direction of groundwater flow with high confidence; (2) to use more than one analytical laboratory to analyze groundwater samples at a disposal facility, thereby raising confidence in the results; (3) to establish background water quality at several well sites in the vicinity of a disposal facility, when the variability of groundwater quality among the monitoring wells was too great to justify a single monitoring well; (4) to establish background water quality by using historical offsite groundwater quality data, since use of the site data alone for this purpose was too uncertain.

Modeling uncertainties encountered in UR reviews are similar to those encountered in other programme areas. These include: (1) uncertainty in the conceptual flow model, related to limited data and site characterization; (2) uncertainties in parameter value estimates; (3) limited validation of site-specific flow and transport model codes, that are developed by the licensee, but which have not been satisfactorily documented and for which there is no track record of validation or application; (4) limited model calibration resulting from sparse historical data.

Staff approaches to modeling uncertainties in the UR programme are generally consistent with the confidence building measures in models used in other regulatory programmes. These include requesting the licensees to: (1) be reasonably conservative, and meet the ALARA requirement; (2) use simple analytical techniques, especially when more complex models are unwarranted by the site conditions or unjustified by the available site data; (3) use known and proven modeling techniques and established codes; (4) verify that the models are validated and calibrated, hence representative of site conditions; (5) conduct bounding and sensitivity analysis when needed; and (6) evaluate constituent concentrations in the groundwater using alternatives to the preferred conceptual model, if these alternatives cannot be ruled out by the available data. The application of these measures is tailored to site-specific conditions.

4.2. *Experiences in HLW*

The NRC staff has evaluated DOE approaches for modeling groundwater flow and transport described in four iterative performance assessments for the proposed Yucca Mountain repository (Barnard, 1992; Wilson, 1994; TRW 1995; U.S. DOE, 1998). In the course of these four iterations, the conceptual model for flow and transport at Yucca Mountain has changed from a model of primarily very slow, steady flow in the unsaturated rock matrix to a model of some rapid, fracture flow penetrating to the repository horizon. Similarly the modeling of infiltration at the surface has evolved from a model of uniform average flow over the repository footprint to a model of variable flow produce by variations in topography and soil thickness. With the movement toward a dose-based standard, transport in the saturated zone has become more important than for the previous release-based standard. A current issue is whether the limited data available to describe the characteristics of the saturated zone will limit the degree to which DOE may be allowed to take credit for the ability of the saturated zone to retard radionuclide migration to the likely receptors. The DOE has made extensive use of expert judgment and expert elicitation in the development of the Yucca Mountain repository performance assessments; NRC has developed guidance for expert elicitation and DOE believes it has generally followed this guidance. However, some areas of improvement have been identified. Although several natural analogues are available for the Yucca Mountain repository, this information is used only to a limited extent in substantiating the safety case.

4.2.1.   How Hydrologic Processes May Affect Yucca Mountain Repository Performance

Groundwater flow and radionuclide migration is believed to have a profound influence on the performance of the proposed Yucca Mountain repository. Except for disruptive scenarios, such as volcanism, groundwater migration is the main pathway for radionuclides to reach the accessible environment. Even for the disruptive scenarios, radionuclide migration in the geosphere may have a significant effect on performance. According to DOE's Viability Assessment (U.S. Department of Energy, 1998).), "A repository would be built about 1 000 feet below the surface and 1 000 feet above the water table in what is called the unsaturated zone. The water table is about 2 000 feet beneath the crest of Yucca Mountain." To evaluate the impact on repository performance, groundwater flow at the proposed Yucca Mountain repository may be divided into flow above the repository horizon and flow below it. Above the repository horizon groundwater flow in the unsaturated zone (UZ) may affect performance in the following ways: (1) as an avenue for meteoric water to reach the repository through infiltration and deep percolation, (2) as an influence on waste package corrosion, and (3) as the means for radionuclide dissolution. Below repository horizon groundwater flow in the UZ and saturated zone (SZ) may affect performance in the following ways: (1) provides the most likely pathway for radionuclide transport from the repository to the critical group; (2) the UZ can reduce radionuclide migration from the repository to the water table by diffusion, dispersion, sorption, and radioactive decay; (3) the SZ can delay radionuclide arrival by: (i) low flow velocities and flow rates in the rock matrix and in the valley fill aquifer and (ii) sorption; and (4) the SZ can reduce dose to receptors by diffusion, dispersion, dilution, sorption, and radioactive decay. Table 1 indicates how certain attributes of the geohydrology at the proposed repository may affect performance.

Table 1. **Attributes of hydrogeologic subsystems that may affect performance**

GEOHYDROLOGIC SUBSYSTEM	HYDROLOGIC ATTRIBUTES AFFECTING PERFORMANCE
CLIMATE	Present climate and future climates, duration and frequency Precipitation rates, average temperatures,
INFILTRATION	Spatial and temporal infiltration rates Percolation rates
FLUXES NEAR REPOSITORY	Spatial and temporal fluxes diverted around disposal drifts and waste packages Spatial and temporal flux in contact with waste Aqueous chemistry in contact with waste Radionuclide flux from repository
UZ RADIONUCLIDE TRANSPORT	Flow velocities in UZ Fracture vs. matrix partition of flow Effective retardation
SZ FLOW AND TRANSPORT	Lateral SZ flux Flow velocities in SZ Fracture vs. matrix partition of flow Effective retardation Radionuclide diffusion and dispersion Groundwater use and dilution Exposure scenario for members of critical group

4.2.2. *Yucca Mountain Performance Assessments*

The primary role of the NRC is to evaluate the safety case for the repository provided by the DOE. To assist in this evaluation, the NRC performs independent quantitative analyses. Therefore this discussion of building confidence in geosphere-transport models relates to: (1) confidence or lack thereof, in DOE's performance assessments and (2) confidence in the independent quantitative tools used by the NRC. NRC will review DOE's PA's at various stages of repository development, such as site characterization, site recommendation, license application, amendment to operate, and amendment to close. DOE indicates that complex computer models will be used to appropriately portray the complex natural and engineered systems that compose the repository.

Over a period of years several performance assessments for the proposed repository at Yucca Mountain have been completed. A hallmark of these performance assessments is that they have been performed **iteratively**; that is, each completed performance assessment provides an indication of areas for improvement, which may be addressed in the subsequent iteration. Areas to be improved may include additional data, modifications of design, and the comprehensiveness and realism of models. In recent years DOE (Barnard, 1992; Wilson, 1994; TRW 1995; U.S. DOE, 1998), NRC (Codell, 1992; Wescott, 1995), and the Electric Power Research Institute (EPRI) (McGuire 1990, McGuire 1992, Kessler 1996) have all completed iterations of a performance assessment for the proposed Yucca Mountain repository. DOE has used these PA's to guide its site characterization programme, give insights for choosing design options, and to test the readiness of its analytical capability. Ultimately DOE must use a performance assessment to demonstrate compliance with regulatory requirements. NRC has used its PA's to: (1) Focus discussions between the DOE and NRC, (2) Define technical issues, (3) Identify the need for regulatory guidance, and (4) Provide input on site-specific regulations (as mandated by the Energy Policy Act of 1992). Ultimately NRC intends to use its PA capability to help review the DOE license application. As the number of iterations increases in time more data become available and are incorporated into the performance assessments. In addition, some aspects of the repository system are modeled with greater refinement; these refinements may be manifested as changes to the performance assessment models or as changes to the more detailed models, which form the basis for abstracted models in the performance assessment.

There has been a significant shift in hydrologic concepts and modeling from the early to the later PA's for the proposed Yucca Mountain repository. The earliest concept of Yucca Mountain repository was that of a "dry" environment; there was expected to be minimal contact of waste by water and subsequent groundwater transport of radionuclides. Most initial performance assessments (DOE, NRC, EPRI) assumed: (1) steady flow (2) vertical flow from repository to water table, and (3) lateral flow in the SZ. For the early PA's the regulatory objective was provided by 10 CFR 60, the NRC's generic regulation for high-level radioactive waste. 10 CFR 60 requires compliance with subsystem and total system performance objectives; the total system objective limits cumulative release of radionuclides to accessible environment (~5 km) for 10 000 years. During this time period (1988-1996), there was a significant difference between DOE and NRC PA's. DOE's PA's used very low infiltration rate, which virtually guaranteed slow, matrix-only flow in the unsaturated zone; NRC's analysis used higher infiltration rates, with some fracture flow in the unsaturated zone. Another significant difference is that DOE took substantial credit for matrix diffusion in the saturated zone, while NRC did not. An interesting aspect of these PA's was the conclusion that cumulative release of gaseous C-14 was a major factor in the determination of compliance. Furthermore the release or transport of C-14 did not depend significantly on water flow; instead air circulation through unsaturated zone facilitated transport to biosphere. Although individual doses from C-14 were small, cumulative releases, the performance measure in the regulation, were large.

The nature of both the NRC and DOE PA's changed as additional site data were acquired and as the repository design evolved. Some of the increased understanding of the site, which changed the hydrogeologic conceptual models, included: (1) evidence of deep percolation from bomb-pulse Cl-36, (2) increased infiltration estimates, which with other information changed the UZ conceptual model, (3) little evidence for matrix diffusion, since isotopes in pore water and rock are out of chemical equilibrium. A significant design change has been to move from small vertical casks to very robust, large containers placed horizontally in drifts. These various differences have reduced the significance of the UZ to repository performance.

In addition to changes to design and to the knowledge about the Yucca Mountain site, the regulatory framework for licensing the repository has changed, as mandated by congressional legislation. In response to this new law, the NRC issued a proposed regulation, 10 CFR Part 63. Some key features of the Part 63 proposed regulation are: (1) Administrative, preclosure, retrievability, and QA requirements are similar to Part 60; (2) The overall measure for post-closure performance is expected annual dose to a member of critical group and is limited 25 mrem; (3) A reference biosphere is specified; (4) The characteristics and location of the critical group for the Yucca Mountain repository are defined; (5) A stylized scenario and performance criterion are specified for human intrusion; (6) DOE must demonstrate that post-closure performance is achieved by system of multiple barriers, although no quantitative requirements are placed on the performance of individual barriers or components; (7) Unnecessary, generic siting and post-closure design criteria are not included; and (8) The all pathway total system performance requirement is considered protective of groundwater. These proposed changes in the regulatory requirements have some profound implications for the performance assessments of the proposed repository. Some of these implications include: (1) Contribution to expected annual individual dose is elevated as the most important factor in determining the significance of an issue; (2) C-14 is eliminated as a substantial contributor to compliance; and (3) Much more emphasis is placed on saturated zone transport and dilution.

In summary, the framework for NRC activities includes these fundaments: (1) the DOE has the burden to demonstrate the safety of the repository system; (2) the primary focus for NRC is the rigorous review of the DOE PA's; (3) to assist in the review of DOE PA's, NRC will conduct independent analyses. NRC enhances confidence in PA tools (both DOE's and NRC's) by evaluating soundness of the model abstractions, robustness of its sensitivity analyses, and thoroughness of the quality assurance processes used in code development with an emphasis on transparency and traceability. The PA tools are developed iteratively. A new version of a code will be based on shortcomings identified in previous iterations, and may involve further development of models and correction of errors. This iterative process also enhances confidence. Complete and clear documentation is another important aspect of enhancing confidence in PA analyses.

4.2.3.   Use of Sensitivity and Uncertainty Analysis

The PA's for geologic repositories are based on conceptual models of physical processes and parameters derived from field and laboratory data or expert elicitations. Because of the variability and sparsity of measured data and the underlying uncertainty involved with modeling physical processes for many thousands of years, the results are uncertain. Therefore, an important aspect of conducting a PA is quantification of the sensitivity of the performance to various input parameters; also important is the degree to which uncertainty in overall performance depends upon the uncertainty in a particular input parameter. Quantification of the degree, to which input parameters and their variability affect overall performance estimates and its variability, identifies from two perspectives (sensitivity and uncertainty, respectively) the significance of parameters. This determination of significance can be used to improve the code and build confidence in the numerical results produced by the code. Also this determination of parameter significance helps indicate where future design, site characterization, and analysis activities should be focused.

The significance of uncertainties in models and parameters are evaluated by observing the effect of these uncertainties on model output, i.e. the total system performance measure. If results demonstrate a narrow range of sensitivities to inputs, then uncertainties are judged insignificant. NRC uses a Monte Carlo method to propagate parameter uncertainty though its total system PA models. Model uncertainty is evaluated by determining the effect on the performance measure of alternative conceptual models and scenarios. The importance of various parts of the engineered and natural systems are evaluated by examination of the parameter sensitivities and alternative conceptual models, and by evaluating the outcome for runs in which the effect of the barrier is nullified (e.g. one-off analysis, importance analyses).

The NRC staff has used primarily two approaches to determining parameter sensitivities:

1. Regression of Multiple Realizations – The total system performance measures, resulting from a series of simulations based on realizations of the input parameters, are treated as a sample of the behavior of the repository. Regression analyses on the results of these simulations provide estimates of sensitivity for each input parameter, that represents the "average" sensitivity over the region of parameter space sampled.

2. Differential Analyses – The total system PA code is run for a specific set of variables, which represent a fixed point in the multidimensional space of the parameters. Then, one at a time, each variable is perturbed a small amount from its nominal value. The sensitivity is then determined for each variable by taking the normalized difference in output results and dividing by the normalized difference in the input variable. Since this analysis is focused at a single point, it is repeated for several different points to represent potentially different ranges of parameter behavior.

It is important to conduct both types of sensitivity analyses to overcome the disadvantages of each. Regression analysis allows the variables to change of a large range of values, but is limited in its ability to extract useful information to only the most significant variables because the results are confounded by the simultaneous variation in all other variables in the sampling procedure. Differential analysis overcomes this problem because only one variable is changed at a time; however, results of differential analysis are limited to a small region of parameter space around the nominal point.

The following example shows how sensitivity analyses can be used to identify dominant variable. Although the NRC code is used for this example, DOE has performed similar analyses for similar purposes. NRC expects DOE to perform this type of analysis in any PA submitted for NRC review. Figure 1 illustrates a typical result from NRC's Total Performance Assessment code. Stepwise regression is used to show which input variables are most important for the 10 000 and 50 000 year time periods of interest. This analysis helped to determine which of 243 sampled variables contributed most to repository performance. Modeling uncertainty in total system performance analysis is usually described by comparing the results of the TPA code under different assumptions about the models. Figure 2 shows the results, in terms of the effects of alternative conceptual models, on the peak of the mean dose predicted by the models for 10 000 years.

Figure 1. **Results of a stepwise regression of peak dose within 10 000 years for the proposed Yucca Mountain repository. Raw variables were transformed logarithmically. The residual sum of squares (RSS) indicates how much of the variation in model output can be accounted for by a linear regression; variables with more influence reduce the RSS by a larger increment.**

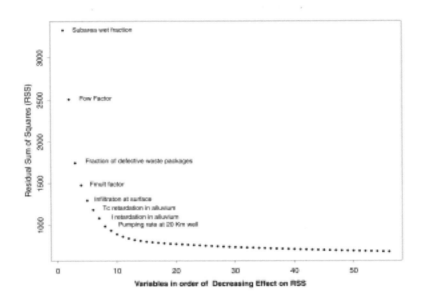

Figure 2. **Evaluation of the effect of various alternative conceptual models on the estimate of peak mean dose over 10 000 years for the proposed Yucca Mountain repository. The larger the difference for the alternative indicated from the "Base" case, the more significant the alternative model might be.**

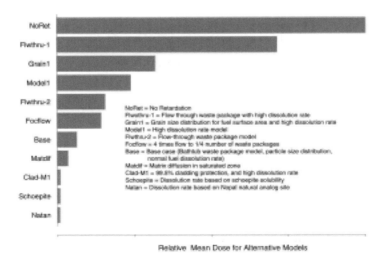

During its interactions with DOE over the past 10-15 years, the NRC staff has drawn several conclusions related to building confidence in geosphere-transport models. Some of the more important "lessons-learned" are:

1. If DOE's safety case depends largely upon a particular hydrologic parameter value(s) being within a certain range of values, then: (1) There must be high confidence that the measurements support the parameter having a value within the required range and (2) Since many hydrologic parameters are inferred, there must be high confidence in the conceptual model used to infer values.

2. As DOE gets more data, their impact on performance may be greater, if the data affect the hydrologic conceptual models, rather than if the data affect parameter values only.

3. As DOE's repository design evolves and DOE's site characterization continues, the role of various repository system components may change; therefore, the performance assessment tools and the detailed hydrologic models, on which they are based, must be flexible.

4. DOE must take care in developing abstracted models for performance assessment, that correlations among variables and models are appropriately preserved, so that estimates of performance are not unjustifiably optimistic.

5.   NRC Research to Treat Uncertainties in Geosphere-transport Modeling

The U.S. Nuclear Regulatory Commission has sponsored numerous grants and contracts to investigate the nature of uncertainties in geosphere-transport models and to identify methods for their treatment. One grant with the National Academy of Sciences (NAS) focused on the scientific and regulatory applications of ground-water models (NAS, 1990). The NAS report identified major sources of uncertainty in ground-water models arising from "(1) the inability to precisely describe the natural variability of model parameters (e.g. hydraulic conductivity) from a finite and usually small number of measurement points, (2) the inherent randomness of geologic and hydrogeologic processes (e.g. recharge rates and erosion) over the long term, (3) the inability to measure or otherwise quantify certain critical parameters (e.g. features of the geometry of fracture networks), and (4) biases or measurement errors that are part of common field methods". The NAS report provided an overview of uncertainty issues in ground-water modeling related to decision making in a regulatory arena. A later NAS study, funded in part by the NRC, on fracture characterization and fluid flow through fractured rock identified the need to explicitly consider uncertainty in decision making and the role fracture characterization plays in reducing uncertainty (NAS, 1996).

In 1983, the NRC funded Massachusetts Institute of Technology (MIT) to utilize stochastic methods to represent the variability of hydraulic parameters, and use the stochastic representation to effectively model partially saturated flow and transport. A principal motivation in funding the MIT research was to develop a stochastic model to quantify uncertainties in outcomes related to the model parameter variability (Mantoglou and Gelhar, 1987a, b and c). The MIT research demonstrated for the first time that large-scale effective hydraulic conductivity of unsaturated soils show tension-dependent anisotropy and hysteresis due to the natural heterogeneity. The MIT stochastic theory provided an explicit method of predicting the "effective large-scale" parameters using practical small-scale measurements interpreted in a stochastic framework. Uncertainties in model outcomes could be traced to parameter variability in a systematic manner.

In 1983, NRC began funding research at the University of Arizona (UAZ) to develop a multidimensional stochastic theory with applications to saturated, fractured rock. The UAZ predictive model focused on estimating the far-field dispersion due to the spatial variability of hydraulic conductivities measured in packer tests conducted in boreholes penetrating fractured granites at the Oracle field site (Neuman and Depner, 1988). Later UAZ field studies of air permeability testing using straddle-packer tests in shallow boreholes at the Apache Leap Research site examined the issue of spatial variability at four different scales (3.0, 2.0, 1.0 and 0.5 meters) (Illman and others, 1998). Geostatistical analysis indicated that the apparent permeability data behave as a stochastic multiscale continuum, and suggest that site characterization be based on hydrogeologic data collected on a spectrum of scales relevant to the performance assessment. Uncertainties related to measurement scale were identified and quantified within a systematic stochastic framework.

A significant contributor to uncertainty in modeling shallow ground-water flow at low-level radioactive waste facilities was estimating infiltration. The NRC funded Pacific Northwest National Laboratory (PNNL) to develop a methodology for estimating infiltration with a focus on quantifying uncertainties. The PNNL research developed generic probability distributions for unsaturated and saturated soil hydraulic parameters based on soil textures, and an approach for applying the Bayesian method for updating these estimates using limited site-specific measurements (Meyer and others, 1997). The uncertainty assessment methodology is being tested using actual site data and "hypothetical" source data from decommissioned sites for evaluating dose assessment models.

Recent research was begun at UAZ to develop a methodology for evaluating and testing alternative conceptual models for ground-water flow and transport, and their associated uncertainties. The methodology will examine the nature of conceptual model uncertainties and will identify appropriate methods to understand, treat, propagate and characterize them. The methodology will be tested using detailed data sets from four distinct hydrogeologic settings: saturated, fractured rock (i.e. Fanay-Augeres mine); unsaturated, fractured rock (i.e. Apache Leap Research site); saturated, porous media (i.e. Maricopa Environmental Monitoring site); and unsaturated, porous media (i.e. Las Cruces Trench site).

Starting in 1987, the NRC funded geosphere related research and technical assistance work at the Center for Nuclear Waste Regulatory Analyses in San Antonio, the NRC's Federally Funded Research and Development Center. Deterministic (Bagtzoglou *et al.*, 1994, Runchal and Sagar, 1993, Ahola and Sagar, 1992, Wu *et al.*, 1992, Wittmeyer, 1995, Mohanty *et al.*, 1995) and stochastic (Bagtzoglu and Muller, 1994) process level modeling methods for flow and transport including sorption processes (Turner, 1995) at various scales were investigated as well as their abstraction suitable for inclusion in the total system performance models. Also investigated was the role of natural analogs (Murphy and Kovach, 1993) and laboratory and field-scale experiments (Green *et al.*, 1993) to garner experience in building confidence in the numerical models. Two computer codes capable of modeling detailed processes were also developed, one for modeling the shallow infiltration process (Stothoff, 1994) in fractured media and the other for modeling the coupled hydrologic, thermal, and chemical processes (Seth and Lichtner, 1996).

6.   **Conclusions and Recommendations**

The discussion above has indicated various means used to enhance confidence in geosphere-transport models. In general the licensee or implementor is required to make a case that uncertainties has been sufficiently characterized and reduced to permit a decision to be made. Various analytical techniques may be used to identify critical uncertainties and focus reviews of licensing actions. Two overriding themes that result from the NRC experience are:

- It is not necessary to resolve all uncertainties related to flow and geosphere-transport to the same degree.

- Performance assessment can help to define what issues are key.

Regardless, the characterization and evaluation of uncertainties in geosphere-transport modeling is an essential aspect of building confidence in a regulatory context.

6.1. *Perspective on Status*

Many of the quantitative methodologies used in conventional flow and transport modeling fit well into the regulatory context of performance assessment and decision making. For the regulator, whose goal is adequate protection of public health, safety, and the environment, the issue is evaluating the relevant uncertainties in the information provided by the applicant or licensee and making an appropriate decision given these uncertainties. Although several methods are available for the definition, evaluation, and propagation of uncertainty, additional consideration might be given to development of methods for quantifying and propagating (1) conceptual model uncertainty and (2) uncertainties introduced when detailed models are abstracted for use in performance assessments.

Based on compliance strategy and availability of computing power, a licensee may choose to include models of geosphere at several levels of detail in the safety case. The most detailed of these are research-level models which explore the effects of various coupled processes, details at fine space- and temporal-scales, and the effect of specific uncertainties. Based on conclusions from the research-level models, the next level of process models may be of lower dimensionality, assume some of the processes as uncoupled and introduce new derived parameters (thus introducing new uncertainties) that are based on more fundamental parameters of the research models. For use in the total system performance assessment, the process-level models may be simplified even further by introducing assumptions that can be reasonably shown to lead to pessimistic results. As noted before, some components of the geosphere models may be more detailed than others depending upon the importance of that component to meeting the system performance goal. From the regulator's perspective, the basic principles for modeling flow and transport are largely established at an adequate level. It is the application of these basic principles to a specific site and licensing context (decommissioning, low-, intermediate- and high-level waste disposal) with uneven and limited data that introduces uncertainties in the results. For the regulator, whose goal is adequate protection of public health, safety, and the environment, the decision rests on evaluating the relevant uncertainties in the information provided by the licensee.

6.2. *Aspects Needing Progress*

Currently, the estimation of parameter uncertainties and their propagation through the model are the most explored and perhaps better understood of all the uncertainties. The distinction between parameter variability and parameter uncertainty is generally well understood although in the upper tier simplified models embedded in the total system performance assessments, this distinction is sometimes obliterated. Quantification and propagation of uncertainties other than parameter uncertainties are less well understood. One of these is the uncertainties introduced during the model abstraction process, i.e. in deriving the simplified model from the research- or process-level model to the total system performance assessment model. Confidence in the simplified flow and transport models and associated parameters can be garnered through comparison to more detailed models that do account for both variability and uncertainty, thereby assuring that the simplified models produce suitably conservative results. Another aspect needing further study is the conceptual model uncertainty. The site-specific data is seldom sufficient to clearly identify a single unique conceptual model for the flow and transport system under all potential future boundary conditions imposed by changing climate and geology. While focus is usually on a preferred conceptual model, it is important that the effect of alternate conceptual models be included in the safety case. From a regulator's perspective, greater confidence in performance assessment results is generated if the results from each of the conceptual models is displayed separately even if they are eventually combined through assignment of weights (or probabilities) into a single risk curve.

Although several methods are available for the definition, evaluation, and propagation of uncertainty, additional consideration might be given to development of methods for quantifying and propagating (1) conceptual model uncertainty and (2) uncertainties introduced when detailed models are abstracted for use in performance assessments.

Specific aspects of hydrogeologic models and how such models treat uncertainty also would benefit from additional work. Some additional questions presently being pursued in NRC-sponsored research activities include:

1. How can current stochastic models be used to determine risk (i.e. combination of consequence of a release event and its probability of occurrence) related to contaminant transport associated with various design and/or remediation options?

2. Which stochastic approaches present opportunities for quantifying and reducing uncertainties?

3. How should stochastic models be applied to realistic hydrogeologic systems for determining uncertainties in conceptual models?

4. Given a scarcity of site-specific data at most sites, which stochastic methods would be appropriate for estimating spatial variability and uncertainty in flow and transport properties at the relevant scale of analysis?

5. In determining compliance with a well-defined performance criteria (e.g. MCL's), how can a stochastic framework be developed to address reliability and uncertainty in modeling predictions?

6. How can stochastic approaches be used to assist in designing characterization and monitoring strategies?

Since parameter measurement and estimation methods, and conceptual model definition and testing are strongly linked to uncertainty, answers to these questions will further advance the ability to identify, characterize, propagate, and treat uncertainty in ground-water modelling in a regulatory context.

7.   References

1.   Ahola, M. and B. Sagar, "Regional Groundwater Modeling of the Saturated Zone in the Vicinity of Yucca Mountain, Nevada," CNWRA 92-001 and NUREG/CR-5890, October, 1992.

2.   Andrews, R.W., T.F. Dale, and J.A. McNeish, "Total-System Performance Assessment – 1993: An Evaluation of the Potential Yucca Mountain Repository," (INTERA, Inc., Las Vegas, Nevada, March 1994).

3. Bagtzoglou, A.C., S. Mohanty, A. Nedungadi, T.J. Yeh, and R. Ababou, "Effective Hydraulic property Calculations for Unsaturated, Fractured Rock with Semi-Analytical and Direct Numerical Techniques," CNWRA 94-007, March, 1994.

4. Bagtzoglou, A.C. and M. Muller, "Stochastic Analysis of Large-Scale Unsaturated Flow and Transport in Layered Heterogeneous Media," CNWRA 94-012, June, 1994.

5. Barnard, R.W. *et al.*, "TSPA 1991: An Initial Total-System Performance Assessment for Yucca Mountain," SAND91-2795, Sandia National Laboratories, Albuquerque, NM, July 1992.

6. Codell, R.B., *et al.*, "Initial Demonstration of the NRC's Capability to Conduct a Performance Assessment for a High-Level Waste Repository," NUREG-1327, (U.S. Nuclear Regulatory Commission, Washington, D.C., May 1992).

7. Green, R.T., R.D. Manteufel, F.T. Dodge, and S.J. Svedeman, "Theoretical and Experimental Investigation of Thermohydrologic Processes in a Partially Saturated, Fractured Porous Medium," CNWRA 92-006 and NUREG/CR-6026, July, 1993.

8. Illman, W., and others, Single- and Cross-Hole Pneumatic Tests in Unsaturated Fractured Tuffs at the Apache Leap Research Site: Phenomenology, Spatial Variability, Connectivity and Scale, NUREG/CR-5559, U.S. Nuclear Regulatory Commission, Washington, DC, November 1998

9. Kessler, J. *et al.*, "Yucca Mountain Total System Performance Assessment, Phase 3," EPRI TR-107191, Electric Power Research Institute, Palo Alto, CA, December 1996.

10. Mantoglou, A. and L.W. Gelhar, Stochastic Modeling of Large-Scale Transient Unsaturated Flow Systems, Water Resources Research, 23 (1), 37-46, 1987a.

11. Mantoglou, A. and L.W. Gelhar, Capillary Tension Head Variance, Mean Soil Content and Effective Specific Soil Moisture Capacity of Transient Unsaturated Flow in Stratified Soils, Water Resources Research, 23 (1), 47-56, 1987b.

12. Mantoglou, A. and L.W. Gelhar, Effective Hydraulic Conductivities of Transient Unsaturated Flow in Stratified Soils, Water Resources Research, 23 (1), 57-67, 1987c.

13. McGuire, R.K. *et al.*, "Demonstration of a Risk-Based Approach to High-Level Waste Repository Evaluation," EPRI NP-7057, Electric Power Research Institute, Palo Alto, CA, October 1990.

14. McGuire, R.K. *et al.*, "Demonstration of a Risk-Based Approach to High-Level Waste Repository Evaluation: Phase 2," EPRI TR-100384, Electric Power Research Institute, Palo Alto, CA, May 1992.

15. Meyer, P.D., M.L. Rockhold, and G.W. Gee, "Uncertainty Analyses of Infiltration and Subsurface Flow and Transport for SDMP Sites," NUREG/CR-6565, U.S. Nuclear Regulatory Commission, Washington, DC, September 1997.

16. Meyer, P.D., M.L. Rockhold, and G.W. Gee, "Uncertainty Analyses of Infiltration and Subsurface Flow and Transport for SDMP Sites," NUREG/CR-6565, U.S. Nuclear Regulatory Commission, Washington, DC, September 1997.

17. Mohanty, S., R.T. Green, and K.A. Meyers-Jones, "Study of Flow in a Fracture under Shear," CNWRA 96-001.

18. Murphy, W. M., and L.A. Kovach, "The Role of Natural Analogs in Geologic Disposal of High-Level Nuclear Waste," CNWRA 93-020, September 1993.

19. Neuman, S.P. and J.S. Depner, Use of Variable-Scale Pressure Test Data to Estimate the Log Hydraulic Conductivity Covariance and Dispersivity of Fractured Granites Near Oracle, Arizona, Journal of Hydrology, 102, 475-501, 1988.

20. NAS, Ground-Water Models: Scientific and Regulatory Applications, National Academy Press, Washington, DC, 1990.

21. NAS, Rock Fractures and Fluid Flow: Contemporary Understanding and Applications, National Academy Press, Washington, DC, 1996.

22. Runchal, A.K. and B. Sagar, "PORFLOW: A Multifluid, Multiphase Model for Simulating Flow, Heat Transfer, and Mass Transport in Fractured Porous Media," CNWRA 92-003 and NUREG/CR-5991, February, 1993.

23. Seth, M.S. and P.C. Lichtner, "User's Manual for MULTIFLO: Two Phase Nonisothermal Flow Simulator," CNWRA 96-005, May, 1996.

24. Stothoff, S., "Breath Version 1.0 - Coupled Flow and Energy Transport in Porous Media," CNWRA 94-020, September 1994.

25. TRW, "Total System Performance Assessment – 1995: An Evaluation of the Potential Yucca Mountain Repository." TRW, Las Vegas, NV, November 1995.

26. Turner, D.R., "A Uniform Approach to Surface Complexation Modeling of Radionuclide Sorption," CNWRA 95-001, January 1995.

27. U.S. Department of Energy, Office of Civilian Radioactive Waste Management, "Viability Assessment of a Repository at Yucca Mountain," DOE/RW-0508. U.S. Department of Energy, Office of Civilian Radioactive Waste Management, Washington, D.C., December 1998.

28. Wescott, R.G., et al. (eds.), "NRC Iterative Performance Assessment Phase 2: Development of Capabilities for Review of a Performance Assessment for a High-Level Waste Repository," U.S. Nuclear Regulatory Commission, NUREG-1464, October 1995

29. Wilson, M.L. et al., "Total-System Performance Assessment for Yucca Mountain – SNL Second Iteration (TSPA-1993)," (Sandia National Laboratory, Albuquerque, New Mexico, SAND93-2675, 2 vols., April 1994).

30. Wingle, W.L., E.P. Poeter, S.A. McKenna, UNCERT User's Guide (Version 1.16), Colorado School of Mines, Golden, CO, 1998.

31. Wittmeyer, G.W., R. Klar, G. Rice, and W. Murphy, "The CNWRA Regional Hydrogeology Geographic Information System Database," CNWRA 95-009, June, 1994.

32. Wu, Y-T., A.B. Gureghian, B. Sagar, and R.B. Codell, "Sensitivity and Uncertainty Analyses Applied to One-Dimensional Radionuclide Transport in a Layered Fractured Rock," CNWRA-92-002 and NUREG/CR-5917, December, 1992.

The Role of Matrix Diffusion in Transport Modelling in a Site-specific Performance Assessment: Nirex 97

S. Norris and J.L. Knight
Nirex Limited, UK

Abstract

Rock-matrix diffusion (RMD) is recognised as a potentially significant process that can act to retard the transport of radionuclides dissolved in groundwater flowing through fractures, and accordingly is included in performance assessments.

Clearly, it is important that confidence can be built in the approach undertaken to modelling a transport process such as RMD. The work undertaken by the Nirex Safety Assessment Research Programme (NSARP) to determine relevant RMD properties is discussed in this paper. This work led to the development of the Nirex 97 [1] model of RMD, and assists in building confidence in the approach adopted in that performance assessment.

The NSARP has adopted two approaches to investigate RMD: laboratory-scale experimental measurements and observations of RMD in natural geochemical systems. Quantification of the RMD properties of rocks has been achieved through a laboratory-scale aqueous phase diffusion experimental programme, which allows precise and accurate measurements of the diffusive properties of rock cores on the scale of 10-100 cm^3. It is recognised, however, that the process of acquiring and preparing the samples might affect their RMD properties: the importance of any such changes needs to be assessed.

Natural analogue studies have been used to support the laboratory data by demonstrating that RMD operates over the time scale of interest in a performance assessment. However, larger uncertainties are associated with the quantification of diffusion coefficients from observations in such natural geochemical systems, because of uncertainties in both the time scale over which the process has operated, and the diffusive mobility and retardation properties of the trace element being studied. Further, it is difficult to prove that diffusion alone is responsible for the observed distribution of radionuclides in such natural analogues. Natural analogue studies therefore are not used *a priori* to quantify rock properties for input to assessment calculations; rather they provide information that, qualitatively, can be used to justify the operation of the RMD process.

Information derived from NSARP studies was used to develop the simple model of RMD used in Nirex 97. This model incorporates various approximations, such as all the fractures in the BVG have the same properties and are arranged regularly. However, in a real system the fractures would be arranged irregularly and their properties would vary from fracture to fracture, and within a single fracture. Variant numerical modelling studies have therefore been undertaken, to assess the adequacy of the Nirex 97 approach. It was found that the approach was acceptable, and conservative.

1.   Introduction

Groundwater flow in fractured rocks takes place predominantly through open fractures in the rock rather than through the rock matrix. This occurs because the permeability of the rock matrix is much lower than the permeability of the interconnected network of open fractures. If a solute being transported by groundwater flowing in fractures can access additional porosity, either in the rock matrix or in any part of the fracture porosity through which groundwater does not flow, then its concentration in the flowing groundwater will decrease and its migration will be retarded relative to the groundwater flow. Yet further retardation could occur if the solute is sorbed onto the additional rock surfaces accessed by such a process. The process by which solutes are transported through the pore water into the low permeability rock is diffusion. In the context of diffusive transfer between fractures and the rock matrix, this process is termed "rock-matrix diffusion (RMD)". It has been invoked in many assessments of contaminant migration, both in nuclear and non-nuclear contexts.

This paper presents an overview of work that has been undertaken by the Nirex Safety Assessment Research Programme (NSARP) to develop, justify and parameterise appropriate conceptual models of RMD for implementation in post-closure performance assessment calculations, such as those presented in Nirex 97 [1]. Several complementary "strands" to the work are summarised: laboratory experiments, natural analogue observations and research model development. Recently, studies have examined the potential for a methodology that could determine *in situ* RMD properties, based on information derived from borehole investigations – this work is also summarised.

2.   Potential Effect of RMD on Radiological Risk

RMD is potentially a process that can lead to significant retardation of both non-sorbing radionuclides and sorbing radionuclides. In both cases, it provides access to greater porosity, and in the latter case, it potentially provides access to more sorption sites. RMD may or may not affect calculated radiological risk, depending on the properties of the site under consideration. Furthermore, its effect may vary from radionuclide to radionuclide.

The effect on the risk from a very-long lived radionuclide will depend on whether the risk is strongly controlled by the near-field source term, as was the case for some probabilistic safety assessment realisations for certain radionuclides in the Nirex 95 performance assessment [2], or by transport in the geosphere, as in Nirex 97. In the first case, the main geological control on risk is the groundwater flow through the repository and, although RMD still operates, it has comparatively little effect on overall calculated radiological risk. In the Nirex 97 case however, the travel time through the geosphere is one of the main controls on risk and RMD has the potential to have a comparatively greater effect on calculated radiological risk. However, this generally would only be the case if the travel time through fractured rocks, assuming access to all of the matrix porosity, is the dominant contribution to the total travel time, and, even in this case, it is necessary for the total travel time to be long enough to allow significant diffusion into the matrix.

In cases in which RMD does have significant effect, it could increase the effective geosphere travel time through the rocks in question, possibly by up to several orders of magnitude. The increased travel time will increase the extent of radioactive decay during geosphere-transport. For short-lived radionuclides, it will probably not have a significant effect on repository performance, given the long groundwater travel times at a site that would be suitable for a radioactive waste repository. However, for long-lived radionuclides, the effect could potentially be important.

RMD may also have more indirect effects on risk. For example, it may affect the transport of salinity through the rocks, which may affect the groundwater flow pattern, and hence the risk. The impact of the effect would depend on the site.

3.   Nirex 97 Performance Assessment

In the Nirex 97 assessment of the post-closure performance of a repository at Sellafield, the potential repository host rock was the Borrowdale Volcanic Group (BVG). This comprises a thick sequence of low permeability fractured volcaniclastic rocks. At that site, groundwater flow in the BVG is predominantly through a subset of the total set of discontinuities – the Flowing Features (FFs). These are discussed further in [1, 3]. FFs can be identified in core samples by the presence of recent (in geological terms) calcite minerals.

Radionuclides transported by groundwater can diffuse into the (relatively) immobile groundwater in the rock matrix between the FFs. This potentially provides an important retardation mechanism, particularly for sorbing radionuclides for which it can provide access to additional sorption sites in the rock matrix away from FFs. In the one-dimensional radionuclide transport calculations undertaken in Nirex 97, RMD was explicitly modelled using the simple model shown in Figure 1. In this model, the fractures are parallel and equally spaced and have constant aperture. This is a considerable simplification of the real system, which might be similar to that shown in Figure 2. The features carrying flow are not parallel or equally spaced, but irregular.

Figure 1. **Simplified model of rock-matrix diffusion**

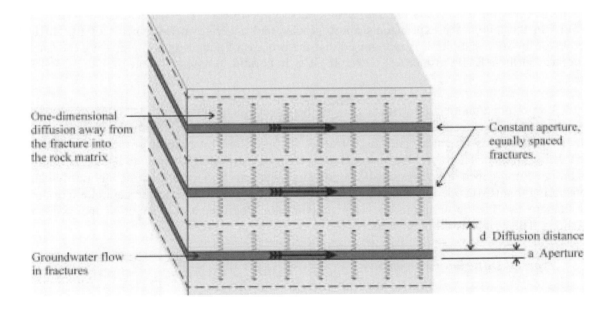

Figure 2. **Schematic showing "real rock"**

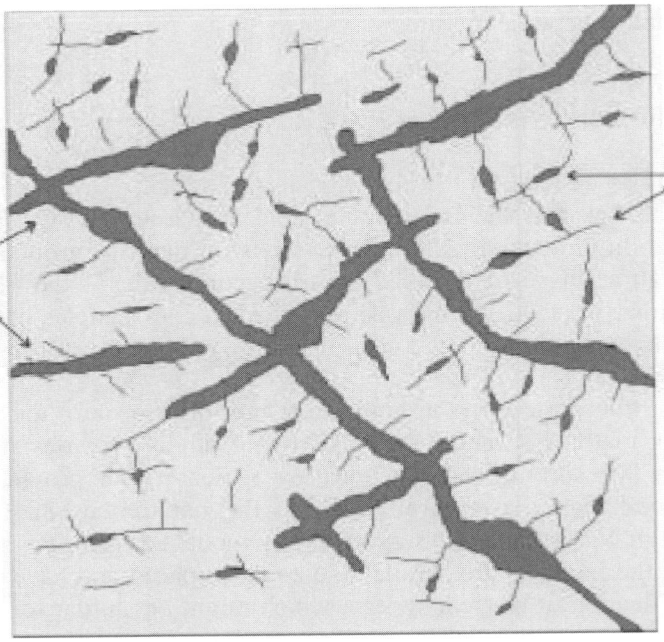

Irregular rough fractures carrying flow

Micro-fractures and micro-pores in rock matrix

4.   Laboratory Experimental Programme

The approach that has been adopted by Nirex to determine RMD parameters for use in performance assessment has been to measure relevant parameter values (e.g. diffusion coefficients and the porosity accessible to the radionuclide) in laboratory experiments. This is because the well-defined conditions in the laboratory experiments allow precise and accurate measurements of the diffusive properties of the rock sample. It is recognised that the process of acquiring and preparing the samples may impact on their RMD properties; therefore, it is important to assess the magnitude of any such changes.

Larger uncertainties are associated with the direct calculation of diffusion coefficients from observations in natural geochemical systems, because of uncertainties in both the time scale over which the process has operated, and the diffusive mobility and retardation properties of the trace element being studied. Therefore, the approach adopted by the NSARP has been to use observations of diffusion in natural geochemical systems ("natural analogue" studies) to support the laboratory data by demonstrating that RMD operates over relevant spatial and temporal scales in the natural environment, rather than to use the observations to calculate diffusion coefficients for input to the assessment calculations. This is discussed in Section 5.

4.1   Types of Diffusion Experiment

There is a range of experimental techniques available for measuring the diffusion properties of a geological sample, such as a granite or a clay [4]. The choice of experimental technique is influenced both by the physical properties of the rock (its diffusivity and porosity) and by the geochemical behaviour of the radionuclide being studied (the extent of sorption).

164

Laboratory-scale diffusion experiments involve measuring one of the following:

(a)   the breakthrough and steady-state flux of a radionuclide that has diffused into and through a rock sample (through-diffusion experiment);

(b)   the depletion of a radionuclide from a reservoir in contact with a rock sample initially free of the tracer (reservoir-depletion experiment);

(c)   the distance that a radionuclide has diffused into a rock sample from a reservoir, of constant concentration, in contact with the sample (in-diffusion);

(d)   the diffusive flux or total activity of radionuclide released from a rock sample that initially contains the radionuclide at a known concentration (out-diffusion experiment or solute accessible porosity measurement).

These experiments either measure steady-state diffusion or transient diffusion and, therefore, are analysed using solutions to Fick's First and Second Laws respectively.

4.2   *Tracers Used in Diffusion Experiments*

The objective of the experiments is to characterise the aqueous phase diffusion properties of a range of rocks of relevance to radioactive waste disposal. Consequently, most experiments involve the direct use of non-sorbing tracers dissolved in water.

The radionuclide tracer most commonly used by the NSARP for diffusion experiments is ^{3}H as tritiated water (HTO). HTO is an uncharged solute, and will access the same pore space that can be accessed by the natural groundwater. Therefore, HTO has been used to obtain baseline parameter values for D_i and ϕ, against which values for other tracers can be compared.

Experimental studies have indicated that, whereas cations and neutral species have access to almost all the pore volume in a particular rock sample, anions have access to significantly less pore space. This is due to charge effects, and is process termed "anion exclusion". This has been investigated using a range of anionic tracers. The following radioactive and stable isotopes of anions have been used.

(a) ^{36}Cl, as NaCl. ^{36}Cl has a half-life of $3.0\ 10^5$ years, and is of direct radiological significance in the Nirex post-closure performance assessments. In radioactive waste, it is produced as a neutron activation product from the irradiation of stable nuclides in reactors or accelerators.

(b) ^{125}I, as NaI. ^{125}I has a half-life of 60 days. It is used as an analogue for ^{129}I (half-life $1.6\ 10^7$ years), which is of direct radiological significance in the Nirex post-closure performance assessment. In radioactive waste, ^{129}I is produced as a medium-yield fission fragment in nuclear reactors. ^{129}I is expected to occur as iodide in the geosphere, and to have similar geochemical behaviour to ^{36}Cl.

(c) ^{127}I, as NaI. ^{127}I is the stable isotope of iodine, and is used as an analogue for ^{129}I.

(d) ^{35}S, as Na$_2$^{35}SO$_4$. ^{35}S has a half-life of 87 days. Sulphur isotopes are not of radiological significance for post-closure performance assessment. ^{35}S, a divalent cation, has been included in the NSARP to assess if anion exclusion increases as a function of the charge of the radionuclide.

In addition to tracers suitable for measuring charge exclusion effects, other tracers are used to measure size exclusion (or steric hindrance) effects. NSARP experiments have mainly used ^{14}C-labelled isosaccharinic acid (ISA), a degradation product of cellulose under alkaline anaerobic conditions. ISA is one of a number of degradation products that are predicted to be generated in a cementitious radioactive waste repository, such as that currently envisaged by Nirex.

Helium gas has been used by the NSARP to scope the diffusivity of dry samples prior to their resaturation and use in through-diffusion experiments. Using a gas tracer allows a rapid measurement of the diffusivity of the material, because the self-diffusion coefficient of a tracer gas within a bulk gas is much higher than that of an aqueous tracer in water. However, it must be assumed that the diffusibility of the rock towards a gas is the same as that towards water. At the detailed level, this might not be the case. Helium diffusivity measurements have therefore not been used by the NSARP to provide quantitative input data for assessments.

4.3   *Microstructural Studies*

The results obtained from diffusion experiments provide quantitative data on the diffusion coefficient of the rock and on the porosity accessible to the tracer. In order to build further confidence in these data, and to ensure that any assessment model of RMD is well-founded, it is valuable to develop a conceptual model of the pore network within the rock. Therefore, microstructural studies have been undertaken as part of the NSARP to provide both qualitative and quantitative data on aspects such as pore structure and pore size and size distribution. The techniques applied to the rock samples have included:

(a)   fluorescence microscopy on rock samples impregnated with fluorescein-doped low viscosity epoxy resin, to provide data on the location, amount and geometry (e.g. tortuosity and constrictivity) of macro-porosity (here defined as pores with dimensions greater than 50 nm) in the rock matrix;

(b)   nitrogen gas adsorption/desorption isotherm measurements on vacuum-degassed rock samples, to provide surface area and pore size distribution (down to 2 nm) data;

(c)   mercury intrusion porosimetry to provide measurements of mercury-accessible porosity and pore size distribution;

(d)   helium gas expansion porosimetry, to measure the matrix porosity accessible to helium gas.

4.4   *Ionic Conductivity Studies*

One of the main problems with aqueous phase diffusion measurements is the limited lengthscale over which tracers can diffuse on laboratory time scales In through-diffusion experiments, the time required to establish steady-state diffusion is proportional to the square of the diffusion distance; given typical properties of the rock and the diffusing solute, such experiments are restricted to measuring diffusion over distances of a few centimetres on realistic time scales. It would require several thousand years to establish steady-state diffusion over a 1m long rock sample, using an aqueous phase tracer such as HTO. Clearly, conventional aqueous phase diffusion experiments on this length scale would not be practicable.

Ionic conductivity measurements provide an alternative technique for determining the diffusive properties of rocks over length scales up to a metre. Further, these measurements are rapid, and can be undertaken as soon as the rock sample is saturated with simulated groundwater solution.

The measurement requires the rock sample to be fully saturated with a solution containing an electrolyte such as NaCl. Once the rock sample is saturated, typically with a simulated groundwater, electrodes are attached to the opposite ends of the core sample and an AC current is passed through the sample and the impedance and phase angle of the electrical circuit are measured. The alternating current flows through the connected pore space in the rock, with electrical charge being carried by the electrolyte in the saturating solution. The crystalline structure of the rock matrix is assumed to act as an insulator.

The property typically measured in ionic conductivity measurements is the ratio of the conductivity of the saturated rock, σ_s, to the conductivity of the free electrolyte in the saturating solution, σ_f. The Formation Factor, F, first introduced by Archie [5], is the inverse of this ratio. The equivalent aqueous diffusion parameter to this ratio is the diffusibility, Ψ. F and Ψ are related as follows:

$$\frac{D_i}{D_o} = \frac{\phi}{\tau} = \frac{\sigma_s}{\sigma_f} = \frac{1}{F} = \psi \tag{1}$$

D_o is the free water diffusion coefficient for the diffusing radionuclide, and D_i is the intrinsic diffusion coefficient (sometimes called the effective diffusion coefficient). ϕ and τ respectively are the porosity accessible to solute and the geometrical factor (incorporating tortuosity and constrictivity) of the rock sample under consideration. Under the experimental conditions considered by NSARP, it has generally been assumed that electrical conductivity is through the free electrolyte, and that surface diffusion can be ignored. This assumption is discussed further in subsection 4.6.

4.5   *Experimental Results*

The earliest diffusion experiments on rocks from the Sellafield site were on samples distant from hydraulically active zones (termed "Flowing Zones") in the boreholes. These experiments were undertaken on a representative selection of lithologies from the most relevant stratigraphic units in the Sellafield area, in order to obtain data for the diffusivity and porosity of the unaltered rock matrix. Subsequent measurements were targeted on core samples where there was visible evidence of alteration close to potential water-conducting fractures; these measurements produced information on the extent to which the RMD properties of the rocks were altered around fractures. Later in the programme, experiments were undertaken solely on rock samples from identified Flowing Zones and from surrounding fractures in the core identified as potential flowing features. These experiments mainly used rock samples from the BVG.

The objective of studying rock around hydraulically active fractures is to build understanding of the way in which the diffusive properties of the rock vary as a function of distance from the fracture. For example, if the porosity and diffusivity of the rock adjacent to fractures were reduced by natural cementation processes, then the diffusive flux into the block of rock would be reduced, and the efficiency of RMD as a retardation mechanism may decrease. Conversely, if porosity and diffusivity increased adjacent to the fractures, as a consequence of mineral dissolution or stress-relief processes, then the diffusive flux into the block (and the equilibrium capacity of the block for "storing" non-sorbing radionuclides in the porewater) would increase and the importance of RMD as a retardation mechanism may be greater.

Changes to RMD Properties around Fractures

Figure 3 presents a compilation of data for similar studies of rock matrix diffusivity around fractures in the BVG. The bars in this figure indicates the distance over which the measurement was made (*i.e.* individual $\pm x$ bar gives the half length of the rock sample) and the uncertainty in the measured diffusivity (*i.e. y* error bars are calculated from experimental errors), which are typically of the order of $\pm 30\%$. Typically, the sample closest to the fracture face shows the highest diffusivity, consistent with the data discussed above. The large uncertainties in some of the diffusivity results are for early experiments, using the standard through-diffusion cell, where samples of BVG had diffusivities similar to that of the sealant. The results from these experiments gave the impetus to develop new cells capable of measuring lower diffusivities.

Figure 3. **Compilation of profiles of diffusivity measurements from probable water-conducting fractures in the BVG**

mbRT is the depth in metres (m) below the rotary table (bRT) of the drilling rig; Longlands Farm Member, Town End Farm Member, Fleming Hall Formation and Bleawath Formation are units within the Borrowdale Volcanic Group (BVG).

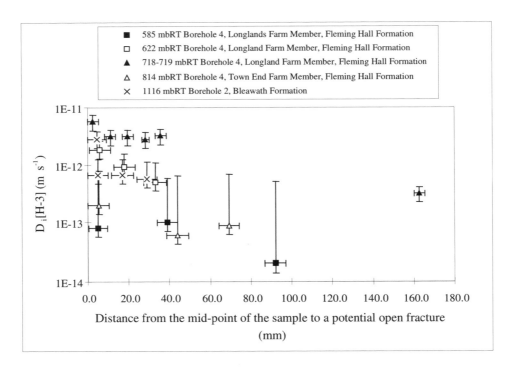

Available data from the Sellafield site have shown that the diffusion properties of rocks may be modified in the altered margins around water-conducting fractures. The intrinsic diffusion coefficient is observed to be increased in many alteration haloes around fractures in the BVG; however, no decreases, suggestive of pore blocking, have been observed.

Heterogeneity

Heterogeneity in diffusion properties has been investigated in a recent study of RMD in a section of BVG from borehole RCF3. The section contains a known Flow Zone, the position of which is given in Figure 4. Results from diffusion experiments undertaken using the modified through-diffusion cell are presented in this figure, together with estimates of diffusivity derived from ionic conductivity measurements. Ionic conductivity measurements were undertaken both on samples used in the through-diffusion experiments (of order 10 mm thick) and on long cores (up to 0.5 m length).

The results in Figure 4 demonstrate that there is significant variability in diffusion properties on small length scales, both within the Flow Zone (which comprises multiple fractures) and in the adjacent wallrock. For 10 mm long samples, D_i (measured either directly by diffusion experiment or calculated from ionic conductivity measurements) varies by up to three orders of magnitude over short distances. In contrast, ionic conductivity measurements on longer core sections show much less variability (calculated D_i values were all within one order of magnitude). For this particular interval of rock, it is also noticeable that there is no significant enhancement in D_i in the Flow Zone relative to that in the surrounding rock.

Figure 4. Compilation of diffusivity measurements for a BVG section in Borehole RCF3 containing a flow zone with associated Potential Flowing Features [1]

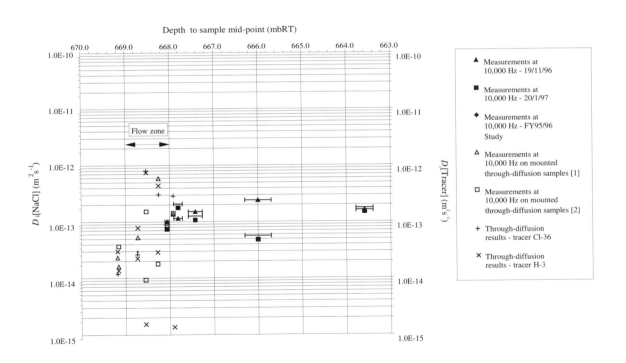

Anion Exclusion

In order to quantify the extent of anion exclusion in a rock sample, it is necessary to compare the diffusion properties of the rock towards two contrasting tracers: one anionic, such as iodide or chloride, and the other uncharged, such as tritiated water. Two types of experiment are used to evaluate anion exclusion effects. Through-diffusion experiments are used to determine values for D_i.

Comparison of the diffusibility of the rock, Ψ, towards the anion and uncharged tracer gives an indication of the extent to which the anion is excluded from the transport porosity in the rock. Out-diffusion experiments, in which the total activity (or mass) of a tracer in a sample is determined, are used to determine values for ϕ. Comparison of the solute-accessible porosities available to anions and uncharged tracers gives a direct measurement of the exclusion of the anion from pores in the rock.

Ideally, the two tracers would be used simultaneously in the experiments, and values of D_i or ϕ would be calculated for each tracer. However, although such "multi-tracer" experiments are preferable, they may not be practicable because of radiochemical analysis considerations. Instead, it may be necessary to undertake experiments sequentially on the same sample (e.g. HTO experiment followed by a ^{36}Cl experiment), by "recycling" the sample.

Figure 5 shows an example of a solute accessible porosity profile. This demonstrates anion exclusion (of iodide) from the porosity in the rock matrix. The solute-accessible porosity adjacent to the fracture is approximately two orders of magnitude higher than that at a distance of 150 mm from the fracture. Higher values of solute-accessible porosity correspond to the zone of haematite alteration around the fracture. Evidence for anion exclusion is seen at distances greater than 50 mm from the fracture face.

Figure 5. **Profile of solute accessible porosity measurements from a probable water-conducting fracture into the rock matrix of BVG sample**

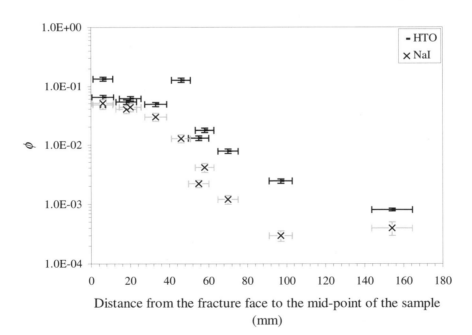

Distance from the fracture face to the mid-point of the sample

(mm)

4.6 *Surface Diffusion*

In addition to the diffusive transfer of radionuclides in solution, it is possible that sorbed radionuclides are mobile on mineral surfaces and hence can diffuse across them in response to concentration gradients of the sorbed radionuclide on the mineral surface. This process is described in the literature as "surface diffusion" and has been invoked by some workers [e.g. 6] to explain results from laboratory experiments. Although theoretically possible, the direct evidence for this process in laboratory diffusion experiments undertaken with saline groundwaters typical of Sellafield is currently ambiguous.

4.7   *Perturbations due to Sample Collection and Preparation*

The rock samples used in the laboratory experiments have been obtained from boreholes, at depths of up to 1 600 m below ground level. It is possible that the drilling process could cause some changes to the rock properties. For example, the pore space in the rock could be reduced as a result of the invasion of drilling muds into the rock sample. Alternatively, additional porosity could be formed as a result of stress relief when the sample is removed from the ground. If laboratory diffusion data are to be used as input to post-closure performance assessment calculations, it is necessary to investigate whether these process could lead to significant changes in the rock properties, such that the laboratory measurements are not representative of *in situ* diffusion conditions.

The impact of some of these effects can be investigated in the laboratory. For example, by returning samples to *in situ* effective confining stresses, it is possible to evaluate the effect of elastic compressibility effects on RMD.

The influence of elastic compressibility effects on diffusivity has been investigated by placing the rock sample in a triaxial pressure vessel, and measuring diffusivity as a function of confining stress. A limited number of through-diffusion experiments have been performed under, first, effective confining pressures representative of *in situ* burial depths (of order 9 MPa for a depth of 0.7 km) and, second, under conditions of low confining stress (0.2 MPa). Only one experiment was carried out using a rock sample from the Sellafield site. In this experiment, the diffusivity of a sample of lapilli tuff towards HTO was insensitive to confining pressures in the range 0.2-9.0 MPa. However, the diffusivity towards iodide decreased by a factor of five over the same pressure range. Similar trends, of increasing ion exclusion effects under higher confining stress, have previously been reported in clays [7]. Further data would be required to demonstrate whether the limited data from rocks at Sellafield were representative.

4.8   *Determination of In situ RMD Properties*

As discussed above, it is possible that the drilling process could cause some changes to the rock properties, affecting the results of subsequent RMD experiments. Recent work published as part of the Swedish Radioactive Waste Management R&D Programme has extended the electrical conductivity approach to the field scale, by inferring RMD properties of rocks using borehole geophysical data. The benefit of this approach is that it would allow RMD properties to be estimated over large areas both quickly and *in situ*. This is valuable, in that information on RMD could then be available early in the site characterisation process. The technique is based on an extension of the methodology to determine RMD parameters using ionic conductivity (as discussed in subsection 4.4):

$$\frac{1}{F} = \frac{R_w}{R_o} = \frac{\sigma_s}{\sigma_f} \tag{2}$$

where R_o is the resistivity of the saturated rock, and R_w is the resistivity of the pore fluid.

In the absence of a rigorous proof, it has been customary in the geophysical literature to assume (implicitly) the validity of Equation (2) as a working hypothesis. This provides the basis for using conductivity measurements to estimate the RMD properties of water-saturated rock.

The correlation between F and Ψ values has been tested, using experimental data produced by the NSARP. There are only a small number of direct comparisons, where both electrical conductivity and intrinsic diffusion coefficient have been measured on the same sample in the laboratory. The limited number of measurements precludes detailed analysis. Instead, the cumulative frequency distributions for diffusibilities and $1/F$ have been compared. Although electrical conductivity and diffusivity have typically been measured on different samples, there has been no systematic difference in the types of samples studied using the different techniques. Results indicate that the cumulative frequency distributions for Ψ[HTO] and $1/F$ are very similar. Thus, based on experimental data for the BVG at Sellafield, it can be concluded that $1/F$ is approximately equal to Ψ[HTO]. To derive values for Ψ[I⁻] and Ψ[Cl⁻] from electrical conductivity data, it would be necessary to establish the relationship between Ψ[HTO] and Ψ[I⁻] and Ψ[Cl⁻].

Formation factor F profiles for the BVG were calculated from both wireline resistivity and porosity logs from a number of the boreholes at Sellafield. Formation factor data calculated from laboratory conductivity measurements on small core samples were also available, collected as part of experiments to determine RMD properties. These data sets can be compared directly. In Figure 6, the laboratory derived F values from core samples obtained from borehole RCF3 are co-plotted with a section of the downhole F profile calculated from the RCF3 deep resistivity log. There is an apparently good agreement between the core and wireline F values over a region of the borehole where there is a significant variation in F values (by up to 2-3 orders of magnitude) over relatively small changes in depth. The laboratory-measured F values appear to follow the trends in the wireline F profile. It should be noted that in Figure 6 the core-derived F values have been plotted at their mean depth, although the core specimens studied varied between 9 and 75 cm in length.

Some deviation between core sample and wireline derived F values is to be expected owing to the marked spatial variability of *in situ* formation resistivity and porosity evident from the wireline logs and the small-scale heterogeneity of the core samples evident from the laboratory measurements. Thus, the observed differences, particularly at higher F values, must partly reflect the small-scale heterogeneity of the bulk rock.

There is good agreement between formation factors derived from wireline and core sample conductivity measurements. This provides confidence that wireline deep resistivity measurements provide data concerning the electrical connectivity of the *in situ* rock which is both qualitatively and quantitatively accurate. On the basis of the correlation found between $1/F$ and Ψ[HTO], it is reasonable to convert the Formation Factor profile calculated from the wireline data into an intrinsic HTO diffusivity profile. Thus, this study has provided evidence that the wireline deep resistivity log may be suitable for use in estimating the rock matrix diffusion properties of the *in situ* rock. However, further testing of the approach, e.g. by comparison of the downhole formation factor profile for other Sellafield boreholes with laboratory data, is required to build confidence in the method.

5.   Natural Analogue Evidence

The laboratory experimental programme has provided quantitative information on the diffusivity and porosity of the rock matrix. However, experimental time scales are short compared with the time scales of relevance to the post-closure performance assessment of a radioactive waste repository. (In the context of RMD, "appropriate time scales" are of the order of 10^3-10^6 years). Therefore, it is desirable to find evidence that RMD operates on these time scales in natural geological systems. A number of natural analogue studies have been undertaken to investigate RMD. These studies address time scales from tens of years (e.g. [8]), through thousands of years (e.g. [9]) to millions of years (e.g. [10-12]), and have involved observations of the diffusive mobility of a wide range of tracers, such as uranium series radionuclides and halides.

In fractured rocks, the most common natural analogue approach is to determine the spatial distribution of an element or radionuclide in the rock matrix adjacent to a natural fracture surface. If RMD has occurred, concentration (or activity) gradients may be observed around the fracture surface as a consequence of diffusion of the radionuclide into, or out of, the rock matrix. These profiles can be interpreted at one of two levels. At the simplest level, it can be assumed that, given sufficient time, repository-derived radionuclides would be able to access at least the volume of rock in which natural diffusion processes have been observed (see below for a discussion as to why the volume of rock accessed by natural diffusion may be a minimum value). At the more complex level, the profile may be analysed quantitatively to derive information on the diffusivity of the rock matrix and the sorptivity of the diffusing radionuclide.

Figure 6. **Comparison of Wireline-derived & Laboratory-derived Formation Factors. *In situ* formation factor profile from BVG in borehole RCF3 (derived from deep resistivity measurements) compared with *F* values derived from laboratory conductivity measurements on core samples**

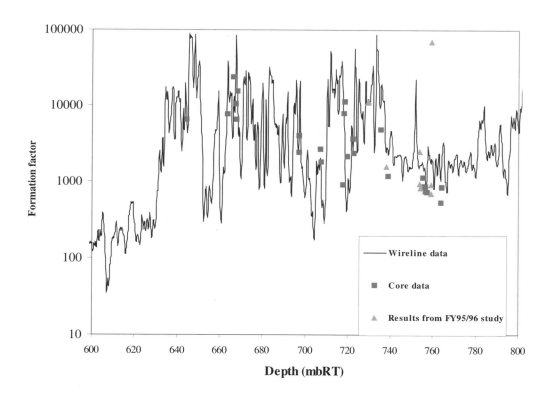

Since 1991, NSARP work has focused on two sites: Sellafield, UK and El Berrocal, Spain. Uranium concentrations in the potential repository host rocks at Sellafield are low, typically ranging from 1-6 ppm. Analytical uncertainty in the radiochemical analyses was therefore often significant; this has limited interpretation of the data.

At the Sellafield site, widths of haematised wallrock margins around open fractures have been measured in three sections of BVG taken from the Potential Repository Zone (PRZ). Where present, the maximum width of the haematised margin is approximately 16 mm; the modal width is approximately 1 mm. This demonstrates a minimum distance from the fracture surface that can be accessed by RMD. It is important, however, to note that the limited depths of penetration are probably due to the limited flux of dissolved oxygen through the fractures, which is in part due to the low permeability of the rock, and the large buffering capacity of the BVG towards oxidation.

Much effort in such rocks has focused on the study of uranium series radionuclides, in part because of their direct relevance to performance assessments. However, the distance over which diffusive transfer has occurred, either "recently" (i.e. as demonstrated by isotopic disequilibrium) or in the geological past, typically does not exceed 10 cm. These short migration distances are consistent with our knowledge of the geochemical retardation characteristics of uranium series radionuclides; it is not necessary to postulate the rock matrix at greater distances from the fracture face is inaccessible to any diffusing radionuclide.

The following point should be emphasised. Provided that the matrix porosity is connected to the flowing fractures, then RMD will occur, and for radionuclides that are not otherwise retarded by interaction with the fracture wall or the rock matrix, it will provide a mechanism for retardation of radionuclide transport that is potentially very significant. Some radionuclides may interact with the fracture walls or the rock matrix. For such radionuclides RMD potentially provides access to many more interaction sites and hence even more retardation.

There is also the possibility that some radionuclides may become irreversibly incorporated into the rock matrix (the process of "mineralisation"). This process has not so far been represented in assessment calculations because it is considered that the appropriate models and necessary data have not been established yet, and it would, at present, be difficult to justify any particular model. However, it is necessary to take the effects of mineralisation into account in analyses of natural analogues in which the objective is to understand the processes that occur in real systems. If there has been significant mineralisation, this may obscure the effects of RMD. Care must therefore be taken in making inferences from natural analogues about the fraction of the matrix porosity that is accessible.

There is incontrovertible evidence for diffusive transfer of non-sorbing tracers, such as chloride, over distances of metres in high porosity rock matrices such as clays and chalk. The chloride profile through a granite block that had been immersed in sea water, as part of a sea defence for 32 years, showed that chloride had accessed the matrix porosity across the block (600 mm) [8]. Evidence of chloride access to matrix porosity in granites over time scales of thousands of years has been obtained from the Gulf of Finland [9]. In this area the land surface is rising in response to unloading following melting of the ice sheet formed around the last glacial maximum. This process, which is believed to have been operating for the last 5 000 years, has resulted in the gradual replacement of saline groundwaters by dilute meteoric waters both by infiltration from the ground surface along fractures and by diffusion of chloride ions from the granite matrix to a depth of at least 40 m. Diffusive transport appears to have operated on a scale of metres away from the fractures.

Finally, indirect evidence for the operation of RMD at the Sellafield site is provided by studies of groundwater residence times, which have been undertaken as part of the Site Characterisation Programme. Palaeohydrogeochemical studies indicate that, within the BVG of the PRZ, the groundwater is derived by mixing of saline brines from the Irish Sea Basin (and possibly from beneath the Lake District) and dilute meteoric waters from recharge to the Lake District Massif. Stable isotope and noble gas data for the groundwaters in the PRZ indicate a predominance of "old recharge", possibly of late Pleistocene age (of order 10^4 years ago). Given the current conceptual hydrogeological model of the Sellafield site [1], in which fracture flow is considered to be dominant in the BVG, these long residence times imply that the rock matrix has been accessed by the solutes.

Natural analogues tend to suggest that diffusive processes operate over a range of length scales in low permeability fractured rocks. However, it is pointed out by critics that it is difficult to prove that diffusion alone is responsible for the observed concentration profiles, as the complete history of the natural analogue is unknown. It is possible that the profile could have developed at some time in the past, when the microfracture network was more favourable to diffusion or when advection may have been more important.

6. RMD in Nirex 97

Section 3 introduced the RMD model used in Nirex 97, as illustrated in Figures 1 and 2. In order to use this simple model, it was necessary to derive appropriate effective parameters to best represent the irregularity and heterogeneity of the real rock in the simple model, and to demonstrate that the simple model was appropriate. Information derived from laboratory and field experiments, and from natural analogues evidence, was utilised in this process. The effective parameters were appropriate averages of the corresponding properties of the rocks on small length scales, and were essentially derived by an upscaling process. They were determined so that the model would behave in the right way at very long times, when all of the accessible matrix porosity is accessed by diffusion, and at very short times, when only the fracture porosity is accessed and radionuclides are diffusing away from the fractures into the matrix at a rate determined by the intrinsic diffusion coefficient. In order to achieve this, the effective fracture aperture and the effective diffusion distance for the simple model were determined so that the fracture porosity in the model matched the effective fracture porosity in the real rock.

6.1 Effective fracture porosity

The effective fracture porosity for the BVG was determined from the effective permeability due to the FFs. For a network of fractures, the effective permeability can be related to the distribution of the transmissivity of the fractures as follows (see [1]):

$$k_{ei} = \frac{C_{Gi}G_i\hat{T}}{b},$$ (3)

where k_{ei} is the effective fracture permeability;

C_{Gi} is a factor that represents the combined effect of modelling uncertainty (see [1]) and the connectivity and variability of the fractures;

G_i is a geometric factor, which accounts for the orientation of the fractures relative to the direction of interest;

$\hat{T}$ is the geometric mean fracture transmissivity;

b is the average spacing of the fractures (for convenience, measured in a vertical borehole).

This can be used to infer the geometric mean fracture transmissivity $\hat{T}$ from the effective permeability. Then the geometric mean fracture aperture $\hat{a}$ can be determined from $\hat{T}$ using

$$\hat{T} = \frac{1}{12}(\hat{a})^3.$$ (4)

The effective porosity on a large length scale due to the network of fractures can then be calculated using

$$\phi_f = \frac{C_F C_A C_a G' \hat{a}}{b},$$ (5)

where ϕ_f is the fracture porosity

G' is a geometric factor that represents the effects of the orientation of the fractures;

C_a is the ratio of the arithmetic mean fracture aperture to the geometric mean aperture (and is usually close to 1);

C_A is a factor that represents the combined effect of modelling uncertainty and the connectivity and variability of the features;

C_F is a 'fracture roughness' factor that accounts for the fact that numerical fracture-network calculations tend to underestimate fracture porosities by a factor somewhere between 1 and 10.

The factor C_a is introduced because the effective porosity is closer to being proportional to the arithmetic mean aperture than the geometric mean aperture, but it is very convenient to use the geometric mean aperture evaluating the uncertainty in the effective porosity.

For models in which the FF clusters are connected, this approach was extended further as follows. First, the effective fracture porosity of an FF cluster was derived from the effective permeability of the FF cluster as described above. Then the overall effective fracture porosity of the rock was estimated using a modification of the above approach in which the porosities of the features in the network were taken to be the effective porosities of FF clusters rather than the porosities of simple fractures, which are determined by their aperture.

6.2 *Effective fracture aperture*

The effective half-aperture $a_{1/2}$ (Figure 1 indicates the effective aperture a) for use in the simple model of rock-matrix diffusion was taken to be given by (see [1]):

$$a_{1/2} = \frac{1}{2} \phi_f b \cos\theta,$$ (6)

where θ is an average dip angle for the fractures.

6.3 *Effective diffusion distance*

The maximum distance for diffusion into the matrix, d, (see Figure 1) was taken to be (see [1]):

$$d = \frac{1}{2} b \cos\theta.$$ (7)

176

6.4   *Effective capacity factor and effective intrinsic diffusion coefficient*

Two additional key parameters in the rock-matrix diffusion model are the effective capacity factor α and the intrinsic diffusion coefficient D_i [1]. It was considered that the appropriate values for the effective capacity factor and intrinsic diffusion coefficients taking account of the heterogeneity would be respectively:

$$\alpha_{eff} = Average[\alpha], \tag{8}$$

$$D_{i,eff} = \frac{\left(Average\left[\sqrt{D_i \alpha}\right]\right)^2}{\alpha_{eff}}, \tag{9}$$

where α is the small-length-scale capacity factor.
         D_i is the small-length-scale intrinsic diffusion coefficient;

In practice, it was adequate, and very convenient, to approximate the arithmetic averages in the above by geometric averages.

The distributions of small-length-scale capacity factor and intrinsic diffusion coefficient for the BVG were derived from the results of laboratory experiments carried out within the NSARP. A significant number of the measured intrinsic diffusion coefficients were below the resolution of the experimental technique used to measure the intrinsic diffusion coefficient, because diffusion through the sealant used to secure the sample into the experimental apparatus was greater than diffusion through the BVG, which has a very low intrinsic diffusion coefficient. This was taken into account in the analyses. There was some uncertainty about the way in which this was done, and this uncertainty was incorporated into the overall uncertainty.

7.   Research Model Development

As discussed above, a simple model of RMD has been used in assessment calculations, [1]. This model incorporates various approximations, such as all the fractures having the same properties and being arranged regularly, whereas in the real system the fractures are arranged irregularly and their properties vary from fracture to fracture, and within a single fracture. In the real system, therefore:

(a)   the flow velocity will vary both from fracture to fracture and within a fracture. Indeed, the flow may even be confined to channels within each fracture, and/or along fracture intersections;

(b)   the distance that radionuclides can diffuse into the rock matrix varies from fracture to fracture and within a fracture;

(c)   diffusion away from the fractures into the matrix is likely to be three-dimensional rather than one-dimensional.

Therefore, research calculations have been undertaken:

(a)   to assess the adequacy of the simple assessment model;

(b)   if necessary to support the development of improved models for use in assessment calculations;

(c)   to support the determination of the appropriate effective parameters to best represent a real system, using either the simple model or suitably modified models.

In order to do this, a realistic network in a suitable block was set up using the fracture-network programme NAPSAC [13]. The parameters for the network model of the BVG in the Nirex 95 [2] analysis were used. A number of transport paths through the network were computed. The transport velocity varied from fracture to fracture along the paths and between the paths. The effect of the variation in velocity was examined by calculating the breakthrough curve for the combined paths. Transport along each path was represented using equations similar to those for the simple model. The equations were solved by Laplace transform techniques. It was concluded that the simple model with appropriate effective parameters adequately represented the effects of the variation in velocity.

The effect of variations in the diffusion distance was examined by considering a model with a number of parallel paths through the blocks, in which the diffusion distance for each path was allowed to vary, and the equations for all the paths were solved together as a coupled system. Thus, the diffusion distance was effectively determined by the transport process, rather than being imposed *a priori*. It was found that the simple model gave an acceptable representation, for the case considered in which the spread of the transmissivity distribution was not large.

In order to examine the effect of three-dimensional diffusion away from highly-channelled flow, calculations were undertaken for a variant of the simple model with radial diffusion away from pipes rather than linear diffusion away from a planar fracture. It was concluded that the simple assessment model was conservative.

8.   Summary

Clearly, it is important that confidence can be built in the approach undertaken, in a performance assessment, to modelling a transport process such as RMD. This paper has presented an overview of work that has been undertaken by the NSARP to develop, justify and parameterise appropriate conceptual models of RMD for implementation in post-closure performance assessment calculations, e.g. Nirex 97. There are several complementary "strands" to the work: laboratory experiments, natural analogue observations and research model development, which were utilised in developing the conceptual model of RMD used in that performance assessment. Recent studies have examined the potential for a further methodology that could determine *in situ* RMD properties; this could provide additional information to justify the current approach to modelling RMD in a performance assessment.

9.   Acknowledgements

This paper is based on information derived from the following NSARP projects:

- *The Role of Rock-matrix Diffusion in Retarding the Migration of Radionuclides from a Radioactive Waste Repository*;

- *The Suitability of Wireline Logs for Evaluating the Matrix Diffusion Properties of In situ Rock*;

- *Behaviour and Geological Evolution of Uranium in the Borrowdale Volcanic Group at Sellafield*;

- *An Assessment of Alteration Zones Related to Fractures in BVG Rocks in Nirex RCF3: Implications for Distance of Rock-Matrix Diffusion from Fractures.*

The following experts worked on these projects, and their contributions are gratefully acknowledged:

- A.J. Baker, C.P. Jackson, N.L. Jefferies, T.R. Lineham, G. Longworth & S.W. Swanton (AEA Technology, Harwell).

- N.R. Brereton, N.J. Fortey, E.K. Hyslop, P.D. Jackson, S.J. Kemp & A.E. Milodowski (British Geological Survey).

10. References

[1] *Nirex 97: An Assessment of the Post-closure Performance of a Deep Waste Repository at Sellafield* (5 Volumes), Nirex Science Report S/97/012, 1997.

[2] *Post-closure Performance Assessment. Nirex 95: A Preliminary Analysis of the Groundwater Pathway for a Deep Repository at Sellafield* (3 Volumes), Nirex Science Report S/95/012, 1995.

[3] *The Treatment of Water-conducting Features in Groundwater Flow and Transport Modelling of the Borrowdale Volcanic Group in Nirex 97*, C.P. Jackson, S. Norris, S.J. Todman & S.P. Watson. Proceedings of 3rd NEA GEOTRAP Workshop on "*Characterisation of Water-conducting Features and their Representation in Models of Radionuclide Migration*", Barcelona, 1998.

[4] *Some Notes on Experiments Measuring Diffusion of Sorbed Nuclides Through Porous Media*, D.A. Lever, UKAEA Report AERE-R.12321, 1986.

[5] *The Electrical Resistivity Log as an Aid in Determining Some Reservoir Characteristics*, G.E. Archie, Trans. Am. Inst. Mech. Eng. 146, 54-62, 1942.

[6] *Diffusion in Crystalline Rocks of Some Sorbing and Non-sorbing Species*, K. Skagius and I. Neretnieks, SKBF/KBS Technical Report 82-12, 1982.

[7] *Mass Transfer Mechanisms in Compacted Clays*, P J Bourke, N.L. Jefferies, D.A. Lever and T.R. Lineham. In "*Geochemistry of Clay-Pore Fluid Interaction*", ed. D.A.C. Manning, P.L. Hall and C.R. Hughes, pp331-350, Chapman and Hall, 1993.

[8] Long Term Solute Diffusion in a Granite Block Immersed in Seawater, N.L. Jefferies. In "Natural analogues and radioactive waste disposal", ed. B. Come and N.A. Chapman, pp249-260; 1987, Graham and Trotman.

[9] *Concentration Profiles of Anions in Granite Bedrock Caused by Postglacial Land Uplift and Matrix Diffusion*, M. Olin and M. Valkiainen, TVO/Ydinjatetutkimumset Technical Report 90-1, 1990.

[10] *Rock Matrix Diffusion as a Mechanism of Radionuclide Retardation: a Natural Analogue Study of El Berrocal Granite, Spain*, M.J. Heath, M. Montoto, A. Rodriguez Rey, V.G. Ruiz de Argandoña, and B. Menendez, Radiochimica Acta 58/59, 379-384, 1992.

[11]   *An Analogue Validation Study of Natural Radionuclide Migration in Crystalline Rocks Using Uranium Series Disequilibrium Studies*, J.A.T. Smellie, A.B. Mackenzie and R.D. Scott, Chemical Geology 55, 233-254, 1986.

[12]   *History of Actinide and Minor Element Mobility in an Archaean Granitic Batholith in Manitoba, Canada*, M. Gascoyne and J.J. Cramer, Applied Geochemistry 2, 37-54, 1987.

[13]   *NAPSAC (Release 4.0) Summary Document*, L.J. Hartley, A.W. Herbert and P.M. Wilcock, AEA-D&R-02771, 1996.

The Effect of Onsager Processes on Radionuclide Transport in the Opalinus Clay

J.M. Soler
Paul Scherrer Institut, Switzerland

1.   Introduction

Coupled phenomena (thermal and chemical osmosis, hyperfiltration, coupled diffusion, thermal diffusion, thermal filtration, Dufour effect) may play an important role in fluid, solute and heat transport in clay-rich formations, such as the Opalinus Clay (OPA), which are being considered as potential hosts to radioactive waste repositories.

Rocks containing large proportions of compacted clays may act as semipermeable membranes due to the existence of ionic double layers on the clay surfaces. Since the structural charge on clay surfaces is negative due to isomorphic substitutions in the mineral lattice, a diffuse double layer of counter-ions (cations) in solution develops right next to the mineral surface. If the rock is sufficiently compacted so the diffuse double layers of adjacent clay platelets overlap, the distribution of electrical charge in solution allows water and non-charged solutes to pass through the pores but prevents ionic species from doing so. Under such conditions, a clay-rich rock can potentially act as a semipermeable membrane, with chemical osmosis, hyperfiltration, and coupled diffusion playing important roles in fluid and solute transport. Also, the existence of temperature gradients in the vicinity of a repository could promote fluid and solute transport by thermal osmosis and thermal diffusion (Soret effect). Coupled heat transport phenomena (thermal filtration, Dufour effect) could also, in principle, contribute to the heat fluxes.

The Opalinus Clay, a shale formation in northern Switzerland, has been selected as a potential host rock for a repository for vitrified high-level radioactive waste (HLW) and spent nuclear fuel (SF). An underground rock laboratory is in operation at Mont Terri, Canton Jura, Switzerland, in order to study the geological, hydrogeological, geochemical, and rock-mechanical properties of this formation. The clay mineral content of the rock ranges from 20 to 75 wt% (NAGRA, 1989; MAZUREK, 1999), and therefore, coupled phenomena could play an active role in fluid and solute transport, including radionuclide transport, through this formation. As part of the effort to characterize the transport properties of OPA, vital to performance assessment studies of a repository, the effects of coupled phenomena have to be taken into consideration. The objective of this study is to provide estimates of the magnitudes of the fluxes associated with these coupled phenomena, and to identify the processes that may have the highest impact on performance assessment studies and possibly require further investigation.

An extended version of the article has been published as a PSI Technical Report by SOLER, 1999.

2.   Direct and coupled transport phenomena

Coupled transport phenomena are described in the reference frame of the thermodynamics of irreversible processes. The theory of irreversible thermodynamics starts from an extension of the second law of thermodynamics, which introduces the concept of entropy (S). It can be shown that the rate of local or internal entropy production of a given system, per unit volume, can be written in terms of

$$\frac{1}{V}\frac{dS}{dt} = \sum_i J_i X_i \tag{2.1}$$

where V is the volume of the system, J_i is a flux (e.g. flux of heat, fluid, solutes, electrical current), and the X_i terms are driving forces (e.g. temperature, hydraulic, chemical potential, or electrical potential gradients). The assumption in the theory of irreversible thermodynamics is that the forces appearing in Eqn. 2.1 are the only forces needed to fully describe the kinetics and evolution of the system (LASAGA, 1998). If this assumptions holds, the second law of thermodynamics will state that

$$\sum_i J_i X_i \geq 0 \tag{2.2}$$

which means that each flux term is a function (although unknown) of all the driving forces. Now, an additional assumption is that this function is linear, and therefore, a flux can be expressed as

$$J_i = \sum_j L_{ij} X_j \tag{2.3}$$

where the L_{ij} terms are the so-called phenomenological coefficients. The term direct or diagonal phenomena is used for the $L_{ii} X_i$ contribution to a flux J_i, and the term coupled or off-diagonal phenomena is used for the $L_{ij} X_j$ contribution when $j \neq i$. Also, the phenomenological coefficients associated with coupled phenomena are related by the Onsager Reciprocal Relations (ONSAGER, 1931)

$$L_{ij} = L_{ji} \tag{2.4}$$

Table 2.1 is the matrix for direct and coupled phenomena for different transport processes. The effects of thermal, hydraulic, and chemical gradients, on heat, fluid, and solute fluxes, have been considered. The contribution of coupled heat transport phenomena to total heat fluxes will not be included in this article. It should be mentioned that the estimates by SOLER (1999) suggest that the heat fluxes associated with thermal filtration and the Dufour effect are negligible compared to the fluxes associated with thermal conduction.

Table 2.1. **Onsager matrix – Matrix of direct (diagonal) and coupled (off-diagonal) transport phenomena (DE MARSILY, 1986; HORSEMAN et al., 1996). Shaded boxes correspond to the processes considered in the study.**

FLUX J	POTENTIAL GRADIENT X			
	Temperature	Hydraulic	Chemical	Electrical
Heat	Thermal conduction	Thermal filtration	Dufour effect	Peltier effect
Fluid	Thermal osmosis	Advection	Chemical osmosis	Electrical osmosis
Solute	Thermal diffus. or Soret effect	Hyperfiltration	Diffusion	Electro-phoresis
Current	Seebeck or Thompson eff.	Rouss effect	Diffusion & Membr. Pot.	Electrical conduction

The effects of possible natural electric potential gradients on the different fluxes have not been considered in the study. To date, there is no information regarding electric potential gradients in the Opalinus Clay, and although it is true that spontaneous potentials are measured in clay formations, it is not clear what the connection is between a potential at a microscopic scale (surface charge on clays, formation of diffuse double layers), which may be the cause of such spontaneous potentials, and one at a macroscopic (metric) scale. Also, only very scarce information is available regarding the coupling coefficients for coupled transport phenomena driven by electric potential gradients.

3. Formulation of solute fluxes associated with coupled transport phenomena

The solute fluxes associated with direct and coupled transport phenomena will not be written here according to proper irreversible thermodynamics (see SOLER, 1999), but in terms of parameters and gradients that are commonly measured in field or laboratory experiments. These fluxes, all of them in units of kg/m^{2}rock/s, will be formulated as follows:

$$\text{Advection} \qquad J_{ADV} = -c_i K \frac{\partial h}{\partial x} \qquad (3.1)$$

$$\text{Chemical diffusion} \qquad J_D = -D_e \frac{\partial c_i}{\partial x} \qquad (3.2)$$

$$\text{Chemical osmosis} \qquad J_{CO} = c_i \sigma K \frac{\partial \Pi_h}{\partial x} \qquad (3.3)$$

$$\text{Hyperfiltration} \qquad J_{HYP} = c_i \sigma K \frac{\partial h}{\partial x} \qquad (3.4)$$

Thermal diffusion $\qquad J_{TD} = -D_e s c_i \dfrac{\partial T}{\partial x}$ $\qquad\qquad$ (3.5)

Thermal osmosis $\qquad J_{TO} = -c_i k_T \dfrac{\partial T}{\partial x}$ $\qquad\qquad$ (3.6)

A detailed explanation of the formulation used for the fluxes is given below.

3.1 *Advection*

The flux of fluid associated with advection (Darcy velocity, v_D), in units of volume of fluid per unit cross-section area of rock per unit time, will be written according to Darcy's law, expressed in terms of hydraulic conductivities and hydraulic heads. In one dimension, Darcy's law has the form

$$v_D = -K \frac{\partial h}{\partial x} \qquad\qquad (3.7)$$

Field tests have indicated that hydraulic conductivities for the Opalinus Clay (OPA) are less than 10^{-11} m/s (NAGRA, 1989). Also, laboratory measurements (HARRINGTON & HORSEMAN, 1999; DE WINDT & PALUT, 1999) have yielded values between 10^{-14} and $8\cdot10^{-14}$ m/s. Based on these measurements, the range of values for the hydraulic conductivity of OPA goes from 10^{-14} to 10^{-12} m/s, with an extended range (less probable values) reaching up to 10^{-11} m/s.

$$K: 10^{-14} \, \Lambda \; 10^{-12} \; (10^{-11}) \; m/s$$

The solute flux associated with advection is given by Eqn. 3.1.

3.2 *Chemical diffusion*

The diffusive solute flux will be written in terms of Fick's first law, given by Eqn. 3.2. D_e is the effective diffusion coefficient for species i, which can be given in terms of (a)

$$D_e = \phi\left(\frac{\chi}{\tau^2}\right)D_0 \qquad\qquad (3.8)$$

where χ and τ are the constrictivity and tortuosity of the porous medium, respectively, and D_0 is the molecular diffusion coefficient of species i in water, or (b)

$$D_e = \frac{D_0}{F} \qquad\qquad (3.9)$$

where F (the resistivity or formation factor) is given by

$$F = \phi^{-m} \qquad\qquad (3.10)$$

and m (the cementation exponent) has reported values between 1.3 and 5.4 (HORSEMAN et al., 1996, and references within), and values around 2 for deeply buried compacted sediments (ULLMAN & ALLER, 1982).

Coupled diffusion (the diffusive flux of one species driven by the chemical potential gradients of other species) will not be considered when estimating the solute fluxes associated with the different transport mechanisms. The goal is to compare the solute fluxes associated with coupled phenomena to the diffusive flux as given by Eqn. 3.2 and see what the additional effect is.

In the case of simple (binary) aqueous electrolyte solutions, coupled diffusion has been shown to be due to the coulombic coupling between cation and anion. When cation and anion have different intrinsic (tracer) diffusion coefficients, the value of the net combined interdiffusion coefficient falls in between the values of the two individual tracer diffusion coefficients (LASAGA, 1998). In the case of a clay-rich rock with overlapping double layers, the negative charge on the surface of the clays will prevent (to some extent) anions from passing through the pores. This phenomenon, known as anion exclusion, will in turn cause a reduction of the cation flux, because of coulombic coupling (charge balance).

Porosity values for OPA determined from weight loss measurements due to evaporation of pore water range from 12% to 19% (MAZUREK, 1999). Porosities determined from Hg injection porosimetry and Cl content, which reflect geochemical or diffusion-accessible porosities, range from 5% to 11% (MAZUREK, 1999). Based on these values, relatively low temperatures (25-60°C), and assuming that the effective diffusion coefficient for OPA can be calculated according to

$$D_e = \phi^m D_0 \qquad (3.11)$$

with a value of m around 2 (ULLMAN & ALLER, 1982), and $D_0 = 10^{-9}$ m²/s at 25°C (see for instance LI & GREGORY, 1974), the range for effective diffusion coefficients goes from 10^{-12} to 10^{-11} m²/s, with an extended range (less probable values) going from 10^{-13} to 10^{-10} m²/s.

$$D_e: 10^{-12} \ \Lambda \ 10^{-11} \ (10^{-13} \ \Lambda \ 10^{-10}) \ \text{m}^2 / \text{s}$$

Laboratory measurements of effective diffusion coefficients for tritiated water and iodine in Opalinus Clay samples have provided values in that same range from 10^{-12} to 10^{-11} m²/s (DE WINDT & PALUT, 1999).

3.3 Chemical osmosis

Chemical osmosis is the flow of fluid (solution) caused by chemical potential gradients. Chemical-osmotic flow across a semipermeable membrane is up the salinity gradient, or equivalently, down the activity of water gradient.

The chemical-osmotic flux of fluid can be expressed in terms of a flow law similar to Darcy's law (KEMPER & EVANS, 1963; KEMPER & ROLLINS, 1966; BARBOUR & FREDLUND, 1989). This flow law, in one dimension, and with units of m³/m²rock/s, has the form

$$v_{CO} = K_\pi \frac{\partial \Pi_h}{\partial x} = \sigma K \frac{\partial \Pi_h}{\partial x} \qquad (3.12)$$

where K_π is the coefficient of osmotic permeability, σ is the coefficient of osmotic efficiency ($0 \leq \sigma \leq 1$), K is the hydraulic conductivity, and Π_h is the osmotic pressure head.

The coefficient of osmotic efficiency (σ) is a measure of how close to ideal a semipermeable membrane is. For an ideal membrane (no solute flux through the membrane is allowed) σ is equal to one. On the other hand, if there is no restriction on the flux of solute through the membrane, σ is equal to zero. The first measurements of osmotic efficiencies of OPA samples in the laboratory (HARRINGTON & HORSEMAN, 1999) give values of σ around 0.1.

The osmotic pressure head, Π_h, is defined as

$$\Pi_h = \frac{\Pi}{\rho g} \tag{3.13}$$

and the osmotic pressure, Π, is given by

$$\Pi = -\frac{RT}{V_w} \ln a_w \tag{3.14}$$

where V_w and a_w are the molar volume and activity of water, respectively. The activity of water can be calculated according to (GARRELS & CHRIST, 1965)

$$a_w = 1 - V_w \sum_i \frac{c_i}{W_i} \tag{3.15}$$

Equation 3.3 describes, in one dimension, the solute flux associated with chemical osmosis.

3.4   Hyperfiltration

Hyperfiltration is the flow of solutes up the hydraulic gradient, caused by the fact that an ideal semipermeable membrane will let water flow by advection through the pores but will prevent solutes from doing so. The hyperfiltration flux can be written as in Eqn. 3.4 (GROENEVELT, ELRICK & LARYEA, 1980) where σ, again, is the coefficient of osmotic efficiency. In reality, this coefficient has a specific value for each different species. However, for the estimates presented in this study, the same average value of σ as in the formulation of the chemical-osmotic flux (Eqn. 3.3) will be used (this would only be strictly true for the case of a solution of a single salt, e.g. a NaCl solution).

3.5   Thermal diffusion

Thermal diffusion promotes the transport of solutes due to temperature gradients. Thermal diffusion coefficients are usually reported in the literature as Soret coefficients (s), which arise from a formulation of thermal diffusion according to Eqn. 3.5. Soret coefficients measured in the laboratory are usually positive, i.e. they indicate solute transport down temperature gradients, and range between 10^{-3} and 10^{-2} K^{-1}, under a whole range of concentrations and temperatures (LERMAN, 1979; THORNTON & SEYFRIED, 1983; DE MARSILY, FARGUE & GOBLET, 1987).

3.6   *Thermal osmosis*

Thermal osmosis causes the flow of fluid (solution) down a temperature gradient. The flux of fluid caused by thermal osmosis has been written as (DIRKSEN, 1969)

$$v_{TO} = -k_T \frac{\partial T}{\partial x} \qquad (3.16)$$

where k_T is the thermo-osmotic permeability (m^2/K/s) of the medium. v_{TO} has units of m^3/m^{2}rock/s. The solute flux associated with thermal osmosis is given by Eqn. 3.6.

A big source of uncertainty in evaluating the potential role of thermal osmosis is the lack of data regarding the thermo-osmotic permeability k_T. Reported values of k_T for Na-saturated kaolinite and Na-bentonite (DIRKSEN, 1969), at an average temperature of 25°C and various temperature gradients and porosities, range between 10^{-14} and 3×10^{-13} m^2/K/s. In another study (SRIVASTAVA & AVASTHI, 1975) thermal osmosis across a kaolinite membrane was investigated. An estimate of k_T from the data in SRIVASTAVA & AVASTHI (1975), corresponding to conditions similar to the ones giving the high k_T values in DIRKSEN (1969), yields a value of 2.6×10^{-10} m^2/K/s, which is three orders of magnitude larger. As a first approximation, the values of k_T mentioned above can be used to define a range of values:

$$k_T: \ 10^{-14} \ \Lambda \ \ 10^{-10} \ \text{m}^2 \, / \, \text{K} \, / \, \text{s}$$

Model calculations (SATO *et al.*, 1998) suggest that temperature gradients near the repository at the time of waste canister failure (t $\approx 1\,000$ y) will be less than 1 K/m, and probably only about 0.25 K/m.

4.   Simple one-dimensional transport simulations including thermal and chemical osmosis, hyperfiltration, and thermal diffusion

In this section the solute fluxes associated with coupled transport phenomena, advection and diffusion will be combined into a simple one-dimensional transport model to obtain information on the combined effect of the different processes and their potential contributions to solute transport.

4.1   *The transport equation*

The solute fluxes (kg/m^2/s) associated with advection, chemical diffusion, chemical osmosis, hyperfiltration, thermal diffusion, and thermal osmosis are given by Eqns. 3.1 to 3.6. If the hydraulic head (∇h), osmotic pressure head ($\nabla \Pi_h$), and temperature gradients (∇T) are assumed to be constant along a one-dimensional section of the Opalinus Clay, and assuming also constant porosity (ϕ), effective diffusion coefficient (D_e), hydraulic conductivity (K), osmotic efficiency (σ), Soret coefficient (s), and thermo-osmotic permeability (k_T), all the fluxes can be incorporated into a transport equation of the form

$$\frac{\partial c_i}{\partial t} = D \frac{\partial^2 c_i}{\partial x^2} - v \frac{\partial c_i}{\partial x} \qquad (4.1)$$

where

$$D = \frac{D_e}{\phi} \qquad (4.2)$$

and

$$v = \frac{-K\frac{\partial h}{\partial x} + \sigma K\frac{\partial \Pi_h}{\partial x} + \sigma K\frac{\partial h}{\partial x} - D_e s\frac{\partial T}{\partial x} - k_T\frac{\partial T}{\partial x}}{\phi} \qquad (4.3)$$

The assumption of constant gradients is intended only to allow the estimation of the relative role of the different transport phenomena under different conditions (sets of parameter values).

The release of a tracer from a repository hosted by the Opalinus Clay can be simulated by making use of Eqn. 4.1 with the following initial and boundary conditions:

$$c(x,0) = 0 \qquad x \geq 0$$
$$c(0,t) = c_0 \qquad t \geq 0$$
$$c(\infty,t) = 0 \qquad t \geq 0$$

The analytical solution to Eqn. 4.1 with the initial and boundary conditions described above was reported by OGATA & BANKS (1961) and is given by

$$\frac{c}{c_0} = \frac{1}{2}\left[\text{erfc}\left(\frac{x - vt}{2\sqrt{Dt}}\right) + \exp\left(\frac{vx}{D}\right)\text{erfc}\left(\frac{x + vt}{2\sqrt{Dt}}\right)\right] \qquad (4.4)$$

4.2   Simulations

A series of simulations have been run with the objective of estimating the effects of the different coupled transport phenomena within the reference frame of a one-dimensional transport calculation. The parameters of the model intend to simulate the conditions in the vicinity of a repository for vitrified high level waste (HLW) and spent nuclear fuel (SF) hosted by the Opalinus Clay at the estimated time of waste canister failure (t $\approx$ 1 000 y).

4.2.1   Model parameters

4.2.1.1   Hydraulic, temperature, and osmotic pressure head gradients

A unit hydraulic gradient $\left(\partial h/\partial x = -1\right)$ has been assumed, based on measured hydraulic heads in Opalinus Clay and in aquifers above and below Opalinus Clay (NAGRA, 1994). Also, based on model calculations by SATO *et al.* (1997) regarding thermal gradients in the vicinity of the repository, a temperature gradient of 0.25 K/m ($\partial T/\partial x = -0.25$ K/m) has been used. Both hydraulic and temperature gradients are negative so they promote transport by advection, thermal diffusion, and thermal osmosis in the direction of increasing x.

Concerning the osmotic pressure head gradient, calculations have been performed with values of $\partial\Pi_h/\partial x = 1$ and $\partial\Pi_h/\partial x = 10$ (see SOLER, 1999), which correspond to salinity gradients equivalent to the change between OPA groundwater and a dilute solution occurring in a distance of 100 m and 10 m, respectively. Positive gradients imply that the chemical osmotic flux will be in the direction of increasing x (the same direction as advection, thermal diffusion, and thermal osmosis).

4.2.1.2   Other parameters

Hydraulic conductivity (K) values of 10^{-13} and 10^{-12} m/s, and effective diffusion coefficients (D_e) of 10^{-12} and 10^{-11} m^2/s have been used, given the range of reasonable values assumed for OPA. A porosity (ϕ) of 0.05 has been used in the calculations. The value of the osmotic efficiency coefficient (σ) has been set to 0.1, based on measurements on OPA samples by HARRINGTON & HORSEMAN (1999). However, results for other values of σ will also be shown. The values of the Soret coefficients and thermo-osmotic permeabilities are based on the range of possible values for OPA. All the results that will be shown correspond to $t = 5\ 000$ y after initial release (waste canister failure). This point in time was arbitrarily chosen, but is in the range of times before temperatures go significantly down in the repository (SATO *et al.*, 1998).

4.2.2   Results and discussion

Figure 4.1 shows concentration (as c/c_0) vs. distance at $t = 5\ 000$ y, for four different values of the thermo-osmotic permeability (k_T). The other parameters characterizing this set of calculations are $K = 10^{-13}$ m/s, $D_e = 10^{-12}$ m^2/s, $s = 0$, $\partial\Pi_h/\partial x = 1$, and $\sigma = 0.1$. Notice that the fact that the hydraulic and osmotic pressure head gradients are -1 and 1, respectively, means that the hyperfiltration and chemical osmotic fluxes cancel each other (see Eqns 3.4 and 3.3), and also that the results are independent of the value of σ.

The results show thermal osmosis will have a significant effect if $k_T > 10^{-12}$ m^2/K/s. Notice that the range of possible values of k_T, based on experimental studies on compacted clays, goes between 10^{-14} and 10^{-10} m^2/K/s. Therefore, there is a potential for thermal osmosis having a strong impact on solute and fluid transport under these conditions.

Figures 4.2(a) and 4.2(b) show the same type of calculation, but for different values of K and D_e. The results in Fig. 4.2(a), which correspond to the case where $K = 10^{-12}$ m/s, show that the effect of the increased hydraulic conductivity is quite minor (the system is not advection-dominated), and also that thermal-osmosis will have a significant effect for $k_T > 10^{-12}$ m^2/K/s. The same conclusion can be drawn from the results shown in Fig. 4.2(b), which corresponds to the case with $D_e = 10^{-11}$ m^2/s.

The results of an evaluation of the potential effect of thermal diffusion on solute transport are shown in Fig. 4.3. The solid line corresponds to the case where the only transport mechanisms are advection and chemical diffusion. The dashed line shows the additional effect of thermal diffusion, characterized by a Soret coefficient (s) of 0.1 K^{-1}. Notice that even with this large value of s, which is about one order of magnitude larger that any reported value, the effect is negligible.

The effect of chemical osmosis can be observed from the results shown in Figs. 4.4(a) and 4.4(b). An osmotic pressure head gradient ($\partial\Pi_h/\partial x$) of 10 has been used, so the chemical osmosis and hyperfiltration fluxes do not cancel each other (see Eqns. 3.3 and 3.4). A value of 10 for the osmotic pressure head gradient is roughly equivalent to the change between OPA groundwater and a dilute solution occurring in a distance of 10 m. Notice that this salinity gradient is quite steep, and that a salinity gradient that were any steeper could only be maintained for very small distances through the rock (<<10 m).

Figure 4.1. **Relative concentration vs. distance at $t = 5\,000$ y. The different curves correspond to different values of the thermo-osmotic permeability k_T. Thermal osmosis will only have a significant effect if $k_T > 10^{-12}$ m^2/K/s**

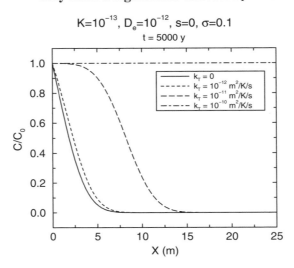

Figure 4.2. **Relative concentration vs. distance at $t = 5\,000$ y, for (a) hydraulic conductivity K equal to 10^{-12} m/s, and (b) effective diffusion coefficient D_e equal to 10^{-11} m^2/s. The different curves in both plots correspond to different values of the thermo-osmotic permeability k_T**

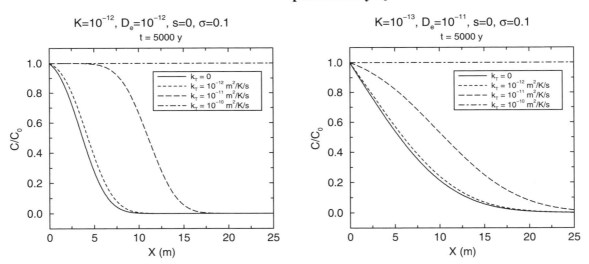

190

Comparing Figures 4.1 and 4.4(a) it can be seen that the additional effect of chemical osmosis on solute transport is quite negligible. Furthermore, the possibility that the Opalinus Clay were characterized by larger osmotic efficiencies ($\sigma > 0.1$) is considered in Figure 4.4(b). The different curves correspond to different values of σ. Thermal diffusion and thermal osmosis are not considered in this case. It can be observed that even in the improbable case of ideal efficiency ($\sigma = 1$) and high prevailing osmotic pressure head gradient ($\partial\Pi_h/\partial x = 10$), the effect of chemical osmosis is rather minor, especially compared to the potential effect of thermal osmosis [see Figure 4.4(a)].

Figure 4.3. **Relative concentration vs. distance at t = 5 000 y, for two different values of the Soret coefficient (s = 0 and s = 0.1 K^{-1})**

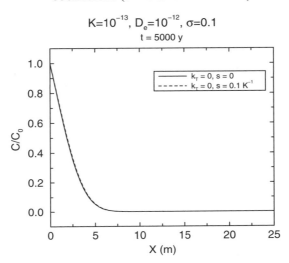

Figure 4.4. **Relative concentration vs. distance at t = 5 000 y, for the case with an osmotic pressure head gradient $\left(\partial\Pi_h/\partial x\right)$ equal to 10. (a) Results for different values of the thermo-osmotic permeability k_T. (b) Results for different values of the osmotic efficiency σ, with $k_T = 0$**

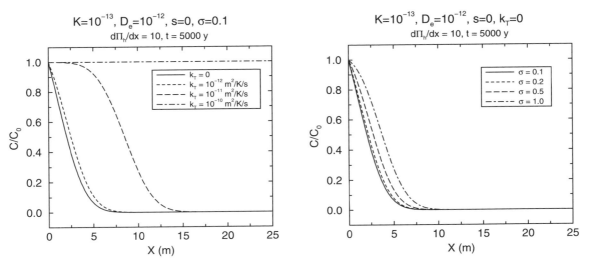

5. Coupling between advection and thermal osmosis: two-dimensional flow calculations

In the previous sections it has been shown that thermal osmosis is the only coupled transport mechanism that could have a strong impact on fluid and solute transport in the vicinity of a repository hosted by the Opalinus Clay. The contribution of thermal osmosis to fluid and solute transport can be significant if its effect is simply added to the other transport mechanisms. However, since thermal osmosis is a flux of fluid, conservation of fluid mass has to be taken into account in order to make any accurate predictions about its role in the performance of a nuclear waste repository. Thermal osmosis promotes the transport of fluid down the temperature gradient, and would therefore cause groundwater to move away from the repository (the heat source) in all directions. It is clear that without an extra source of solution, transport will be limited by the available amount of fluid in the system (conservation of fluid mass).

Two- and three-dimensional flow models including advection and thermal osmosis have been developed. The results from model simulations allow the evaluation of the effect of thermal osmosis when conservation of fluid mass and conservation of energy are taken into account. Additional two-dimensional simulations including temperature-dependent fluid density and viscosity terms have also been run. Only results from the two-dimensional simulations with constant fluid density and viscosity will be shown here. The conclusions from these results apply to the other simulations as well (SOLER, 1999).

5.1 Model formulation

The model that will be used solves numerically the equations of conservation of fluid mass and conservation of energy at steady state, for constant porosity and fluid density, in an homogeneous and isotropic porous medium. Observations from ten motorway and railway tunnels in which the Opalinus Clay is exposed have shown no water flow through the formation, including fractured zones, for sections with more than 200 m of overburden (GAUTSCHI, 1997). Also, field tests at the Mont Terri Underground Rock Laboratory have shown that there is no significant contrast in terms of hydraulic properties between a major fault zone and the wall rock (WYSS, MARSCHALL & ADAMS, 1999). These observations are consistent with the assumption of an homogeneous and isotropic medium for the flow model.

The model intends to simulate the conditions near the repository at the time of waste canister failure. Previous model calculations (SATO *et al.*, 1998) suggest that temperature gradients in the vicinity of the repository will be rather low at the time of canister failure (gradients less than 1 K/m, and probably of the order of 0.25 K/m, at $t \approx 1\,000$ y). These results led to the assumption of steady state and constant fluid density.

The equations of conservation of fluid mass and conservation of energy are written as

$$\nabla \cdot v = 0 \tag{5.1}$$

and

$$\nabla \cdot \kappa \nabla T - \rho c_f \nabla \cdot (vT) + A = 0 \tag{5.2}$$

respectively, where v, the total specific discharge or total flow velocity, which has units of $m^3/m^2 rock/s$, is given by

192

$$v = -K\nabla h - k_T \nabla T \qquad (5.3)$$

The first term on the right-hand-side of Eqn. 5.3 corresponds to Darcy's law (advection), and the second term describes the contribution to the total specific discharge from thermal osmosis. κ, K, and k_T are the thermal conductivity, hydraulic conductivity, and thermo-osmotic permeability of the porous medium, respectively. They are all assumed to be constant. ρ and c_f are the density and heat capacity at constant pressure of the fluid, and are also assumed to be constant. A is the heat source term, and h and T are the hydraulic head and temperature, respectively.

The system composed of Eqns. 5.1 and 5.2 is numerically solved for the hydraulic heads (h) and temperature (T), and fluid velocities are then calculated according to Eqn. 5.3. The spatial discretization of the system is done according to a cell-centered finite difference scheme. The resulting system of algebraic nonlinear equations is solved using Newton's method.

The flow domain (Figure 5.1) is a square, where constant hydraulic heads and temperatures are specified at the nodes on the exterior side of all the boundaries. Hydraulic head and temperature on the left- and right-hand-side boundaries are constant along the boundary, defining overall gradients from one side to the other of the domain. Hydraulic heads and temperatures along the top and bottom boundaries change linearly with space between their values on the left- and right-hand side boundaries.

5.2   *Results and discussion*

Two different sets of results are shown in Figs. 5.2 and 5.3, corresponding to two different overall temperature differences between left- and right-hand side boundaries. In the first case (Fig. 5.2) the temperature is the same in both sides, and in the second one (Fig. 5.3) the temperature on the left-hand-side is about one degree higher than on the right-hand-side. In both cases there is a unique (only one node) heat source at the center of the domain. The value of the heat source was arbitrarily chosen to produce relatively low temperature gradients.

The values of the hydraulic conductivity (K) and thermo-osmostic permeability (k_T) are certainly too high compared to the values that would apply to the Opalinus Clay (see sections 3.1 and 3.6). These values were used because they provided good convergence of the numerical calculations. It should be emphasized that the goal of the calculations in this section is to study how the coupling between advection and thermal osmosis works, rather than obtaining specific values for the flow velocities.

Figure 5.1. **Schematic diagram showing the geometry and boundary conditions of the flow domain used in the model. The flow domain is enclosed by a thick solid line. External nodes, where the boundary conditions are defined, are enclosed by dashed lines.**

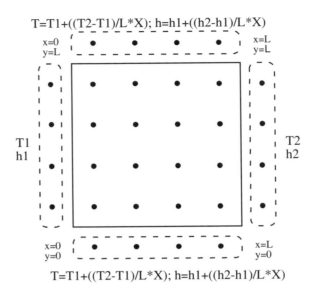

The parameters for both simulations are:

L = 16 m; 1 m spacing between nodes

h1 = 25 m; h2 = 9 m (overall unit hydraulic gradient from left to right)

T1 = T2 = 298.15 K (case shown in Fig. 5.2)

T1 = 299.22 K; T2 = 298.08 K (case shown in Fig. 5.3)

$K = 10^{-9}$ m/s; $k_T = 10^{-6}$ m^2/K/s; $\kappa = 2.6$ W/m/K

$A = 20$ J/m^3/s (only in one node at the center of the domain)

$\rho = 1000$ kg/m^3; $c_f = 4180$ J/kg/K

5.2.1   Case 1 (T1 = T2 = 298.15 K)

Figure 5.2(a) shows the calculated temperature distribution in the system. Figures 5.2(b), 5.2(c), and 5.2(d) show the total flow velocity, the advective component of the total flow velocity ($-K\nabla h$), and the thermal-osmotic component ($-k_T\nabla T$), respectively. Notice that thermal osmosis promotes fluid transport down the temperature gradient (from the center of the domain outwards). However, and due to the fact that there is no extra source of fluid in that region of the flow domain, the resulting hydraulic gradient causes the advective component ($-K\nabla h$) to oppose thermal osmosis. The result is that there is no net effect of thermal osmosis. The total flow velocity field is homogeneous (no component of velocity in the y direction), and the magnitude of the total velocity at all points equals the hydraulic conductivity (K) times the overall hydraulic gradient ((h2-h1)/L).

194

Figure 5.2. **Results for case 1 (T1 = T2 = 298.15 K). (a) Temperature, (b) total specific discharge, (c) advective component of the total specific discharge, and (d) thermal-osmotic component of the total specific discharge. Notice how the advective component of flow cancels the thermal-osmotic component, and there is no net effect of thermal osmosis on the flow field.**

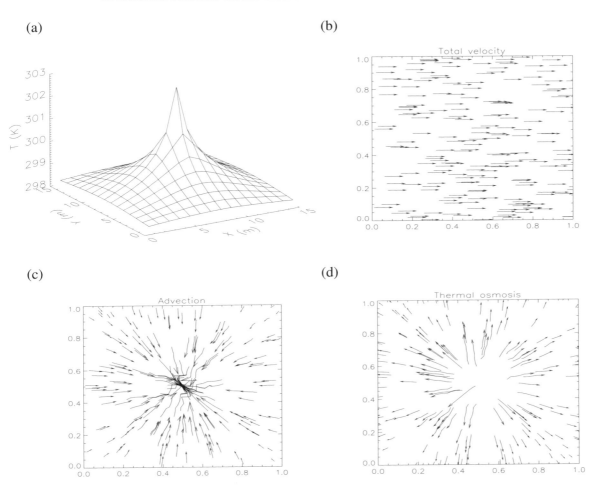

5.2.2 *Case 2 (T1 = 299.22 K, T2 = 298.08 K)*

Figure 5.3(a) shows the calculated temperature distribution in the system. Figures 5.3(b), 5.3(c), and 5.3(d) show the total flow velocity, the advective component of the total flow velocity $(-K\nabla h)$, and the thermal-osmotic component $(-k_T\nabla T)$, respectively. Notice again that thermal osmosis promotes fluid transport down the temperature gradient (from the center of the domain outwards), and also, that there is an extra component of thermal osmosis in the x direction, caused by the temperature difference between the left- and right-hand-side boundaries (the streamlines curve to the direction of increasing x near the top and bottom boundaries). As in the previous case, and due to the fact that there is no extra source of fluid in the interior of the flow domain, the resulting hydraulic gradient causes the advective component $(-K\nabla h)$ to cancel the disturbance in the flow field caused by thermal osmosis, and the end result is that there is no net effect of thermal osmosis arising from the presence of a heat source in the interior of the domain. The total flow velocity field is homogeneous (no component of velocity in the y direction). The only effect of thermal osmosis is to change the magnitude of the total velocity. The total velocity at all points is now given by

$$v = -K((h2-h1)/L) - k_T((T2-T1)/L)$$

Figure 5.3. **Results for case 2 (T1 = 299.22 K, T2 = 298.08 K). (a) Temperature, (b) total specific discharge, (c) advective component of the total specific discharge, and (d) thermal-osmotic component of the total specific discharge. Notice how the advective component of flow cancels the thermal-osmotic component, and there is no net effect of thermal osmosis arising from the presence of a heat source in the interior of the domain on the flow field.**

(a)   (b)

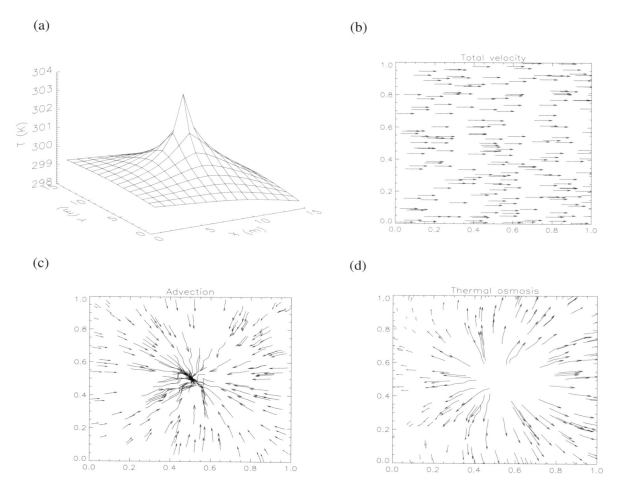

(c)   (d)

6.   Conclusions

The objective of the study was to place some constraints on the potential effects of coupled phenomena on fluid, solute and heat transport in the vicinity of a repository for vitrified high-level radioactive waste (HLW) and spent nuclear fuel (SF) hosted by the Opalinus Clay (OPA), and in the context of a fully-saturated OPA at the time of waste canister failure (t ≈ 1000 y).

- Firstly, estimates of the solute fluxes associated with chemical osmosis, hyperfiltration, thermal diffusion and thermal osmosis were calculated. These estimations, which are not included in this article, are reported in SOLER (1999). Available experimental data concerning coupled transport phenomena in compacted clays, and the hydrogeological and geochemical conditions to which the Opalinus Clay is subject, were used for these estimates.

Hyperfiltration can only be as large in magnitude as advection, and it always acts in the opposite direction, suggesting that any contribution from such a transport mechanism would be beneficial for repository performance (against the release of radionuclides from the OPA).

Chemical-osmotic fluxes could be significant under some conditions but, for conditions similar to those in the Mont Terri Underground Rock Laboratory (high salinity in the OPA; low salinity outside), any chemical-osmotic flux would be directed towards the OPA, and therefore against any release of radionuclides from the OPA.

Thermal diffusion does not seem to contribute in a significant manner to solute fluxes, given the low temperature gradient conditions believed to apply to the vicinity of a HLW/SF repository at the time of waste canister failure.

Finally, and without taking conservation of fluid mass into account, thermal osmosis seems to be the only coupled transport mechanism that could potentially have a strong effect on fluid and solute transport in the vicinity of a repository hosted by the Opalinus Clay. However, there is a significant lack of data regarding thermo-osmotic permeabilities of clays, causing a high degree of uncertainty in such estimates.

- Secondly, estimates of the heat fluxes associated with thermal filtration and the Dufour effect in the vicinity of the repository were calculated (again, reported in SOLER, 1999). Due to the lack of experimental data, the phenomenological coefficients for these heat transport mechanisms were obtained by means of the Onsager Reciprocal Relations and the values for their conjugated coupled phenomena (thermal osmosis and thermal diffusion). The calculated heat fluxes are absolutely negligible compared to the heat fluxes caused by thermal conduction.

- As a further step to obtain additional insight into the effects of coupled phenomena on solute transport, the solute fluxes associated with advection, chemical diffusion, thermal and chemical osmosis, hyperfiltration and thermal diffusion were incorporated into a simple one-dimensional transport equation. Under the assumption of constant hydraulic, temperature and osmotic pressure head gradients, constant porosity, effective diffusion coefficient, hydraulic conductivity, osmotic efficiency, Soret coefficient and thermo-osmotic permeability, the equation has a simple analytical solution, which can be used to estimate the effects of direct and coupled transport phenomena. By making use of parameters intended to simulate the conditions in the vicinity of a repository for vitrified high-level waste and spent fuel hosted by the Opalinus Clay, the equation shows that thermal osmosis is the only coupled transport mechanism that could potentially have a strong effect on solute and fluid transport in the vicinity of the repository, supporting the results of the previous estimates.

- Finally, the results of two- and three-dimensional flow models incorporating advection (Darcy's law) and thermal osmosis, under conditions (temperature and hydraulic gradients) simulating the vicinity of the repository at the time of waste canister failure ($t \approx 1\ 000$ y) and taking conservation of fluid mass and conservation of energy constraints into account, have shown that the advective component of the flow velocity cancels the thermal-osmotic component arising from the presence of a heat source in the interior of the flow domain. These simulations were run under the assumption of constant fluid density and viscosity.

Additional two-dimensional simulations with temperature-dependent density and viscosity have shown the same type of solution (SOLER, 1999), with the advective component of flow cancelling the thermal-osmotic component (no net effect of thermal osmosis), at least under the low temperature gradients believed to apply in the vicinity of the repository at the time of waste canister failure. The fact that taking three-dimensional flow into account did not make any difference regarding the coupling between advection and thermal osmosis in the simulations with constant fluid density and viscosity suggests that the same type of coupling will apply to three-dimensional flow with temperature-dependent fluid density and viscosity.

- The main conclusion arising from this study is that coupled phenomena will only have a very minor impact on radionuclide transport in the Opalinus Clay, at least under the conditions at the expected time of waste canister failure (times equal to or greater than 1 000 years). It would be advisable to guarantee the performance of the canisters up to these times through quality control. The effects of coupled phenomena on solute, fluid and heat transport in the near field of the repository (bentonite buffer) and at time scales less than 1 000 years will have to be addressed to achieve a full understanding of the role of coupled phenomena on the whole repository system.

7. Acknowledgements

Partial financial support by the Swiss National Cooperative for the Disposal of Radioactive Waste (Nagra), and the valuable comments from Urs Berner, Andreas Gautschi, Jörg Hadermann, Walter Heer, Andreas Jakob, Martin Mazurek, Joe Pearson, Jürg Schneider, and Piet Zuidema, are gratefully acknowledged. Acknowledgements are also due to Dr. G. de Marsily and Dr. P. Jamet for their detailed review of the original manuscript (SOLER, 1999) and helpful discussions.

8. References

BARBOUR, S.L. & FREDLUND, D.G. (1989): Mechanisms of Osmotic Flow and Volume Change in Clay Soils. Can. Geotech. Jour. 26, 551-562.

DIRKSEN, C. (1969): Thermo-Osmosis Through Compacted Saturated Clay Membranes. Soil Sci. Soc. Amer. Proc. 33, 821-826.

GARRELS, R.M. & CHRIST, C.L. (1965): Solutions, Minerals, and Equilibria. Harper and Row, New York.

GAUTSCHI, A. (1997): Hydrogeology of the Opalinus Clay – Implications for Radionuclide Transport. Nagra Bulletin 31, 24-32.

GROENEVELT, P.H., ELRICK, D.E. & LARYEA, K.B. (1980): Coupling Phenomena in Saturated Homo-Ionic Montmorillonite: IV. The Dispersion Coefficient. Soil Sci. Soc. Am. J. 44, 1168-1173.

HARRINGTON, J.F. & HORSEMAN, S.T. (1999): Laboratory Experiments on Hydraulic and Osmotic Flow. In: M. Thury & P. Bossart, eds., Mont Terri Project: Results of the

Hydrogeological, Geochemical and Geotechnical Experiments Performed in the Opalinus Clay (1996-1997). Swiss National Hydrological and Geological Survey Report, Bern (in preparation).

HORSEMAN, S.T., HIGGO, J.J.W., ALEXANDER, J. & HARRINGTON, J.F. (1996): Water, Gas and Solute Movement Through Argillaceous Media. NEA-OECD Report CC-96/1.

KEMPER, W.D. & EVANS, N.A. (1963): Movement of Water as Effected by Free Energy and Pressure Gradients III. Restriction of Solutes by Membranes. Soil Sci. Soc. Am. Proc. 27, 485-490.

KEMPER, W.D. & ROLLINS, J.B. (1966): Osmotic Efficiency Coefficients Across Compacted Clays. Soil Sci. Soc. Am. Proc. 30, 529-534.

LASAGA, A.C. (1998): Kinetic Theory in the Earth Sciences. Princeton University Press, Princeton.

LERMAN, A. (1979): Geochemical Processes. Water and Sediment Environments. John Wiley and Sons, New York.

LI, Y.H. & GREGORY, S. (1974): Diffusion of Ions in Sea Water and in Deep-Sea Sediments. Geochim. Cosmochim. Acta 38, 703-714.

DE MARSILY, G., FARGUE, D. & GOBLET, P. (1987): How Much Do We Know About Coupled Transport Processes in the Geosphere and Their Relevance to Performance Assessment? Proc. Geoval 1987, 475-491.

MAZUREK, M. (1999): Mineralogy of the Opalinus Clay. In: M. Thury & P. Bossart, eds., Mont Terri Project: Results of the Hydrogeological, Geochemical and Geotechnical Experiments Performed in the Opalinus Clay (1996-1997). Swiss National Hydrological and Geological Survey Report, Bern (in preparation).

NAGRA (1989): Sedimentstudie – Zwischenbericht 1988. Möglichkeiten zur Endlagerung Langlebiger Radioaktiver Abfälle in den Sedimenten der Schweiz. Nagra Technischer Bericht 88-25. Also executive summary in English, Nagra Technical Report 88-25E.

NAGRA (1994): Sedimentstudie – Zwischenbericht 1993: Zusammenfassende Uebersicht der Arbeiten von 1990 bis 1994 und Konzept für weitere Untersuchungen. Nagra Technischer Bericht 94-10.

OGATA, A. & BANKS R.B. (1961): A Solution of the Differential Equation of Longitudinal Dispersion in Porous Media. USGS Professional Paper 411-A.

ONSAGER, L. (1931): Reciprocal Relations in Irreversible Processes, II. Physical Review 38, 2265-2279.

SATO, R., SASAKI, T., ANDO, K., SMITH, P.A. & SCHNEIDER, J.W. (1998): Calculations of the Temperature Evolution of a Repository for Spent Fuel in Crystalline and Sedimentary Host Rocks. Nagra Technical Report 97-02.

SOLER, J.M. (1999): Coupled Transport Phenomena in the Opalinus Clay: Implications for Radionuclide Transport. PSI Bericht 99-07. Also published as Nagra Technischer Bericht 99-09.

SRIVASTAVA R.C. & AVASTHI, P.K. (1975): Non-equilibrium Thermodynamics of Thermo-Osmosis of Water Through Kaolinite. Journal of Hydrology 24, 111-120.

THORNTON, E.C. & SEYFRIED JR., W.E (1983): Thermodiffusional Transport in Pelagic Clay: Implications for Nuclear Waste Disposal in Geological Media. Science 220, 1156-1158.

ULLMAN, W.J. & ALLER, R.C. (1982): Diffusion Coefficients in Nearshore Marine Sediments. Limnology and Oceanography 27, 552-556.

DE WINDT, L. & PALUT, J.M. (1999): Tracer Feasibility Experiment (FM-C, DI). In: M. Thury & P. Bossart, eds., Mont Terri Project: Results of the Hydrogeological, Geochemical and Geotechnical Experiments Performed in the Opalinus Clay (1996-1997). Swiss National Hydrological and Geological Survey Report, Bern (in preparation).

WYSS, E., MARSCHALL, P. & ADAMS, J. (1999): Hydro- and Gas Testing (GP): Hydraulic Parameters and Formation Pressures in Matrix and Faults. In: M. Thury & P. Bossart, eds., Mont Terri Project: Results of the Hydrogeological, Geochemical and Geotechnical Experiments Performed in the Opalinus Clay (1996-1997). Swiss National Hydrological and Geological Survey Report, Bern (in preparation).

Uranium Migration in Glaciated Terrain: Implications of the Palmottu Study, Southern Finland

David Read
Enterpris, University of Reading, UK

Runar Blomqvist, Timo Ruskeeniemi
GTK, Finland

Kari Rasilainen
VTT Energy, Finland

Carlos Ayora
CSIC, Spain

Abstract

Demonstrating confidence in the long-term performance of a spent fuel disposal site is largely dependent on developing an adequate understanding of the geological barrier system and its evolution. However, performance assessment modelling is usually based solely on estimates of present-day groundwater fluxes that cannot be assumed to be representative of time scales on the order of tens to hundreds of thousands of years. The uncertainties associated with changing boundary conditions are particularly important at high latitudes and in coastal regions owing to the effects of climate change. Palaeohydrogeology, using mineralogical, isotopic and hydrochemical evidence of past events provides a firmer scientific basis for predicting future trends. In this respect, the Palmottu study is unique among natural analogue investigations in that it seeks to establish the effects of glaciation on an ancient uraninite deposit hosted in crystalline, igneous rock. Owing to its geological and geographical setting, the results are of direct relevance to repository projects in Finland, Sweden and Canada.

Alteration of the uraninite at Palmottu has led to substantial re-mobilisation of uranium under both reducing and, more recently, near-surface, oxidising conditions. A proportion of the original ore body has been altered hydrothermally resulting in the formation of U(IV) silicate (coffinite) around 1 billion years ago. During the last one million years, the site has experienced multiple glaciations in connection with which uranium has been mobilised under more oxidising conditions in the upper parts of the bedrock, leading to subsequent precipitation of U(VI) silicates (uranophane) at shallow depths. Volumetrically significant amounts of U are also found in association with fracture coatings, mainly impure calcites, clays and iron oxy-hydroxides. Finally, at the surface, post-glacial peat constitutes an additional sink as is typical of U deposits in northern Europe.

The U mineralisation extends from the surface to a depth of more than 400 m where the environment is substantially reducing and, consequently, there is ample scope for testing hypotheses of mobilisation and fixation processes. A variety of modelling approaches has been applied ranging from the simple advection-dispersion-matrix diffusion representation currently employed in performance assessment to a fully coupled chemical transport model. These studies have confirmed a number of the processes responsible for U mobilisation and fixation observed at other natural analogue sites. Additionally, Palmottu includes some unique features owing to its glacial and post-glacial history. By demonstrating very limited dispersion of U over millions of years, the study has greatly increased confidence in the safety of spent fuel disposal in crystalline, igneous rocks. However, it has also highlighted the limitations of our modelling tools in describing U migration in a complex natural geochemical system.

1. Introduction

One of the main purposes of the Palmottu study was to develop an understanding of the mechanisms via which uranium migrates through crystalline bedrock and the time scales on which such processes operate [1]. In this respect, it shares a number of features with other natural analogue investigations, for example Poços de Caldas [2], which also addressed oxidation of an uraninite-pitchblende deposit in fractured, igneous rocks. There are similarities also with Koongarra [3], in terms of the formation of secondary silicate mineralogy, and Broubster [4], where emphasis was placed on organic complexation by peat. However, Palmottu is unique in making the influence of palaeoclimatic changes a central aspect of the investigation. Since Palmottu experienced continental ice margin conditions, a major outcome of the project is increased knowledge of the effects of glaciation and related phenomena (e.g. permafrost) on the hydrogeochemistry of a fracture-controlled system.

In broad terms, our interest centres on:

- the length of time over which U migration has been occurring;

- the amount of U lost from the system and the direction of U movement now and in the past;

- the depth to which oxidising waters, capable of mobilising U have penetrated the system.

In order to address these questions, it is first necessary to establish the phases that control uranium concentrations at Palmottu, where the role of groundwater composition in maintaining aqueous levels is obviously an important factor. This is the focus of the current paper. The following discussion draws on previous mineralogical investigations [5-7] and work aimed at reconstructing the evolution of groundwater chemistry [8].

To date, there has been no structured attempt to use analogue data to critically examine the fundamental concepts underlying models used routinely in nuclear waste research. This may account for the relative absence of natural analogue information in published repository performance assessments. Within the Palmottu Project an attempt has been made to integrate the study with the requirements of PA by:

- assessing and comparing diverse modelling approaches for simulating the migration-fixation of U at the site, with emphasis on the representation of hydrogeologic and geochemical processes;

- testing alternative hypotheses of U behaviour as a means towards constructing a defensible, conceptual model of site evolution.

The models range from a state of the art coupled chemical transport code [9] to a standard advection-dispersion-matrix diffusion model currently employed in assessment calculations [10]. The outcome of the modelling exercise is described below following a brief description of the system and the model domain.

2. Site Characterisation

Petrology and Mineralogy

The deposit is hosted within a Proterozoic (1,800 Ma) meta-sedimentary sequence of rocks in south west Finland, comprising mica gneiss with granitic and granite pegmatite veins (Figure 1). The main U-Th rich ore is associated with the veins and forms a vertical structure extending to a depth of more than 400 m. Estimates suggest around one million tonnes of ore grading at 0.1% U [11]. Uraninite [UO_{2+x}, (0.01<x<0.25)] and monazite [$(Ce,La,Nd,Th)PO_4$] are the primary U-Th minerals. Coffinite [$U(SiO_4)_{1-x}(OH)_{4x}$] is a widespread alteration product of uraninite formed under reducing conditions soon after the precipitation of its precursor. An important consequence of the coffinitisation process is that large amounts of uranium and radiogenic lead are released. Remobilised uranium phases with chemical compositions resembling that of coffinite are found in association with fracture infillings, mainly impure calcites, clays and iron oxides. Minor amounts of thorite [$ThSiO_4$] and thorianite [ThO_2] have also been observed in some weathered samples [5-7].

Figure 1. Geological map of the Palmottu site (500m×600m), after A. Kuivamäki, Geological Survey of Finland. Modified from [12].

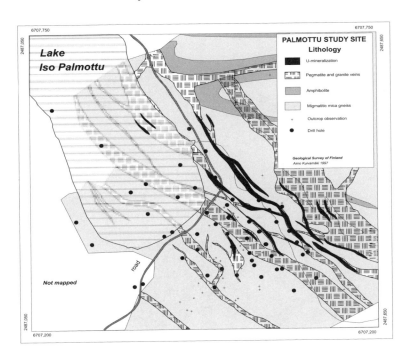

The only U(VI) phase definitely identified to date is uranophane ($Ca(UO_2)_2Si_2O_7 \cdot 5H_2O$). It occurs as a bright yellow crystalline mass close to the ground surface in the uppermost, weathered part of the eastern sector (Eastern Granite [6,7]) or as rosettes in fracture infillings. Many of the uranophane-bearing fractures are above the present groundwater table. Microprobe analyses show that it contains 67-75% UO_2, but, in contrast to the U(IV) minerals, very little Th or the rare earth elements (REE). Two uranophane samples were dated by the U-series disequilibrium (USD) method (J. Suksi, University of Helsinki) giving values of 90-120ka and 189-240ka, respectively [7]. These fit well with the interglacial periods of the two most recent ice ages [13].

Quaternary Geology

Over the last million years the area has been subjected to several glaciations, the last of which (Weichselian Ice Age) ended only 10 000a ago [13]. It is estimated that the maximum thickness of the continental ice cover in the central part of the Fennoscandian Shield was about 3 km. The weight of the ice depressed the underlying crust by up to 800 m at the centre of the glaciated area [14]. At Palmottu, the depression was less, of the order of 300-400 m. Isostatic rebound is continuing at a rate of 4-5 mm/a [15].

Three ice-marginal formations (Salpausselkä I to III) are, together with the esker formations, the most distinctive signs of the retreating ice (Figure 2). They can be followed for hundreds of kilometres from SW Finland eastwards to Russian Karelia and reflect periods when the retreat of the ice cover was temporarily halted by slightly cooler climatic conditions. At Palmottu, a large ice-marginal delta was deposited next to the Salpausselkä III formation at the Yoldian Sea level [16].

Figure 2.   **Eskers and ice-marginal formations in Southern Finland (and part of Russian Karelia). The NE trending formations are the Salpausselkä I to III ice-marginal formations. Palmottu is located immediately SE from the Salpausselkä III formation (marked with the black dot). The size and shape of the individual glacial lobes are clearly reproduced by the pattern of the ice-marginal formations and the eskers. From [16] based on data by [17].**

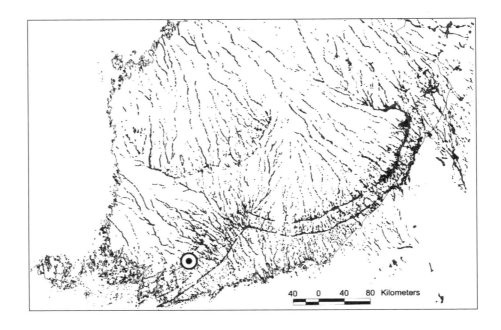

Palmottu is located in a bedrock block surrounded by prominent fracture zones and is topographically higher than its surroundings with a present elevation of 106-138 m. The highest points emerged from the proto-Baltic immediately after the retreat of the ice sheet during the Yoldian Sea stage, 10 000a B.P. (Figure 2). Based on shore-line displacements [13], the separate islands of the Palmottu area soon formed an integrated part of the mainland during the Ancylus Lake stage of the proto-Baltic some 9 200a B.P.

During the Yoldian stage substantial meltwater streams flushed the Palmottu site. However, it was during the previous Salpausselkä II stage, at the time of deglaciation, that glacial melt water flowing beneath the ice sheet will have intruded into the bedrock, caused by the enormous hydrostatic pressures below the ice cover. The depleted $\delta^{18}O$ values of some of the fracture groundwaters sampled at Palmottu [1,8,18] may be due to glacial melt water intrusion. At present, the bedrock is largely covered by glaciogenic tills or glaciofluvial formations. Soon after deglaciation, peat started to form on depressions of bedrock attaining a thickness of up to 3-4 m, though more commonly, 2 m or less.

Groundwater Chemistry

An integrated hydrogeological and hydrochemical model of the Palmottu site is presented in [1, 19] and only a brief summary is given here. Palmottu is thought to comprise an oxidising, Dynamic Upper Flow System (<100 m) dominated by young Ca-HCO$_3$ waters, a Dynamic Deep Flow System (mainly Na-HCO$_3$), and a Stagnant Deep Hydrogeological System containing waters with a Na-SO$_4$ or Na-Cl composition (Figure 3). The $\delta^{18}O$ values of the latter indicate a substantial glacial recharge component introduced some 10 000a ago and both hydrochemistry and isotopic data indicate long residence times. High dissolved U concentrations (100-500 ppb) are associated with the HCO$_3$ groundwaters in the vicinity of the uranium mineralisation down to depths of 130 m. This region forms the focus of the migration modelling study. At greater depths, clearly reducing conditions prevail and U concentrations rarely exceed 10 ppb.

Figure 3. **The hydrogeological model of Palmottu as modified from [1].**

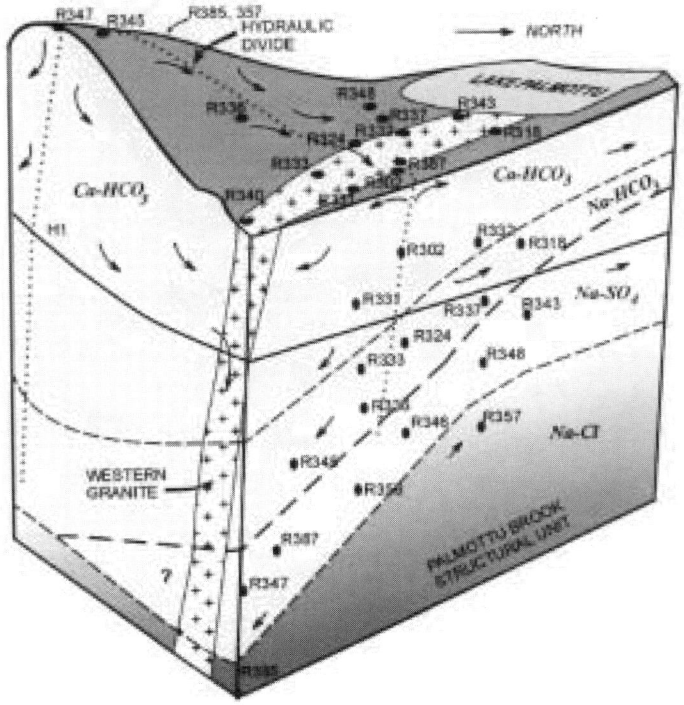

3.   Description of the Modelling Domain

Migration modelling has focused on the well-defined Eastern Flow System, for which a complete data set is available. The postulated flow route, identified on the basis of tracer tests and supporting information, is in the upper, oxidising part of the bedrock and is intersected by six boreholes. Comprehensive analyses of packed-off sections from these and for surface recharge provided seven reference points for the calculations. The data comprise a structural model of the bedrock [20], the results of groundwater flow modelling [21], groundwater and fracture mineral chemistry [22,5,6], *in situ* video logging [23] and the interpretation of a hole-to-hole tracer test [24]. Recent analyses of samples taken from two shallow boreholes, drilled in autumn 1998, have led to a greatly improved understanding of uranium behaviour at the site.

The transport domain is shown in cross-section in Figure 4. Rainwater falling on the topographic high to the north east is assumed to constitute the only source of recharge after which water and dissolved constituents migrate towards borehole R318 along a series of interconnected fractures. The modelling study consisted of two steps, the first involving prediction, the second allowing model refinement. The flow route was assumed to constitute a closed system and to reflect essentially post-glacial events, as it is probable that this latest signal dominates.

Figure 4. **Cross-section of the postulated migration route [25].**

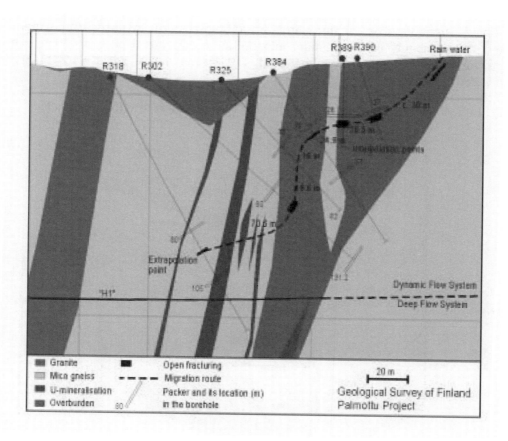

206

4. Mobilisation and Fixation of Uranium

Dominant Processes

As summarised in [25], previous workers have considered a number of processes to explain the means by which U concentrations in Palmottu groundwater are maintained, including:

- diffusion from the rock matrix
- adsorption/desorption onto/from Fe oxyhydroxides
- dissolution of primary U(IV) minerals
- dissolution of fracture calcites
- mixing of water bodies

To the above could be added the dissolution of separate secondary U(VI) minerals as observed at Koongarra [3], though none had been identified at the start of the exercise.

An additional complication is the marked effect of seasonal changes in recharge, which strongly influence uranium concentrations, the latter increasing rather than decreasing after rainfall. Consequently, a field experiment was established to monitor dissolved concentrations of uranium in the uppermost part of the Eastern Granite. A borehole packer was installed in borehole R390 (Figure 4) at 10.4m depth, during a dry period [7]. A video camera survey of the borehole showed that sub-horizontal fracturing was common and seven fractures with small apertures (1-2 mm) were detected above the water table. Some of these fractures contained uranophane, often accompanied by calcite. After approximately 3 weeks following the start of a rainy period, the water table rose above the packer level so that the first groundwater samples of the "unsaturated zone" could be taken. Uranium concentrations in these samples were around 100 ppb. After an additional week, the water table had risen a further 5 m, to a level 5.1m below ground surface and it remained there for at least 3 weeks. Dissolved uranium concentrations increased to 300-350 ppb [7], attributable to the introduction of low-pH water and the resulting higher solubility of uranophane.

Each of the above hypotheses was tested in a two-phase modelling programme incorporating both interpolation (prediction of concentrations for R389 and R390) and extrapolation (R318) of measured data [25]. Initial simulations were carried out "blind" as no hydrochemical or mineralogical information was available for the two shallow boreholes (R389 and R390) at that time. The modelling tools employed were RETRASO [9], a 2D coupled heat, flow and reactive solute transport code, and FTRANS [10], a finite element code employing conventional equilibrium partitioning of solutes to account for adsorption onto fracture surfaces and the rock matrix. These studies helped considerably to constrain the range of feasible interpretations that could be placed on the Palmottu data. The results of the modelling phase are summarised below.

Diffusion from the rock matrix

In paper [26] the potential effect of rock matrix diffusion on U concentrations in flowing groundwater was examined using the FTRANS code. Assuming U to be reversibly sorbed onto (unspecified) phases in the matrix and taking an earlier measured K_d of 0.03 m^3kg^{-1} these authors undertook a sensitivity study varying K_d, groundwater velocity, and matrix diffusion parameters. In addition, the initial U concentration in the pore water was varied from 1 500 ppb to 300 ppb. The simulations provided predictions of U concentration in flowing groundwater as a function of migration distance. Figure 5 gives the results for an initial U concentration of 1 500 ppb.

The results appear reasonably close to the measured values. However, there is a major uncertainty concerning the initial U concentration in the matrix water for which no direct measurements are available. This was assumed to be an order of magnitude higher than levels in the fractures. Taking a more probable 300 ppb as the initial U concentration equates to a factor of 5 reduction in the concentrations shown in Figure 5 which substantially reduces the model fit. It would also of course reduce the tendency for outward diffusion from the matrix. When viewed in conjunction with the rapid response of groundwater concentrations to recharge, it was concluded that rock matrix diffusion could not exert a dominating influence on aqueous uranium in the system studied.

Figure 5.   **Calculated U concentrations in Palmottu ground-waters [26]. Initial U concentration in pore water is 1500 ppb. "Base" means base case, "Lv" means low groundwater velocity, "Hv" high groundwater velocity, "HKd" high Kd, "LKd" low Kd, "Lde" low diffusivity, "Lal" low all parameters, and "Mix" high Kd and low diffusivity. Measured U concentration ranges are shown in the order: R390, R389, R384, R302, R318.**

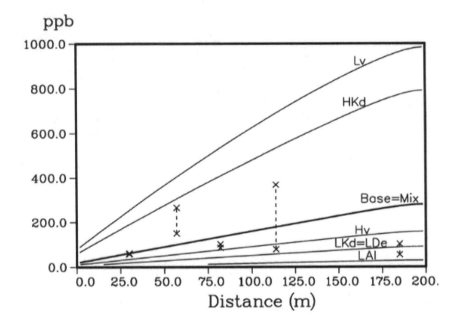

Adsorption/desorption onto/from Fe oxyhydroxides

This study, described in [27], also addressed equilibrium sorption but differs from the above in that the adsorbent was specified (amorphous $Fe(OH)_3$ on fracture surfaces) and the adsorption-desorption reactions were represented explicitly by chemical mass action laws.

The model adopted was that proposed in [28], encompassing strong and weak sorption sites with estimated concentrations of 1.8×10^{-3} mol/mol Fe and 0.2 mol/mol Fe, respectively. Assuming a $Fe(OH)_{3(am)}$ volume fraction of 0.01 in Palmottu fractures gives site concentrations of 5.4×10^{-4} mol dm^{-3} and 6.0×10^{-2} mol dm^{-3}. Since the proportion of $Fe(OH)_{3(am)}$ in Palmottu fractures is not known, sensitivity calculations were performed for scoping purposes. The results of the calculations after 1 year of elapsed time are plotted in Figure 6.

Figure 6. **Distribution of U concentration in solution: Surface complexation modelling [27].**

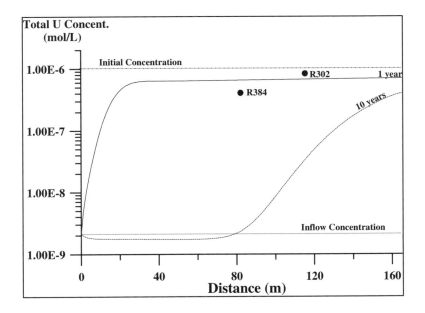

The amount of U(VI) in solution is about 7.0×10^{-7} mol kg^{-1}, a value within the same order of magnitude as those measured in boreholes R384 and R302. The infiltration water, containing around 2.0×10^{-9} mol kg^{-1} U, flushes out the U sorbed initially. Subsequently, U is desorbed from the Fe(OH)$_{3(am)}$ surfaces to the new equilibrium concentration. As the infiltration water percolates through the column, U(VI) species are progressively desorbed from the Fe(OH)$_{3(am)}$ surface and a front of dilute water advances to greater depths. Without replenishment from an additional source, most of the sorbed U species will have been removed after 10 years and the U concentration in the fracture will be similar to the inflow [27].

The time required for the U(VI) species to completely desorb from the Fe(OH)$_{3(am)}$ surface is linearly dependent on the number of initial sites (or Fe(OH)$_{3(am)}$ molar fraction). Therefore, the authors concluded that unreasonably high amounts of Fe(OH)$_{3(am)}$ would be required for adsorption to be responsible for the U concentrations measured over periods longer than about 10 years.

Dissolution of primary U(IV) minerals

Uraninite is the ultimate source of U in the Palmottu system and the dominant ore mineral. There is no doubt that dissolution of UO$_2$ has been active for a considerable length of time and that this process may well be responsible for aqueous levels at depth. However, the extent to which its oxidation influences concentrations in the relatively shallow zone chosen for migration modelling is less certain. This issue has been addressed from both an experimental [29] and theoretical viewpoint [27].

A series of models has been constructed based on differing assumptions concerning the assemblage of minerals undergoing dissolution along the transport path and the kinetics of dissolution. Each model considers the role played by potential redox buffers; Fe^{2+}-Fe^{3+}, U^{4+}-UO$_2$$^{2+}$, H$_2$S-SO$_4$. The outcome, as described in [27], is a sequence whereby Eh control evolves from dissolved O$_2$ in surface waters, through the iron couple in an intermediate zone to the U^{4+}-UO$_2$$^{2+}$ pair at greater depths. Dissolution of uraninite is predicted to occur at a sharp front that moves downstream as the source is exhausted. According to the calculations, aqueous U concentrations should peak around 80m depth before decreasing to a constant value (Figure 7). This matches qualitatively the trend observed. However, predicted U levels are some three orders of magnitude lower than those measured in boreholes R384 and R302; a fact that led the authors to postulate the existence of an additional uranium source.

Figure 7. **Solute concentration resulting from dissolution of uraninite [27]**

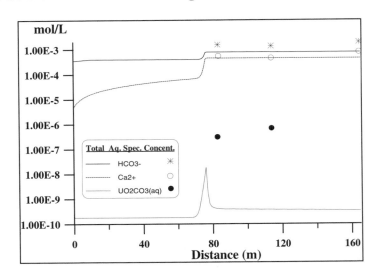

Dissolution of fracture calcites

Fracture coatings at Palmottu, mainly impure calcites, contain a substantial inventory of uranium. Concentrations decrease with depth, however, and higher values, some in excess of 1 000 ppm, are confined to the upper 100 m [5]. Rare earth element patterns indicate at least two generations of calcite, only one of which appears to be U-rich. In terms of major element chemistry, it was concluded in [27] that dissolution of calcite (and uraninite) "seems to account for the general trend of pH, Eh, HCO$_3$ and Ca values measured in the R384, R302 and R318 boreholes".

When examined in detail it is apparent that calcite, itself, cannot constitute a significant uranium source. The U is located in microcrystalline phases associated with the calcite that, unfortunately, are too fine-grained to allow quantitative analysis. Nevertheless, recent mineralogical studies have provided an indication of the dominant oxidation state [30], albeit on a limited sample set, and the relative isotopic abundance of ^{234}U/^{238}U. The latter is fairly constant and close to unity in the upper part of the flow route (R390, R389, R384) but shows a significant increase towards R302 and decreases again towards R318 [31]. This U-series information forms an important reference for the modelling studies. In addition, it has important implications for building an overall conceptual understanding of the site, as discussed in the final section.

Mixing of water bodies

The well-developed stratification of water bodies at Palmottu militates against extensive mixing as discussed by [8]. It is clear, at least for the transport domain studied (Figure 4), that simple mixing cannot explain either the distribution of aqueous U concentrations with depth or the variation in isotopic ratios. Contrary to expectation, at Palmottu flushing with fresh recharge does not cause U dilution but U enrichment, a feature that can only be accounted for by mineral dissolution. Given relatively constant major element composition, mixing should also result in a monotonic increase in U levels with depth. This is obviously not the case.

The use of numerical models to test hypotheses has played an important role in the development of a conceptual model for Palmottu. Current interpretations and the lessons learnt from this approach are outlined below.

5.   Discussion

The results of the Palmottu project have significant implications for performance assessment (PA), in particular with respect to development of evolutionary scenarios, identification of key processes and model testing. The presence of glacial water signatures at depths exceeding 350 m implies that, under the high hydraulic pressures prevailing at the ice margins, melt water can be forced down to close to repository depths. However, there is no evidence of oxidation below 100 m and, as the bulk of the uranium present has remained in the U(IV) state since its emplacement some 1.8 billion years ago, the host rock must exhibit a substantial redox buffering capacity.

When considering the processes responsible for U migration and fixation, it is important to note that at least three and, possibly, four main phases control uranium signatures in Palmottu groundwater. These comprise "amorphous" readily soluble U-phase(s), crystalline uranophane, U-rich fracture fillings, which may have the composition of uranophane or coffinite, and finally uraninite and coffinite. The near surface is a highly active zone in which aqueous U levels vary in response to a fluctuating water table and subsequent repeated dissolution and re-precipitation of amorphous U(VI) minerals. Uranium concentrations in groundwater rise rapidly from close to zero at the surface to around 150-250 ppb. Penetration is limited, however, corresponding to the solubility versus pH relationships of uranophane described in [32] and, subsequently, U levels fall slightly with increasing pH. At depths below ~100 m, reducing conditions persist and U concentrations of a few ppb reflect solubility control by uraninite.

Modelling of this system using a coupled chemical transport code [9] accounted for the direct and indirect effects of Eh-pH variations on U concentrations [27]. The models also predicted the development of a sharp dissolution front, which seems to be borne out by analysis of waters and fracture minerals, together with the existence of an additional uranium source, subsequently shown to be uranophane. On the basis of surface complexation modelling, it was concluded that insufficient amounts of Fe oxides are present for equilibrium adsorption to play a major role in determining aqueous concentrations. Essentially physical models, encompassing sorption, could not provide an explicit representation of the processes believed to control U distributions. It would appear from these studies that rock matrix diffusion could not exert a dominating influence on aqueous uranium in the present-day rapid groundwater system studied.

One of the planning criteria used in repository performance assessment is that the results obtained should be "conservative", i.e. key performance indicators, for example the source term and far-field concentrations, should be over-estimates. At Palmottu, as with other sites, it is evident that, once mobilised by breakdown of UO_2, uranium can be fixed and concentrated by a variety of mechanisms. These include the formation of secondary minerals and organic complexation. Such "sinks" act to limit U migration but also constitute secondary U sources. The resulting source terms tend to be much greater than those associated with the primary mineralisation, in the case of Palmottu by around two orders of magnitude. Performance assessments have (so far) ignored secondary sources downstream, the implications of which may not be conservative over at least part of the modelling domain.

The Palmottu study has made a substantial contribution to our understanding of uranium behaviour in a glaciated environment. By demonstrating very limited dispersion of U over millions of years, it has also greatly increased confidence in the safety of spent fuel disposal in crystalline, igneous rocks. However, shortcomings in the modelling tools currently used to quantify repository performance were identified and will need to be addressed in future investigations.

6. Acknowledgements

The financial support of the European Commission (Contract FI4W-CT95-0010) is gratefully acknowledged.

7. References

[1] Blomqvist, R., Kaija, J., Lampinen, P., Paananen, M., Ruskeeniemi, T., Korkealaakso, J., Pitkänen, P., Ludvigson, J.-E., Smellie, J., Koskinen, L., Floría, E., Turrero, M.J., Galarza, G., Jakobsson, K., Laaksoharju, M., Casanova, J., Grundfelt, B. and Hernan, P., 1998. The Palmottu Natural Analogue Project – Hydrogeological evaluation of the site. European Commission, Nuclear Science and Technology Series, Luxembourg, EUR 18202 EN, 95 p. + 1 Appendix.

[2] Chapman, N.A., McKinley, I.G., Shea, M.E. and Smellie, J., 1991. The Poços de Caldas Project: Summary and implications for radioactive waste management. Swedish Nuclear Fuel and Waste Management Company (SKB), Stockholm, Sweden. SKB Technical Report 90-24.

[3] Murakami, T., Ohnuki, T., Isobe, H. and Sato, T., 1997. Mobility of uranium during weathering. Am. Min. 82, 888–899.

[4] Read, D., Bennett, D.G., Hooker, P.J., Ivanovich, M., Longworth, G., Milodowski, A.E. and Noy, D.J., 1993. The migration of uranium into peat-rich soils at Broubster, Caithness, Scotland, UK. J. Contam. Hydrol. 13, 291–308.

[5] Ruskeeniemi, T. (comp.), 1998. Mineralogical and geochemical database of the Palmottu site. The Palmottu Natural Analogue Project. Technical Report 98-10.

[6] Ruskeeniemi, T., Nissinen, P. and Lindberg, A., 1998. Mineralogical characterisation of major water-conducting fractures in boreholes R302, R318, R332, R335, R373, R384, R388, R389 and R390. The Palmottu Natural Analogue Project, Technical Report 98-06.

[7] Ruskeeniemi, T., Lindberg, A., Pérez del Villar, L., Blomqvist, R., Suksi, J., Blyth, A. and Cera, E., 1999. Uranium mineralogy with implications for mobilisation of uranium at Palmottu. In "Eighth EC Natural Analogue Working Group Meeting." Proceedings of an international workshop held in Strasbourg, France, March 1999 (Eds. H. von Maravic and R. Alexander). European Commission, Nuclear Science and Technology Series, Luxembourg, EUR 19118 EN.

[8] Pitkänen, P., Kaija, J., Blomqvist, R., Smellie, J.A.T., Frape, S., Laaksoharju, M., Negrel, P., Casanova, J. and Karhu, J., 1999. Hydrogeochemical interpretation of groundwater at Palmottu. In "Eighth EC Natural Analogue Working Group Meeting." Proceedings of an international workshop held in Strasbourg, France, March 1999 (Eds. H. von Maravic and R.

Alexander). European Commission, Nuclear Science and Technology Series, Luxembourg, EUR 19118 EN.

[9] Saaltink, M.V., Ayora, C. and Carrera, J., 1998. A mathematical formulation for reactive transport that eliminates mineral concentrations. Water Resour. Res. 34, 1649-1656.

[10] Intera, 1983. FTRANS: A two-dimensional code for simulating fluid flow and transport of radioactive nuclides in fractured rock for repository performance assessment. ONWI Technical Report 426.

[11] Räisänen E., 1986. Uraniferous granitic veins in the Svecofennian schist belt in Nummi-Pusula, Southern Finland. Technical Committee Meeting on Uranium Deposits in Magmatic and Metamorphic Rocks, Salamanca 1986. Report IAEA-TC-571, 12 p.

[12] Niini, H., Vesterinen, M., Kuivamäki, A. and Blomqvist, R., 1993. Groundwater balance in a Precambrian fractured area: Palmottu, SW Finland. In "Selected Papers on Environmental Hydrogeology from the 29th International Geological Congress (I.G.C.)": Kyoto (Japan), August 24 – September 3, 1992 (Ed. Yasuo Sakura). International Association of Hydrogeologists. Volume 4, Hydrogeology, Selected Papers, 143-152.

[13] Donner, J., 1995. The Quaternary history of Scandinavia. World and Regional Geology 7. Cambridge University Press, Cambridge, U.K., 200 p.

[14] Mörner, N.-A., 1979. The Fennoscandian upplift and Late Cenozoic geodynamics: Geological evidence. GeoJournal 3, 287-318.

[15] Ekman, M., 1996. A consistent map of the postglacial uplift of Fennoscandia. Terra Nova 8, 158–165.

[16] Blomqvist, R., Ruskeeniemi, T. and Smellie, J.A.T., 1999. Palmottu natural analogue project – Geological setting and overview. In "Eighth EC Natural Analogue Working Group Meeting." Proceedings of an international workshop held in Strasbourg, France from 23 to 25 March 1999 (Eds. H. von Maravic and R. Alexander). European Commission, Nuclear Science and Technology Series, Luxembourg, EUR 19118 EN.

[17] Kujansuu, R. and Niemelä, J. (eds.), 1984. Quaternary deposits of Finland: 1:1 000 000. Geological Survey of Finland.

[18] Smellie, J.A.T., Blomqvist, R., Frape, S.K., Pitkänen, P., Ruskeeniemi, T., Suksi, J., Casanova, J., Gimeno, M.J. and Kaija, J., 1999. Palaeohydrogeological implications for long-term hydrochemical stability at Palmottu. In "Eighth EC Natural Analogue Working Group Meeting." Proceedings of an international workshop held in Strasbourg, France from 23 to 25 March 1999 (Eds. H. von Maravic and R. Alexander). European Commission, Nuclear Science and Technology Series, Luxembourg, EUR 19118 EN.

[19] Blomqvist, R., Paananen, M., Korkealaakso, J., Pitkänen, P., Kattilakoski, E., Smellie, J. and Ludvigson, J.-E., 1999. Evaluation of local groundwater flow conditions at the Palmottu natural analogue site: An integrated approach. In "Use of Hydrogeochemical Information in Testing Groundwater Flow Models." Technical Summary and Proceedings of a NEA Workshop, Borgholm, Sweden, 1-3 September 1997. Nuclear Energy Agency, Organisation for Economic Co-operation and Development (OECD/NEA), Paris, France, 271–280.

213

[20] Paananen, M., Blomqvist, R., Kaija, J., Ahonen, L., Ruskeeniemi, T., Suksi, J. and Rasilainen, K., 1998. The Palmottu Natural Analogue Project, Progress Report 1998. Geological Survey of Finland, Nuclear Waste Disposal Research, Report 100, 29 p.

[21] Koskinen, L. and Kattilakoski, E., 1997. Modelling of deep groundwater flow under natural conditions at the Palmottu site. The Palmottu Natural Analogue Project. Technical Report 97-03, 31 p. + 22 Appendices.

[22] Kaija, J. (comp.), 1998. The hydrogeochemical database of Palmottu. The 1998 version. The Palmottu Natural Analogue Project. Technical Report 98-08.

[23] Paulamäki S., Lindberg A., Paananen M. and Blomqvist, R., 1997. Structural modelling of water-conductive fractures and fracture zones of the Eastern Granite in the Palmottu study site, based on hydraulic data and TV-logging of open fractures. The Palmottu Natural Analogue Project. Technical Report 97-17, 15 p. + 3 Appendices.

[24] Gustafsson, E., Carlsson, A-C., Wass, E. and Ahonen, L., 1998. Tracer Test in the Eastern Granite, between boreholes R302, R335 and R384, Palmottu Natural Analogue Study Site. Palmottu Natural Analogue Project. Technical Report 98-02, 29 p. + 5 Appendices.

[25] Read, D., Ruskeeniemi, T., Rasilainen, K., Blomqvist, R., Kaija, J. and Paananen, M., 1999. Experimental database for the migration modelling exercise at Palmottu: Phase 2. The Palmottu Natural Analogue Project. Technical Report 99-10, 21 p.

[26] Nordman, H. and Rasilainen, K., 1999. Migration modelling exercise of the Palmottu Project – PA approach. The Palmottu Natural Analogue Project, Technical Report 99-18, 47 p.

[27] Salas, J. and Ayora, C., 1998. Migration modelling exercise at Palmottu. Preliminary results. The Palmottu Natural Analogue Project. Technical Report 98-09.

[28] Waite, T.D., Davis, J.A., Payne, T.E., Waychunas, G.A. and Xu, N., 1994. Uranium (VI) adsorption to ferrihydrite: Application of a surface complexation model. Geochimica et Cosmochimica Acta 58, 5465–5478.

[29] Cera, E., Ahonen, A., Rollin, C., Bruno, J., Kaija, J. and Blomqvist, R., 1999. Redox processes at the Palmottu uranium deposit. In "Eighth EC Natural Analogue Working Group Meeting." Proceedings of an international workshop held in Strasbourg, France from 23 to 25 March 1999 (Eds. H. von Maravic and R. Alexander). European Commission, Nuclear Science and Technology Series, Luxembourg, EUR 19118 EN.

[30] Suksi, J., Ervanne, H., Ruskeeniemi, T. and Blomqvist, R., 1998. Study of U-series disequilibria and U redox speciation in calcites in borehole R346. The Palmottu Natural Analogue Project. Technical Report 98-12.

[31] Suksi, J., Rasilainen, K., Casanova, J., Ruskeeniemi, T., Blomqvist, R. and Smellie, J.A.T., 1999. U-series disequilibria in a groundwater flow route as an indicator of uranium migration processes. Submitted to Radiochemica Acta.

[32] Bruno, J., Cera, E. and Grivé, M., 1999. Uranophane solubility evolution with pH. The Palmottu Natural Analogue Project. Technical Note, January 1999, 8 p.

An Unsaturated Zone Transport Field Test in Fractured Tuff

G.Y. Bussod

Los Alamos National Laboratory, United States

An underground transport test facility has been sited, designed, and constructed at Busted Butte, in Area 25 of the Nevada Test Site (NTS), approximately 160 km northwest of Las Vegas, Nevada, and 8 km southeast of the potential Yucca Mountain repository area. The site was chosen based on the presence of a readily accessible exposure of unsaturated rocks of the Topopah Springs/Calico Hills formations and the similarity of these units to those beneath the potential repository horizon.

The principal objectives of the test are to evaluate fundamental processes and uncertainties associated with flow and transport in the unsaturated zone site-scale models for Yucca Mountain. These include but are not restricted to:

1. The effect of heterogeneities on flow and transport under unsaturated and partially saturated conditions in the Calico Hills. In particular, the test aims to address issues relevant to fracture/matrix interactions and permeability contrast boundaries.

2. The migration behavior of colloids in fractured and unfractured Calico Hills rocks.

3. The validation, through field testing, of laboratory sorption experiments in unsaturated Calico Hills rocks.

4. The evaluation of the 3-D site-scale flow and transport process model (i.e. equivalent-continuum/dual-permeability/discrete-fracture-fault representations of flow and transport)used in performance assessment abstractions.

5. The effect of scaling from lab scale to field scale and site scale.

The test involves the use of a mix of conservative and reactive tracers and polystyrene microspheres and is subdivided into test Phases 1, 2 and 3 (Figure 1). Phase 1 initiated April 2, 1998 is a scoping test which involves 6 injection and 2 collection boreholes each 2m in length and located in the hydrologic Calico Hills and in the Topopah Spring formation. Phase 2 initiated July 23, 1998, involves two horizontal planes each containing four subparallel 7.5-8.0 meter injection boreholes, and twelve 8.5-10.0 meter horizontal collection boreholes located at right angles to the injection boreholes in a large *in situ* block 10 m x 10 m x 7 m and. Phase 3 is under consideration at this time, and may include the use of radionuclide tracers. Electrical Resistance Tomography (ERT), Ground Penetrating Radar (GPR) and neutron logging are used to determine the saturation state of the block prior to and during testing.

Parallel laboratory tests of Busted Butte core samples are underway to characterize the hydrologic and transport properties of the lithologies. Critical evaluation and iterative improvement of the flow and transport conceptual and numerical models awaits the collection of data, which is currently in progress. Field observations from overcoring and mineback of Phase 1 testing lead to several key conclusions of relevance to performance assessment.

The results from both observation and modeling indicate that strong capillary forces in the rock matrix of the hydrologic Calico Hills (CHn) are likely to modulate fracture flow from overlying units, thereby dampening pulses of infiltrating water and providing a large degree of contact between radionuclides and the rock matrix. Several modeling approaches, from deterministic to Monte Carlo to stochastic models, were used to simulate the Phase-1 experiments prior to direct observation. All yielded results similar to observation so that we may conclude tentatively that the deterministic modeling approach taken at the site scale may be adequate. The parameterizations used in performing these calculations however must be evaluated after quantitative data from the UZTT are available.

Preliminary "blind" predictions of the behavior of the Phase-2 block have been made to test the current modeling concepts and tools available to the integrated site-scale model and their abstractions for performance assessment. Modeling results for fluorescein, indicate that we expect tracer breakthrough at several sampling locations within the first year of testing. For some sampling locations, tracer breakthrough is predicted for travel times of less than a month. Tracer breakthroughs could be even quicker than predicted if the ECM assumption does not hold. Fracture flow through the Topopah Springs formation (Tptpv2) could result in faster travel times. The fracture parameters for the van Genuchten model are not known to a high degree of accuracy. Sensitivity analyses on these parameters will be performed to determine how sensitive travel times are to these fracture parameters. Another caveat in these modeling results is the effect of physical heterogeneities within each layer. Small-scale heterogeneities could result in preferential flow paths, which results in faster flow paths in some parts of the block and slow flow paths in other parts of the block. In the future, Monte Carlo simulations and more elegant stochastic techniques will be employed to attempt to capture the uncertainty in the travel times.

So far these predictions use parameters from the available Yucca Mountain hydrologic and geochemical databases. To date no calibrations have been performed using information from Busted Butte. As more data become available from the UZTT, they will be incorporated into refined versions of the models employed in these preliminary predictions and documented

Figure 1. Schematic of the Busted Butte Phase-2 blocks

This schematic shows the relative locations of the different test phases and barehole locations

Southern Busted Butte
UZ Transport Test

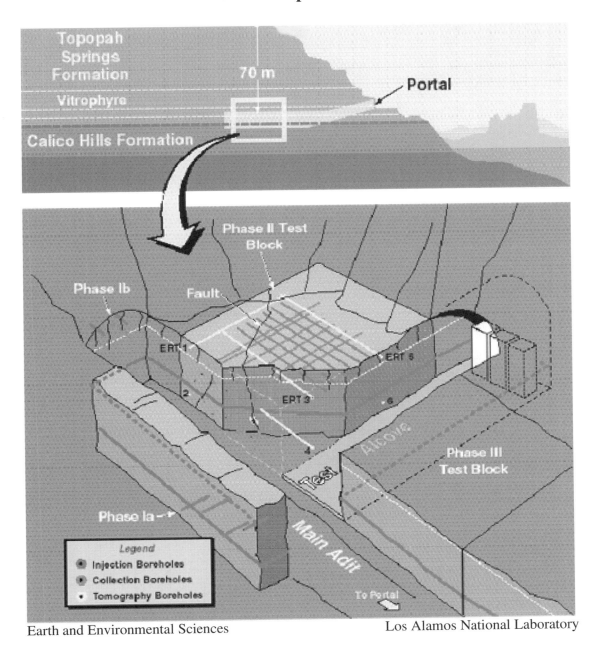

Earth and Environmental Sciences

Los Alamos National Laboratory

The Impact of Performance Assessment on the Experimental Studies at Äspö Hard Rock Laboratory and on Future Site Characterization and Evaluation in Sweden

P. Wikberg, A. Ström and J.O. Selroos
Swedish Nuclear Fuel and Waste Management Company (SKB), Sweden

1. Introduction

The SKB method for disposal of spent nuclear fuel is denoted the KBS-3 method. The repository system, located at about 500 m depth in crystalline rock, includes engineered and natural barriers which co-function to isolate the waste from the biosphere during the time span in which the waste constitutes a potential hazard, roughly 100 000 years. The spent fuel elements will be encapsulated into a copper-coated steel canister surrounded by a layer of bentonite clay. At existing favourable chemical and mechanical conditions the canister will remain intact for several million years. Despite this, an important scenario for the safety assessment is the malfunction of the system, which could cause a leakage of radionuclides from the repository, and migration of some of them up to the biosphere. For the assessment of that scenario, models are made to describe how corrodants could be transported to the canister and subsequently, how radionuclides could be transported from a leaking canister. However, the corrosion process of the canister itself has the largest impact on the outcome of this scenario [SR 97].

Below we discuss the data needed for the modelling of radionuclide transport and possible artefacts in the data, the possibility of using qualitative understanding where hard data may be missing, and upscaling of investigation results from laboratory and field scale to scales relevant for Performance Assessment (PA) applications. However, first we discuss how the early safety assessment exercises have influenced the presently on-going research at the Äspö Hard Rock Laboratory, and how the present knowledge is implemented into the planning of future characterization of repository candidate sites.

Feedback from previous performance assessment exercises, confidence building approaches and reviews.

The early KBS 1, 2 and 3 safety assessment studies were carried out at the end of the 1970s and beginning of the 1980s. Their role was to prove that it is possible (= safe) to dispose of spent nuclear fuel in an excavated repository at roughly 500 m depth in Swedish bedrock. The KBS-3 study was reviewed by a large number of international scientific organisations that confirmed the outcome of the study. It is safe to dispose of the spent nuclear fuel in the Swedish bedrock. An other outcome of the assessment and the review process was that there were fields where the uncertainty was large and where focussed efforts could enhance the knowledge and decrease the uncertainty. The research fields identified were related to the dissolution of the waste form, the corrosion of the copper canister, the

219

(thermo) mechanical properties of the bentonite, the chemistry of deep groundwater and of the relevant radionuclides, and the groundwater flow and transport of radionuclides. Natural analogues were also identified as a means to study (confirm) the effects of the very slow migration processes.

At the same time it was also obvious that it is not enough to prove the safety from a strictly scientific point of view. It is also necessary to convince the public that neither they nor their grandchildren will ever suffer from the radioactive waste in the repository. The only way to do that is to be able to tell them what will happen in the future. That takes more skill than just to calculate a dose-rate as a function of time. Therefore a major effort was started to convert the scientific results into statements that were understood by the public. The experts were also trained to present their results to ordinary people.

2.   What is the rationale for new underground experiments related to transport?

The rationale behind all presently on-going experiments and newly planned ones in the Äspö Hard Rock Laboratory is to:

- Develop and test methods for site characterisation, detailed investigations and repository construction.

- Increase the scientific understanding of the safety margins and provide data for safety assessment.

- Develop techniques which can reduce the costs and simplify the disposal concept.

- Demonstrate technology and educate people for final repository development.

Since the start of pre-investigations of the site in late 1986, up to the end of the construction of the 3 600 m long tunnel, characterisation and evaluation methods have been developed [Rhén and Bäckblom (Eds) 1997, Banwart et. al., 1997]. On-going research, organised as international projects are carried out to increase the understanding of the barrier functions of the rock. The aim is to decrease the uncertainty of important conditions (presently and in the future) and the processes (physical, chemical and biological) that affect the engineered barriers and radionuclide solubility and speciation. Further development of the models that describe the groundwater flow and nuclide migration is focussed within the TRUE projects where tests are carried out with nonsorbing and sorbing tracers in different geometrical scales. The processes that are studied are advection-dispersion in the flow streams, diffusion into the micro fractures of the rock, reversible and irreversible sorption and effects of gauge material in the fractures. In addition the configuration and the flow pattern between the fractures and within the fractures is even more important since this configuration will determine the available surface for interaction with the transported tracers.

The retention properties of the most relevant radionuclides have been investigated in laboratories. Since these nuclides cannot be included in tracer tests in the rock, we have constructed a borehole laboratory where the chemical *in situ* conditions prevail and the other conditions are controlled. This so-called CHEMLAB probe gives the opportunity to verify the results from ordinary laboratory experiments.

The redox conditions of the groundwater are the single most important factor for the Swedish KBS 3 concept. Large efforts have been, and are still made to prove that reducing conditions prevail at different stages of the repository lifetime. Under proved reducing conditions the copper canister will remain intact for millions of years and many of the most important radionuclides will remain in a reduced insoluble form. Two different experiments have been carried out to assess the fate of oxygen in a final repository [Banwart (ed), 1995, Puigdomenech et al., 1999]. The first one showed that the oxygen in infiltrating surface water is rapidly reduced by bacteria, and that there is no risk to have oxygen entering to the repository. The second experiment shows that the oxygen entrapped at closing the repository will be consumed within a few years after closure.

3.   What are the open issues to solve in future site characterisation activities with particular focus on the needs of transport modelling for performance assessment?

The Äspö HRL project co-operates internationally with a number of organisations, all in the field of nuclear waste management. An important part of this co-operation is the work within the Äspö Task Force on Modelling of Groundwater Flow and Transport of Solutes which was initiated by SKB in 1992. Each organization supporting the Äspö HRL was invited to form or appoint a Modelling Team that performs modelling of HRL experiments.

Key issues regarding among other things radionuclide transport, influence of heterogeneity on radionuclide transport predictions and modelling aspects have been compiled within the Äspö Task Force group. A table called Issue Evaluation Table has been produced; the name reflects the table´s intention to provide a basis for identification and evaluation of key issues in performance assessment of a deep geological repository. The table reflects the understanding (at the time of its formulation) of the key issues related to performance of the geological barrier in hard rocks, availability of reliable data and the way in which the issues can or are being addressed by different organizations at different underground laboratories. One of the major aims has been to produce a document which is condensed, informative, but not necessarily complete. The following issues are addressed:

- radionuclide transport issues;
- influence of heterogeneity on radionuclide transport predictions;
- long term stability of geological environment;
- modelling aspects;
- geosphere changes due to repository presence.

The table is named *Issue Evaluation Table* in order to reflect that it intends to provide a basis for identification and evaluation of key issues in performance assessment of a deep geological repository. One part of the table is provided in Table 1, influence of heterogeneity on radionuclide transport predictions/site scale.

One of the objectives of the table is to provide a basis for prioritisation of experimental work to be performed at the Äspö HRL. Many of the issues also apply for future site characterisation. However, SKB has availability to Äspö HRL, which provides crystalline rock at typical repository depth, where a number of transport related experiments are on-going at different geometric scales.

Transport models in PA obtain much of their input from the geoscientific description of hydrogeology and geochemistry. On the other hand, transport modelling makes new demands on site specific data that are not automatically satisfied by the hydrogeological and geochemical description. Site specific information is needed for sorption, porosity, diffusivity etc but the most essential transport related information to collect during site characterisation is the distribution of groundwater flow over different flow paths, as well as the flow wetted surface. Work related to the field characterization of groundwater flow and flow wetted surface, and the application of this information in PA, is briefly discussed below in the section on upscaling.

4.   Artefacts in available data

For the mass balance calculations it is necessary to know the groundwater flow rate and the concentration of corroding constituents in the groundwater, i.e. oxygen and sulphide. It is also necessary to know all the processes affecting the concentration of these constituents in the deep repository, e.g. biological oxygen consumption and sulphate reduction. There are no artefacts that would severely influence the conditions that provide the durability of the canisters. No chemical or biological processes could cause a corrosion that penetrates the 5 cm thick copper coating in a time shorter than a million years.

For nuclide transport modelling it is important to prove that the groundwater is reducing. Furthermore, the total salinity has an influence on the weakly sorbing nuclides. The underlying field data for these two parameters are not considered to be affected by artefacts that would influence the slow migration of radionuclides. However, values for matrix diffusion parameters are primarily obtained from laboratory analyses (with possible modifications in order to account for site-specific water chemistry). The laboratory analyses are typically based on through diffusion experiments where a formation factor of the rock matrix is obtained. Based on the formation factor and free water diffusivity (diffusion coefficient) of all relevant radionuclides, the nuclide specific effective diffusivities are estimated. A possible artefact is thus that the laboratory obtained formation factor (a function of porosity, tortuosity, constrictivity) does not apply to field scale conditions. For example, the porosity of altered fracture surfaces could be different from those of the intact rock matrix samples used in the laboratory. Furthermore, if the K_d value is different in the field than in the used laboratory sample, an additional artefact may prevail.

In evaluation of field experiments, tailing in breakthrough is sometimes attributed to matrix diffusion when, in fact, the dominant process may be e.g. diffusion into gouge material, diffusion into stagnant water in the fracture plane, or simply advective velocity variations. Thus, evaluated field scale values on matrix diffusivities may be overestimated if the dominant underlying process yielding the tailing is some other process than diffusion into the matrix. When such estimates on matrix diffusion parameters subsequently are used in PA, the predictions of breakthrough will be inaccurate. This is due to the fact that the prevailing processes will have different characteristics than matrix diffusion does on the temporal and spatial scales of interest.

To summarise, the main artefacts in data for radionuclide transport modelling are believed to prevail for matrix diffusion. Two possible sources for erroneous data are identified. The first one is related to the applicability of laboratory derived diffusivities in the field, the second one is related to the derivation of field scale diffusivities based on a wrong conceptual model.

5.   Use of qualitative understanding in Performance Assessment

The knowledge of processes affecting the radionuclide transport in the geosphere has increased substantially during the last two decades. Still there is a need to improve the models used to quantify that knowledge. In the SKB radionuclide transport model, the use of qualitative under-standing is primarily limited to the formulation of the conceptual model for mass transfer. To a certain extent all processes within the matrix are treated based on a general understanding. For practical reasons process models utilizing site specific data are not employed. Specifically, the K_d concept is based on a general understanding of several sorption processes. Furthermore, some retention processes that may be important from a process understanding point of view (e.g. co-precipitation) but that are hard to unequivocally show present on the spatial and temporal scales of interest are not included in the PA analysis. Finally, diffusion processes such as diffusion into gauge material and stagnant water are also neglected in the PA analysis. This is primarily done due to the fact that the impact of these processes is limited compared to the effect of matrix diffusion, and that the chosen approach is conservative.

It should be remembered that the formulation of the simplified conceptual model has to be substantiated by more elaborate process models. If the process models accurately enough can describe field experiments, and the PA models are based on all or some of the same processes, the practise is deemed defensible. The possible problem with upscaling of parameters is discussed in the next section.

Thus, based on a more thorough process understanding, PA models which only include a subset of all known processes are abstracted. The reason is not that the needed data cannot be obtained, but that the approach based on a general and simplified understanding of the retention processes is believed to be sufficient for PA purposes. On the contrary, the flow description is based on site-specific data and strives for more realism.

6.   Upscaling of laboratory and field scale data to Performance Assessment models

As discussed above, it is an open issue if laboratory derived diffusivities and K_d values apply in the field. Furthermore, it is not clear if parameters derived on the field scale can accurately be applied for the spatial and temporal scales of interest in PA applications.

The first issue is currently being investigated in the TRUE-1 experiments (Winberg et al., 1999) at SKB's Äspö Hard Rock laboratory. TRUE-1 consists of tracer experiments in a single feature on length scales of approximately 5 m using conservative and sorbing tracers. Laboratory derived K_d values and diffusivities are employed in the evaluation of the tests. The evaluation indicates that laboratory data cannot be used directly for accurate process predictions (Winberg et al., 1999). The mass transfer obtained in the field is underestimated using the laboratory derived parameters. However, for PA purposes the laboratory data seem conservative since mass transfer is underestimated.

The evaluation of the TRUE-1 tests also yields estimates of the flow wetted surface on a detailed field scale. An issue of great importance is how to upscale this flow wetted surface to PA scales. The present development work is based both on TRUE-1 results and the ongoing TRUE Block Scale experiment where tracer tests on scales up to 50 m will be performed. Furthermore, numerical modelling using continuum and discrete fracture network models are used in supportive efforts. The basic idea in the discrete models is to integrate the F-factor (flow wetted surface normalised with flow rate) along flow paths. This entity can subsequently be used for radionuclide transport modelling (Andersson et al., 1998). Questions addressed within the TRUE programme are how the flow wetted surface is conceptualised in single fractures and fracture networks, and how the F-factor depends on scale (upscaling).

The combined field and modelling work related to the determination of the flow wetted surface is believed to provide strategies for future site characterisation and subsequent PA studies.

7. References

Andersson, J., Elert, M., Hermansson, J., Moreno, L., Gylling, B., and Selroos, J.O., 1999. Derivation and treatment of the flow wetted surface and other geosphere parameters in the transport models FARF31 and COMP32 for use in safety assessment, SKB Report, R-98-60.

Banwart S (ed.), 1995. The Äspö redox investigations in block scale. Project summary and implications for repository performance assessment. SKB TR 95-26.

Banwart S., Wikberg P., Olsson O, 1997. A testbed for underground nuclear repository design. Environmental Science & Technology, 31, N°11, 510A.

Puigdomenech I., Banwart S., Bateman K., Milodowski A., West J., Griffault L., Gustafsson E., Hama K., Yoshida H., Kotelnikova S., Pedersen K., Lartigue J.-E., Michaud V., Trotignon L., Morosini M., Rivas Perez J., Tullborg E.-L., 1999. Redox experiment in detailed scale (REX) First project status report. SKB International Co-operation Report ICR-99-01

Rhén I (ed), Bäckblom (ed), Gustafson G, Stanfors R, Wikberg P, 1997. Äspö HRL – Geoscientific evaluation 1997/2. Results from pre-investigations and detailed site characterization. Summary report. SKB TR 97-03.

SKB, 1999 SR 97 – Post closure safety for the deep repository for spent nuclear fuel, main report, SKB Technical Report TR 99-06.

Winberg, A. (ed.), Andersson P., Hermansson J., Byegård J., Cvetkovic V., Birgerson L., 2000. Final Report of the First Stage of the Tracer Retention Understanding Experiments, Äspö Hard Rock Laboratory, SKB TR 00-07.

Table 1 **Influence of heterogeneity radionuclide transport predictions – site scale heterogeneities**

Issues	Relevance for input to PA	Availability of reliable data	Feasibility of – experiments (EX) – test of model concepts (MC)	Project	Comments Remarks – conceptual or data	Reference(s)
Identification of "major" discontinuities (which require separation distance)	Determination of repository layout Important to define distance to nearest major water-conducting zone (fast pathway)	– Yes, from site characterization	EX: Yes, by integrated geologic, geophysical, and hydrologic investigations	POSIVA: Site investigations	Conceptual issue Related to repository concept Criteria and procedure for classification needs to be specified	Nagra (1994a) (S) SKB (1992, 1995) (S) Thury et al. (1994) (S) TVO (1992) (S)
Effect of major fracture discontinuities on site scale flow and transport	Determination of repository layout	– Yes, from site characterization	EX: Yes, by large scale pumping tests MC:Yes, by e.g. MTF Task 3	LPT2 test Äspö tunnel drawdown POSIVA: Site investigations	Data issue	Axelsson et al. (1990) AECL (1994)
Separation ("respect") distance from waste containers to major discontinuities	Repository layout Critical path length Specification of criteria	– Site specific. Issue of "repository-fitting"	MC: Yes, but related to chosen PA model	TRUE Block Scale POSIVA: Site investigations	Conceptual issue Depends on repository concept and site specific PA	SKB (1983) (S) TVO (1992) (S) AECL (1994) (S) SKI (1996) (S)
Boundary conditions to flow model	Validity of data input to PA Change of boundary conditions with time important (see 3.1)	– Limited for recharge and flow through model boundaries – Yes for head	EX: Yes, by large scale pumping tests MC: Yes, by e.g. MTF Task 3	LPT2 Äspö tunnel drawdown Hydrochemistry CRIEPI: Geochemistry Rokkasho	Conceptual issue Data issue Validity of "standard" assumptions? (e.g model boundaries at water divides)	Olsson et al. (1992) (S) Clark et al. (1997)
Identification of recharge and discharge areas	Post closure scenarios; Groundwater flow regime Also important to site selection /characterisation	– Yes, by head monitoring, recharge is determined indirectly	EX: Yes, e.g. MTF Task 3	LPT2 Äspö tunnel drawdown Grimsel: MI	Data issue Necessity/possibility to identify discharge areas?	SKB (1995) TVO (1992) Nirex (1995)
Effect of fresh/saline water boundary on flow and transport	Affects the regional flow field See 3.3, may change with time	– Limited	EX: Yes, by measurement of flow velocities and salinities MC:?	Hydrochemistry POSIVA: Site investigations at Olkiluoto	Depends on geology, topography Stability of interface in long term?	Voss & Andersson (1993) Pirhonen (1994) Claesson (1992) Follin (1995)

Future Prospects for Site Characterization and UndergroundExperiments Related to Transport Based on the H12 Performance Assessment

Y. Ijiri, A. Sawada, K. Sakamoto, H. Yoshida, M. Uchida, K. Ishiguro, H. Umeki
Japan Nuclear Cycle Development Institute, Japan

E.K. Webb
Sandia National Laboratories, USA

Abstract

The paper briefly overviews the methodology and the results of radionuclide migration analysis performed in the H12 performance assessment to be published by the year 2000. From the analysis, it was found that large water-conducting features, which transect the region of interest, have significant impacts on the radionuclide release from the geosphere, while low-permeability domains contribute to the retention of nuclides in the geosphere. We conclude that the locations and characteristics of large water-conducting features as well as the extent of low-permeability domains need to be identified during site characterization. We also place emphasis on the need for large-scale tracer experiments in order to verify the characteristics of large water-conducting features. The interactive relationship between site characterization and performance assessment is discussed. The treatment of large water-conducting features in a repository design in order to enhance the performance credit of the geosphere is presented. The requirements for the future Japanese programme are also discussed.

1.   Introduction

The Japan Nuclear Cycle Development Institute (JNC) is currently preparing the "Second Progress Report on Research and Development for the Geological Disposal of HLW in Japan", referred to as the H12 report, to demonstrate the technical reliability of deep geological disposal of high-level radioactive waste in Japan. Since no potential geological formation and disposal site have yet been nominated, the H12 performance assessment incorporates a wide range of geological, hydrogeological and geochemical conditions observed in Japan. JNC investigated hydrogeological characteristics of rock formations at many outcrops and galleries throughout the country, and established that most rock formations can be conceptualized as fractured media dominated by discrete fracture flow. Since fracture characteristics are heterogeneous and have been treated statistically, a three-dimensional discrete fracture network (DFN) model in which fracture characteristics can be treated as statistical parameters was adopted to represent this conceptual model of the fractured media. Then, radionuclide migration was solved on a channel network model derived from the DFN model [1, 2].

In this paper, we briefly overview the radionuclide migration analysis performed in the H12 performance assessment. Based on the lessons learned from the analysis, we discuss the issues that need to be clarified through site characterization. Additionally, we discuss the linkage between site characterization and performance assessment. The requirements for the future Japanese programme are also discussed.

2. Overview of the H12 Performance Assessment Analysis

2.1 Methodology

First, three-dimensional DFN models are constructed by generating disk-shaped fractures in a 200 meters cubic-block region based on the fracture parameter values determined from field data. Then, the DFN models are converted to channel network models by linking the fracture intersections on each fracture plane with one-dimensional pipes. Radionuclide migration in the channel network models is numerically solved using the Laplace transform Galerkin method taking into account processes including advection-dispersion in a fracture network, diffusion into the rock matrix, sorption in the rock matrix and radioactive-chain decay [2].

Granitic rock is considered the archetype fractured media. Consequently, a large amount of fracture data is available in literature for granitic materials. Furthermore, granitic rock has been treated as a potential host rock in many overseas programmes, and has JNC performed a number of field observations and experiments in granodiorite at the Kamaishi Mine in northern Japan. Thus, a large amount of data on fracture characteristics as well as rock characteristics related to matrix diffusion and sorption is available for granitic rock. Therefore, granitic rock was selected as a reference material for the H12 analysis.

2.2 Model Parameters

Parameter values, transport processes as well as conceptual models for fractured rock are commonly evaluated at less than several meters scale in laboratories and in the field. On the other hand, performance assessment is commonly carried out at more than several hundreds meters scale. Therefore, upscaling of parameter values, processes and conceptual models from the laboratory / field experiment scale to the larger scale of performance assessment models is necessary. In the H12 performance assessment, radionuclide migration is evaluated in a 200 meters cubic block region, whereas most fracture characteristics have been obtained at the scale of less than several meters at outcrops and in experimental underground galleries. We considered that trace length observed in the field as the most scale dependent fracture characteristic, and thus one focus was upscaling the parameters used for fracture size from observed values in fields.

The Log-normal distribution has been often adopted for the probability density distribution of trace length in the literature [e.g. 3]. One of the reasons for this is that trace lengths longer than the window of observation tend to be missing from field data. There are few places in Japan where fractures can be observed at the hundreds of meters scale, because most rock formations in Japan are covered by thick Quarternary sediments and/or deep forests. We investigated fracture distributions obtained at three underground oil storage sites, Kikuma, Kuji and Kushikino [4]. Fracture distributions can be estimated based on data obtained over observation windows of several hundred square meters observed in seven to ten large excavated caverns. From the fracture distributions obtained at these sites, it was found that there are large fractures and faults which transect the entire observation window. Separately, Ohno and Kojima [5] showed the fractal nature of trace-length distribution using fracture maps observed at different scales from meter to kilometer. Based on the above studies, we adopted a power-law distribution for the probability density distribution of fracture radius (size) and consequently we generated large fractures in the DFN model.

Fracture parameters and the probability density distributions for these parameters are displayed in Table 1. One realization of the DFN model constructed from these fracture parameters is shown in Figure 1. More detailed descriptions of the fracture parameters can be seen in [1]. All transport parameters including longitudinal dispersivity in a fracture, diffusion coefficient in rock matrix, flow-wetted surface for matrix diffusion, matrix diffusion depth, rock porosity, dry rock density and distribution coefficients are assumed to be uniform in the analysis.

Table 1. **Fracture parameters used for the DFN model**

Fracture parameter	Probability density distribution
Orientation	Fisher
Radius	Power law
Frequency	Uniform
Spatial distribution	Uniform
Transmissivity	Log-normal
Aperture*	Log-normal

* Aperture is correlated with transmissivity.

Figure 1. **DFN model where 25% of fractures are displayed**

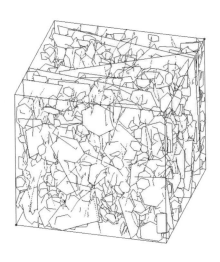

2.3   *Results and Lessons Learned*

Our current fracture network model contains large fractures, which transect the entire model region as shown in Figure 1. The model also yields a wide range of migration velocities along pathways within the system. More specifically, if large fractures are water-conducting, they tend to form fast migration pathways providing for both quicker early-time radionuclide arrival and larger total release from the geosphere than complicated networks of small fractures. Additionally, the results of radionuclide migration analysis vary significantly among realizations of the stochastic DFN model [1, 2]. This results from the presence or absence of large water-conducting features in various realizations. The realizations which contain large water-conducting features yield large release from the geosphere, while those which do not contain large water-conducting features tend to retain nuclides as a natural barrier. Since the total radionuclide release from the entire repository can be estimated by the combined release from a suite of realizations [2], we conclude that it is important to identify the locations and characteristics of large water-conducting features as well as the extent of low-permeability domains within a repository region.

3.   Key Aspects of Site Characterization

In this section, we discuss the key aspects to be clarified through site characterization and underground experiments based on the lessons learned from the H12 performance assessment analysis.

3.1   *Identification of Hydrogeological Structure*

(i)   Large Water-Conducting Features

Large water-conducting features have significant impacts on the performance of geosphere barriers as described in the previous section. Therefore, from the site evaluating point of view, it is crucial to identify the locations and characteristics of large water-conducting features within a repository region. The locations of large features will be detected by geophysical surveys through site characterization. The size of large features may change with the scale of interest as described later. The characteristics of large features can be estimated using hydraulic and tracer tests within the boreholes drilled in the features. It is, however, common that the data vary in space due to heterogeneity. For example, the permeability values of fracture zones at the URL site in Canada ranged over several orders of magnitude within small spatial scales of a few meters [6]. In this kind of situation, one of the conventional approaches for estimating characteristics of large features from field data is numerical simulation using heterogeneous models generated by such numerical approaches as geostatistical simulators [e.g. 7]. Another approach often found in performance assessment analyses is choosing the most conservative value for a homogeneous field of transport characteristics. This conventional approach, however, tends to yield overly pessimistic results.

Davison *et al.* [6] performed a series of tracer tests within the large fracture zones at the URL site in Canada over distance scales ranging from 17 meters to 700 meters in order to quantify the transport characteristics at large scale. From the results of the tracer tests, they concluded that;

> "There were no upscaling trends to enable the smaller scale hydraulic tests and
> tracer tests to be used to forecast the large scale transport properties of these
> fracture zones. Therefore, the tracer tests need to be performed at the same scale

as the modelling or averaging scale to yield values that are appropriate to represent radionuclide transport in the models. "

According to their conclusion, hydraulic tests and tracer tests at the scale of interest seem to be necessary in order to evaluate large-scale characteristics of features. Following this line of reasoning, large-scale tracer experiments over distances of approximately 100 meters were also performed at the Finnsjön site [8] and the Äspö Hard Rock Laboratory in Sweden [9]. Thus, large-scale tracer experiments have become more commonly applied as part of site investigation programmes.

(ii)   Low-Permeability Domains

As noted earlier, it is also important to identify the extent of low-permeability domains that contribute to the retention of nuclides within a repository region, since the overall performance of the geosphere barrier is determined by the combination of rapid migration along large water-conducting features and slow migration within low-permeability domains as shown in Figure 2. Therefore, it is important to design the repository layout based on the hydrogeological structure identified through site characterization in order to enhance the performance of the geosphere barrier as much as possible.

Figure 2. **Schematic view of the hydrogeological structure within a repository region**

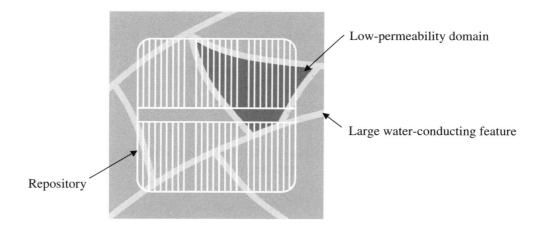

3.2   *Identification of Characteristics Related to Matrix Diffusion*

In our current performance assessment model, diffusion into rock matrix adjacent to fractures is the only mechanism which contributes to the retention of nuclides. Therefore, the characteristics of the matrix diffusion process have a large impact on the performance of the geosphere barrier. The parameters controlling the matrix diffusion process in our current model are the diffusion coefficient within the rock matrix, flow-wetted surface area of fractures representing the area across which diffusion occurs into the matrix, matrix diffusion depth, rock porosity, dry rock density and the distribution coefficients. These parameter values have been investigated using different experiments. However, it is difficult to assign appropriate parameter values to the performance assessment models because:

- Parameter values obtained from field data are not unique and vary significantly in space.
- There are strong disagreement among various investigators concerning the parameter values obtained from different sampling or experimental techniques. For example, matrix diffusion depths obtained from natural analogue studies tend to be smaller than those measured by laboratory diffusion experiments. Distribution coefficients obtained from batch experiments are considered to be larger than those directly measured for rock matrix.
- There is no established technique for evaluating flow-wetted surface area.

In addition, there is no guarantee that models based on these parameter values identified through various experiments will yield a correct answer. Thus, it may be more accurate and more practical to evaluate processes, for example matrix diffusion, through long-term experiments. In the case of matrix diffusion, parameter values could be inversely calculated based on the breakthrough results of a long-term tracer experiment.

4. Approaches for Site Characterization and Performance Assessment

4.1 Optimum Repository Design for Increasing the Performance of Geosphere Barriers

Site characterization, performance assessment as well as repository design are closely related with each other as shown in Figure 3. As noted in the previous section, repository designs should be based on the hydrogeological structure identified through site characterization in order to enhance the performance of geosphere barriers. Therefore, repository design should be revised interactively as site characterization proceeds, and performance assessment should be iteratively updated whenever repository designs are changed or significant new information is derived from site characterization. Additionally, repository design should be determined from the performance assessment point of view.

Figure 3. **Relationship between site characterization, repository design and performance assessment**

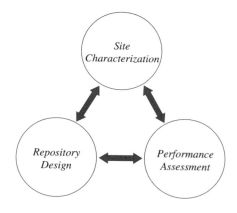

232

Rock formations in Japan are relatively fractured and are covered by thick Quarternary sediments or weathered rock, because Japan is located in the circum-Pacific belt and receives a lot of precipitation annually. Thus, it will be difficult to detect all water-conducting features within a repository region through site characterization. While major, kilometer-scale features can be detected by geophysical surveys at the ground surface and within exploratory boreholes, a large number of minor features, of from hundred meter to kilometer scale, can only be identified during the excavation of repository drifts. Therefore, it is possible and important to treat minor features (large water-conducting features) appropriately within the repository design in order to enhance the performance of geosphere barriers. Possible repository designs incorporate the following principles:

- Emplacement of waste packages should avoid large features, which are likely to move causing collapse of the engineered barrier system (EBS) even though these features are not water-conductive (see Figure 4).

- Plugging and grouting of large water-conducting features should be properly performed so that nuclides released from the EBS do not flow directly into large features.

- Design of the repository should avoid large water-conducting features with extensive widths (resulting in areas without waste packages) so large they cause the repository to be uneconomical (see Figure 4).

Additionally, the repository drifts should be excavated starting at the downstream edge of the repository region and proceeding to the other outer boundary of the region in order to provide information for design of the final configuration of repository as part of the earlier stages of construction. This pattern of excavation provides one approach for identifying and rapidly characterizing large water-conducting features before the full repository layout design is complete.

Figure 4. **Schematic view of repository design accounting for large features**

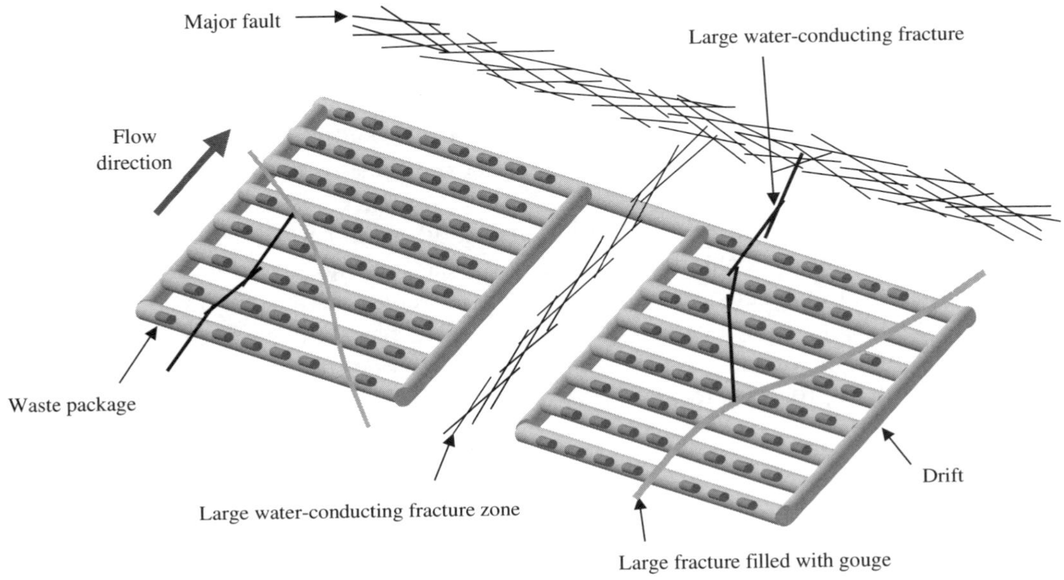

4.2   *Interactive Relationship between Site Characterization and Performance Assessment*

The H12 performance assessment found that the locations and characteristics of large features seem to be more important for the performance assessment of geosphere than complicated networks of small fractures. In addition, large features will be more easily detected by geophysical surveys through site characterization, while small fractures will not be easily detected until repository drifts are excavated. In a manner similar to conventional geological surveys, site characterization can proceed from a large scale to smaller scales. Therefore, performance assessment can also focus first on large features and progressively add more detail at a smaller scale as shown in Figure 5. As site characterization proceeds, smaller features are detected by higher resolution surveys and more complex performance assessment models can be adopted so as to increase the confidence in the results. Thus, site characterization and performance assessment should be implemented interactively.

At an early stage of site characterization, geophysical surveys such as airborne geophysics can be performed at a regional scale in order to detect major faults and fracture zones which should be avoided during construction of the repository. At this stage, the performance assessment of major features will be performed using simple one-dimensional models. Then, surface geophysics such as seismic reflection surveys will be performed at site scale in order to detect minor features within a repository region. The length of these minor features can be longer than several hundreds of meters. One-dimensional models as well as three-dimensional network models of minor features can be generated at this stage. The characteristics of minor features can be identified through borehole surveys. At later stages, geophysical surveys with higher resolution such as seismic tomography can be performed at block scale or even smaller in order to detect dominant migration pathways in a complex fractured rock. The length of large features at this stage can be a few hundreds of meters. The characteristics of large features will be identified through many kinds of field experiments including hydraulic tests and tracer tests. At the construction stage, a lot of fracture data obtained along repository drifts will be available, and thus the detailed three-dimensional DFN models can be applied to confirming the post-closure safety assessment. In this manner, the size of large features may change with the scale of interest.

5.   Requirements for the Future Japanese Programme

5.1   *Development of Approaches for Site Characterization and Performance Assessment*

Although a large amount of subsurface data on rock and fracture characteristics obtained at tunnels and underground open spaces is available in Japan, most of these data were obtained at near subsurface and were not collected to support performance assessment. Therefore, fracture characteristics as well as characteristics related to matrix diffusion need to be identified systematically at depths and geologic settings consistent with deep nuclear repositories. In this sense, the approaches for site characterization and performance assessment presented in the previous section need to be demonstrated and developed for the deep underground. JNC has started a deep underground research laboratory in granite at Mizunami in central Japan in order to investigate rock and fracture characteristics as well as groundwater chemistry at depth [10, 11].

Figure 5 **Interactive relationship between site characterization and performance assessment**

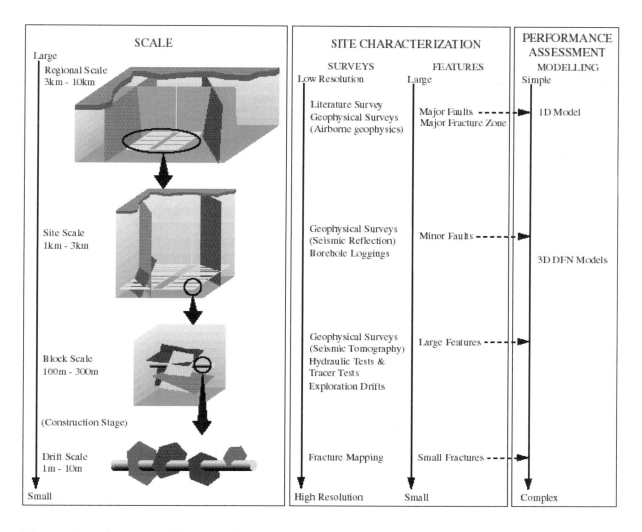

5.2   *Development of Fracture Transport Model*

Based on investigating rock and fracture characteristics at depth, it is important to develop more realistic transport model for water-conducting features. For example, according to fracture observations conducted in various rock types at more than ten outcrops and mine drifts, fracture fillings and fault gouge are more or less observed in all rock types. Ota *et al.* [12] performed tracer experiments within a single fracture filled with clay-rich minerals at the Kamaishi mine and found that sorbing tracers of strontium and cesium are significantly retarded by fracture fillings, while non-sorbing tracers of uranine and chlorite are less retarded as shown in Figure 6. Umeki *et al.* [7] performed tracer experiments within a water-bearing shear zone at the Grimsel Test Site, Switzerland, and found that a suite of tracers were significantly retarded due to diffusion into adjacent fault gouge and sorption within the gouge. According to these experiments, it is obvious that fracture filling and fault gouge have significant impacts on radionuclide migration. Therefore, it seems to be reasonable to incorporate fracture filling and fault gouge into future performance assessment models.

Figure 6. Tracer breakthrough curve obtained from dipole tracer experiments at the Kamaishi mine [12]. X- and y-axis designate elapsed time and normalized concentration, respectively

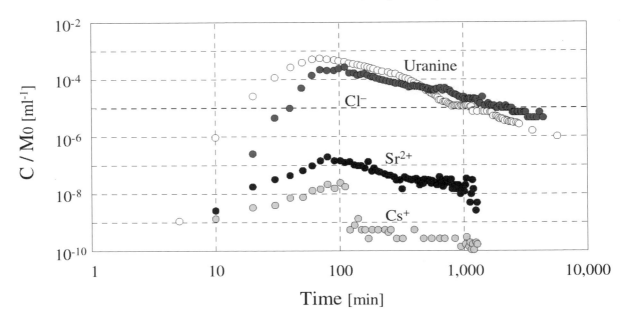

6. Conclusions

The radionuclide migration in heterogeneous fracture network systems has been solved in the H12 performance assessment. The lessons learned from the analysis are:

- It is important to identify the locations and characteristics of large water-conducting features, which have significant impacts on the radionuclide release from the geosphere.

- It is also important to identify the extents of low-permeability domains that contribute to the retention of nuclides in the geosphere.

Therefore, from the point of view of site evaluation, it is very important to identify the hydrogeological structure within a repository region in order to assess the performance of the geosphere. In other words, repository layout should be determined from the hydrogeological structure of a disposal site.

It is erroneous to estimate the characteristics of large water-conducting features from the data obtained through small-scale experiments or tests that stress only several meters. Upscaling of parameter values obtained from small-scale laboratory experiments and field experiments to the larger scale performance assessment models has been one of the key issues that challenge various national nuclear waste programmes. JNC has paid attention to the scale effects in both transport processes and parameter values, and has developed the ENTRY project in which flow and transport experiments are performed at intermediate scales (between laboratory core sample scale and field experiments scale) [1]. In addition to these efforts, large-scale tracer experiments seem to be more straightforward in providing answers on this issue. Another key issue is related to matrix diffusion. Since matrix diffusion is the only mechanism for the retention of nuclides in our current model, parameter values relevant to matrix diffusion are crucial. Each parameter value has been identified from a different experiment. There is, however, no guarantee that the model using the parameter values identified from different experiments yield a correct answer. Long-term tracer experiments might be desirable in order to confirm whether the parameter values are appropriate.

In a manner similar to conventional geological surveys, site characterization can proceed from large scale to smaller, and thus performance assessment will also initially focus on large features progressing to smaller ones. It is difficult to detect all large water-conducting features during site characterization, and many of them will be found during the excavation of repository drifts. Therefore, repository design should be flexible and easily modified in order to acquire the desired performance of the geosphere barrier. The repository design concept presented in this paper should helps us to build confidence in the feasibility of geological disposal in Japan.

It is very important for Japan to demonstrate and further develop the approaches for site characterization and performance assessment presented in the paper at JNC's Mizunami underground research laboratory. The incorporation of fracture fillings and fault gouge into the performance assessment models seems to be reasonable and further efforts are necessary to fully understand and incorporate these features and processes into future waste repository assessments.

7.   Acknowledgements

The authors are grateful to Shingo Watari of CRC Research Institute Inc. for his contribution to the H12 performance assessment analyses. We would also like to thank for Kuniaki Akahori, Kunio Ota, Toshihiro Seo, Kazuhiko Shimizu, Kaname Miyahara and Hitoshi Makino for their useful comments and suggestions.

8.   References

[1]   Ijiri, Y., Sawada, A., Hatanaka, K., Webb, E.K., Uchida, M., Ishiguro, K. and Umeki, H.: Strategy for characterisation and radionuclide migration modelling in block-scale geological media, Water-conducting Features in Radionuclide Migration, 3rd GEOTRAP Workshop, Barcelona, Spain, 1998.

[2]   Ijiri, Y., Sawada, A., Webb, E.K., Watari, S., Hatanaka, K., Uchida, M., Ishiguro, K., Umeki, H. and Dershowitz, W.S.: Radionuclide migration analysis using a discrete fracture network model, MRS 1998 Fall Meeting, 1998 (*in printing*).

[3]   Bridges, M.C.: Presentation of fracture data for rock mechanics, The 2nd Australian-New Zealand Conference on Geomechanics, pp.144-148, 1976.

[4]   Hayashi, N., Tanaka, K., Kitamura, M. and Kamiyama, K.: Construction of large underground caverns – Examples from underground oil storage facilities – Electric Power Civil Engineering, No.244, pp.344-440, 1993 (*in Japanese*).

[5]   Ohno, H. and Kojima, K.: Fractal on the spatial distribution of fractures in rock mass, Part 2: Fractal characteristics and variability of fractal distribution, Journal of the Japan Society of Eng. Geology, 34(2), pp.12-26, 1993 (*in Japanese*).

[6]   Davison, C.C., Kozak, E.T., Frost, L.H., Everitt, R.A., Brown, A., Gascoyne, M. and Scheier, N.W.: Characterizing and modelling the radionuclide transport properties of fracture zones in plutonic rocks of the Canadian shield, Water-conducting Features in Radionuclide Migration, the 3rd GEOTRAP Workshop, Barcelona, Spain, 1998.

[7]   Umeki, H., Hatanaka, K., Alexander, W.R., McKinley, I.G. and Frick, U.: The NAGRA/PNC Grimsel Test Site radionuclide migration experiment: Rigorous field testing of transport models, Proc. Mat. Res. Soc., vol.353, pp.427-434, 1995.

[8]   Gustafsson, E. and Nordqvist, R.: Radially converging tracer test in a low-angle fracture zone at the Finnsjön site, central Sweden. The Fracture Zone Project – Phase 3, SKB Technical Report, 93-25, 1993.

[9]   Gustafsson, E., Andersson, P., Ittner, T. and Nordqvist, R.: Large scale three-dimensional tracer test at Äspö, SKB Technical Report, 92-32, 1992.

[10]   Sakuma, H., Sugihara, K., Koide, K., Mikake, S. and Bäckblom, G.: The Mizunami underground research laboratory in Japan – Programme for study of the deep geological environment – Proc. of The 3rd Äspö International Seminar, Oskarshamn, SKB Technical Report TR-98-10, pp.15-24, 1998.

[11]   JNC: Master plan of Mizunami Underground Research Laboratory, Tono Geoscience Center, JNC, JNC TN7410 99-008, 1999.

[12]   Ota, K., Amano, K. and Ando, T.: Brief overview of in situ contaminant retardation in fractured crystalline rock at the Kamaishi In situ Test Site, Proc. of an International Workshop for the Kamaishi In situ Experiments, JNC Technical Report, JNC TN7400 99-007, pp.67-76, 1998.

Current Understanding of Transport at the Meuse/Haute-Marne Site and Relevant Research and Development Programme in the Planned Underground Research Laboratory

P. Lebon
ANDRA, France

Abstract

The Meuse/Haute-Marne site, on the eastern border of the Paris Basin, presents a very simple geological context of alternating carbonated and clay formations. The host formation (Callovo-Oxfordian argillites lying at a depth of 400-500 m) is uniformly thick (130-135 m) with a constant dip of 2° NW.

The conceptual model for transport has been established based on the data of the preliminary reconnaissance phase (1994-96) and can be summarised as follows. Radionuclides will be transported within the host formation by diffusion. Transport in the carbonated formations over and underlying the argillite will be advective with significant dispersion, but the velocity will remain low. The opening and ventilation of disposal vaults will create an EDZ. It is assumed that the damage mechanisms will only lead to changes in the order of magnitude of the transport time in a zone a few metres thick adjacent to the underground workings, and will not create new pathways for radionuclide migration.

A preliminary performance assessment exercise focusing on the geological barrier was carried out based on these assumptions, and supposing a horizontal single-level disposal facility in the mid-plane of the formation. The exercise carried out with these conservative simplifications offers a number of important and useful lessons for the next performance assessment exercises and for defining the Research and Development Programme strategy.

The underground laboratory programme has to deal with three main questions:

- Could EDZ characteristics change the transfer pathway (effective permeability at large scale and after re-saturation)? Are conceptual models of argillite THM behaviour capable of modelling the EDZ?

- Will diffusion remain the dominant transport phenomenon throughout the host formation in the future (persistence of overpressure in argillite, role of yet undetected discrete fractures at a regional scale, possible significant contribution of coupled phenomena)?

- What constitutes "necessary and sufficient" characterisation of transport within the Oxfordian formation for evaluating transfers 100 000 years from now?

1. Current understanding of transport at the Meuse/Haute-Marne site

1.1 Geological context

The Meuse/Haute-Marne site is located on the eastern margin of the main French sedimentary basin (i.e. Paris Basin). Its geological context is very simple and consists of alternating carbonated and argillaceous formations exceeding 1 000 m in depth at the site location. ANDRA is targeting its investigations on Callovo-Oxfordian argillites situated at a depth of 400-500 m.

Reconnaissance work has demonstrated the highly constant nature of the formation's thickness (130-135 m over a distance of 10 km) and dip (2° NW). No faults were highlighted by the 2D-seismic reflection surveys (with a 2 km profile spacing and a 10 m vertical resolution). Sedimentation cycles are well delineated, determining three scales of variability:

- The first, and largest scale, corresponds to three successive transgression/regression cycles;

- The second corresponds to sequential layers at a decametric scale (for example, there are five in the intermediate cycle, including one very homogeneous 14-m sequence in the mid-plane of the formation);

- The third corresponds to alternating 1-m thick layers, in certain of the second order horizons.

These scales of variability are also found regarding such petro-physical characteristics as total porosity (12-20%) and carbonate content.

On the other hand, the nature of the argillaceous minerals is not correlated at these scales. High-illite-content mixed-layer clays (R1) prevail in the lower half of the formation (up to the maximum intermediate flooding level), while the upper half is dominated by high-smectite-content clays (R0). The same applies to the average pore size, that is $20 \cdot 10^{-9}$ m in the lower half, and $40 \cdot 10^{-9}$ m in the upper half, resulting from compactness and recrystallisation due to secondary calcite formation. This pore structure implies specific strong interactions between water molecules and mineral phases (physisorption on clay minerals, capillary forces).

Figure1 **Geological and petro-physical log of Callovo-Oxfordian (HTM102 borehole)**

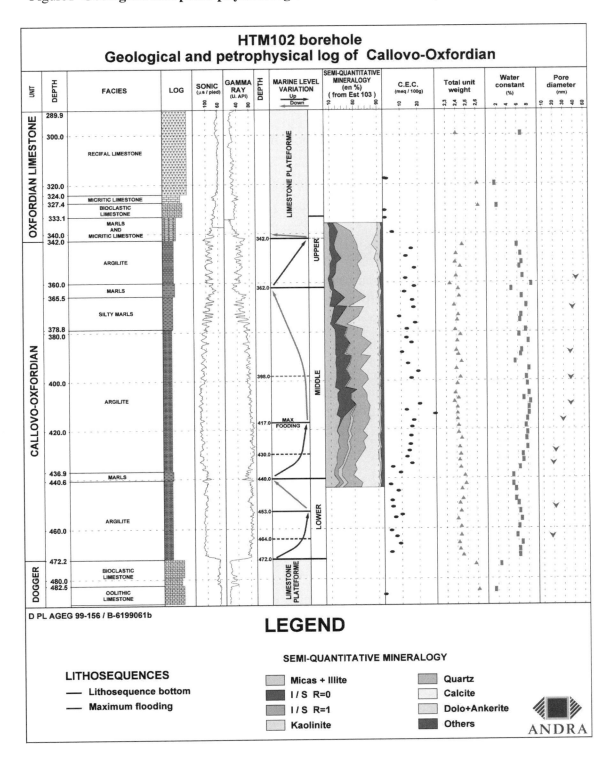

1.2   *Transport-mechanism parameters in the host formation*

The present level of information concerning the formation's physical and chemical properties indicates that the transport parameters of the rock (permeability, diffusion coefficient) are quite constant throughout the formation. Certain other parameters show an obvious relationship with petro-physical properties (e.g. the capacity of cationic exchanges and the salinity of interstitial waters appear to be associated with the nature of the argillaceous minerals).

Permeability is very low throughout the formation making measurements difficult. For this reason, a variety of evaluation techniques involving different phenomena based on various space and time scales were used, such as:

- Permeability measurement in boreholes by pulse tests: $1.3 \cdot 10^{-13}$ to $1.7 \cdot 10^{-11}$ m/sec [1];

- Deconvolution of imbibation profiles using adsorption/desorption isotherms: 0.9 to $6 \cdot 10^{-14}$ m/sec [2];

- Tritiated-water tracing in cells under a pressure gradient: $4.7 \cdot 10^{-15}$ to $3.3 \cdot 10^{-14}$ m/sec [3];

- Reverse analysis of the transient re-charge phase with interstitial pressure in a borehole using an Electronic Pressure Gauge sealed in the intermediate argillaceous horizon.

Except for the very short first method (a few hours), all methods yield relatively comparable values, thus reinforcing confidence in these very low figures.

Diffusion measurements were made in cells using a through-diffusion method and 1 cm thick samples; both tritiated-water and sodium-iodide tracers were used. Measured diffusion coefficients ranged from 0.4 to $2 \cdot 10^{-11}$ m^2/sec for tritium [3] and from $5 \cdot 10^{-13}$ to $3.6 \cdot 10^{-12}$ m^2/sec for the iodide anion. This difference is most probably due to an anion exclusion effect due to the surface electrical charge of the argillaceous minerals.

1.3   *Transport-mechanism parameters in the geosphere outside the host formation*

Familiarity with the hydrogeological context is important for evaluating transport mechanisms for two reasons:

- For establishing the hydraulic and geochemical boundary conditions for the Callovo-Oxfordian formation;

- To aid in defining the potential discharge zones into the biosphere.

From this standpoint, the two carbonated formations surrounding the Callovo-Oxfordian formation (i.e. Oxfordian and Dogger formations), which are known as aquifers in other parts of the Paris Basin, are the most interesting.

The permeability of the carbonated formations surrounding the argillite is relatively low (inferior or equal to 10^{-8} m/sec) [4], because they have also been submitted to an intense recrystallisation of diagenetic origin. The current hydrogeological map shows a general northwesterly run-off direction, consistent with our knowledge of the recharge areas and dip of the formations making up the Paris Basin's aquifer systems, and current data on regional piezometry [4]. On site, hydraulic overpressure is measured in the Callovo-Oxfordian formation (a phenomenon which has also been observed elsewhere, but not explained yet). If it spreads over the whole of this area, it should block off any vertical flow exchanges between the over- and underlying aquifer formations.

2. Preliminary performance assessment exercise

2.1 Conceptual model of the transport parameters at the Meuse/Haute-Marne site

Based on the preceding data, the following conceptual model for radionuclide transport was developed for the Meuse/Haute-Marne site:

- Taking into account the very low permeability and the overpressure measured in the Callovo-Oxfordian formation, diffusion is assumed to be the dominant process for radionuclide migration within the formation for the entire period considered, giving due consideration to possible changes in hydrogeological conditions;

- Diffusion coefficients for each chemical species are considered to be invariant throughout the whole formation and over time. Chemical retention processes (mainly sorption) are controlled by the succession of argillaceous minerals present. It will therefore be necessary to take into account three sub-units: a lower zone dominated by R1 mixed-layer clays, an upper zone dominated by R0 mixed-layer clays, and an intermediate transition zone;

- Disturbances induced by mechanical decompression during the excavation of the disposal structures and their subsequent ventilation (desaturation, chemical oxidation, etc.) will significantly modify the transport parameter values in a an adjacent zone of relatively small extension (a maximum of a few metres); radionuclide migration pathways are assumed not to be significantly modified;

- Transport in the adjacent carbonated formations will take place in relatively thin porous levels which are considered to extend laterally to the formation limits. Radionuclide migration would occur by convection/dispersion.

For the purpose of this preliminary performance assessment exercise, simplifications were used:

- The over- and underlying carbonated formations were considered as zero-concentration limits for radionuclide species because of their much higher permeability relative to that of argillites.

- Radionuclide transport in the adjacent carbonated formations was not taken into account. An instantaneous transfer to Tithonian aquifer and a pumping in that aquifer as route for radionuclide to the biosphere were considered.

2.2 Results of the preliminary performance assessment exercise

A preliminary performance assessment, focusing on the Callovo-Oxfordian formation as the geological barrier, was carried out based on the main assumptions of the conceptual model and on a preliminary design of the repository:

- A horizontal structure located on a single level in the mid-plane of the formation;

- An underground disposal facility divided into separate units containing different waste types, and (high performance?) sealing of each disposal unit.

Figure 2. **Conceptual model scheme of the Meuse/Haute-Marne site**

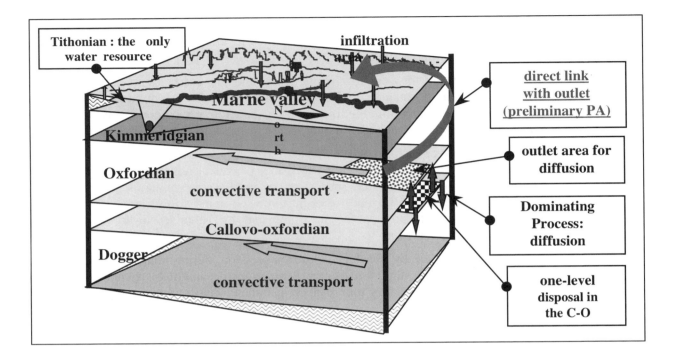

As regards the prevailing transport mechanism, the preliminary performance assessment exercise was carried out using a deterministic mass transport simulation software (OASIS) which represents different radionuclide pathways in terms of a multi-1D network. Each scenario was modelled as an assembly of different modules containing elementary transfer functions for each component (from the waste containers to the biosphere), each module being characterised in terms of its assumed hydraulic and geochemical properties.

Two scenarios were considered:

- A reference-case scenario in which the dose rates is directly estimated from the radionuclide-release rate at the limit of the Callovo-Oxfordian formation. In the exercise, the engineered barrier system is only taken into account as a 500-year corrosion shield for the waste containers. Radionuclide release begins after that period and occurs at a constant rate (e.g. 10^{-5} year^{-1} for vitrified waste, 10^{-3} year^{-1} for waste in a cement matrix).

- A disruptive event scenario represented by a short-circuit of the Callovo-Oxfordian formation in order to design the engineered barrier system (not presented here).

The following table shows for each waste type the radionuclides capable of reaching the top of the Callovo-Oxfordian formation.

Table 1. **Radionuclides of interest for the long term safety from Results of the preliminary performance assessment exercise**

Waste type	Radionuclides (activity > 1 Bq/year at the top)
Vitrified waste	^{94}**Nb*** – ^{99}**Tc*** – ^{79}Se – ^{147}**Sm***
Bitumen-waste packages	^{94}**Nb*** – ^{99}**Tc*** – ^{129}I – ^{79}Se** – ^{147}Sm*/** – ^{14}C**
Hull and end caps packages	^{94}Nb – ^{99}Tc – ^{129}**I*** – ^{79}Se – ^{147}**Sm*** – ^{14}C – ^{93}**Mo*** – ^{36}Cl

* Main contributors.

** Radionuclide activity between 10^{-3} and 1 Bq/year at the top.

Figure 3. Breakthrough curves at the top of the Callovo- Oxfordian

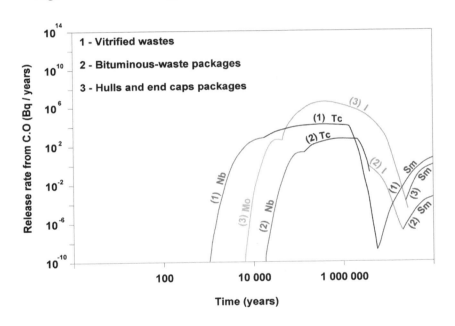

Figure 3 shows the activity release rate of the most important radionuclide contributors from the Callovo-Oxfordian formation for ILW and HLW. The release peak should arrive over a period of roughly 100 000 years according to current transport parameters, but ^{147}Sm is not calculated to appear before 10^{7} years.

Even without taking into account transfers within the carbonated formation, i.e. from the upper limit of the host rock to the biosphere, the individual dose rate for an hypothetical critical group (local self-sustaining farming community) generally remains very low (<10^{-9} Sv/year for a 1m^3/h dilution flow rate) except for that due to the ^{129}I contained in "hulls and end caps" waste packages (10^{-5} Sv/year for a 1m^3/h dilution flow rate after 100 000 years). It is worth noting that the activity at the upper limit of the host formation is due to fission products only, and that actinides are not released into the biosphere during the calculation period because of retardation due to low solubility and sorbtion in the host formation.

The main contributors have high solubility, long half-lives and very low chemical retention.

Groundwater flow within the carbonated formations (Oxfordian and Dogger) were also simulated using numerical hydrogeological models. Water flow trajectory simulations at the regional scale highlight the constancy of the flow directions, even when hypotheses vary. The nearest natural outlet to the site for run-offs in the carbonated formations would be the Marne Valley topographically lower and corresponding to a crustal fault.

The travel time to the biosphere by pure convection takes a minimum of approximately 150 000 years for pathways starting from the top of the Oxfordian host formation, and more than 10 million years for those starting from its base, i.e; within the Dogger formation. It is worth noting that the travel time is at least in the same order of magnitude as the average breakthrough time through the Callovo-Oxfordian formation. In addition, consideration of convective/dispersive radionuclide transport along a transport path in the carbonated formations in the site-performance assessment would lead to a "spreading out" of the exit signal at the outlet, thus reducing the calculated dose significantly.

2.3   *Lessons learnt from the preliminary performance assessment exercise*

- The conceptual transport model in the host formation seems to be robust, but it will be necessary to verify that disturbances due to structures will not modify the transfer pathway, and also that convection cannot prevail locally, or at other time scales.

- Confidence in the assessment of the transport performance in the host formation can be increased by attempting to confirm the diffusion coefficients measured on samples and the conservative approach of the retardation coefficients used for actinides.

- The host formation constitutes the principal containment barrier and, by applying an extreme conservatism, it could be possible to disregard the role of carbonated formations in the performance assessment. The research programme should therefore be limited to a "necessary and sufficient" hydrogeologic characterisation of those formations for the global performance assessment at the scale of 1 million years.

3.   Relevant Research and Development Programme in the planned underground research laboratory

The underground research laboratory (URL) architecture was designed to respond to major issues on disposal possibilities for long-lived HLW in the Callovo-Oxfordian formation, taking into account the very tight deadline mandated by the Law of 31 December 1991. That is the reason why certain scientific activities have already been prepared by ANDRA at the Mol Laboratory (Belgium) and are being subjected to continual development at the Mont-Terri Laboratory (Switzerland) where the geological environment is more similar to the Meuse/Haute-Marne argillites in France.

The competition between diffusive and convective processes represents the major element to be analysed, in particular:

- The orders of magnitude of the transfer parameters (permeability, diffusion coefficient).
- Driving factors (pressure and concentration fields).

We shall now examine how the research programme in the underground laboratory can back up the preliminary performance assessment and provide answers to the above questions.

3.1   *To verify that disturbances due to structures does not modify transfer pathways*

While the Excavation Disturbed Zone (EDZ) study was not a topic of this workshop, it is, on the other hand, a very important part of the URL programme. A point of particular interest is determination of the effect of mechanical and thermo-mechanical damage on the transport and retention properties of the argillites (development of micro-cracks or of organised mesoscopic fractures).

It is clear that radionuclide migration from a repository implies that the host formation has been resaturated around the structures. It is therefore reasonable to question whether or not transmissivity measurements of the disturbed zone that will be possible in the underground laboratory will not be too conservative since they may not take fully into account the fluid/rock interactions occurring along the cracks during rehydration (swelling, calcite deposit). This issue has already been investigated in partnership with the *CNRS* (French Science and Research council), but remains difficult due to the differences in time scales between what is available in the laboratory and what geological processes can offer (several thousand years).

In addition, if an organised system of mesoscal fractures develops, it could vary depending on the spatial distribution of the petro-physical properties of the argillite, thus having repercussions on the longitudinal continuity of that fracture. Systematic measurements will therefore be made along the different drifts of the underground laboratory with a view to determining its variability.

3.2   *To verify that convection continues to prevail even locally or at other time scales*

It is quite possible that some discrete fractures exist (around 2-km long max.) that were below the detection limits of the means implemented during the preliminary surface reconnaissance work. The question is: can they play a significant role in the transfer of radionuclides to the biosphere?

In order to detect them, the research programme includes:

- The implementation, before the actual excavation of underground installations, of a high resolution 3-D-seismic-reflection campaign on a 2 square km area centred on the URL in order to obtain a detailed 3D geological model.

- The addition, at the main experimental level situated in the mid-plane of the formation, of two orthogonal drifts, upwards and downwards respectively, oriented according to the large regional faults. Those drifts will make it possible to compare direct geological observations with the previous model.

In situ measurements of the transport parameters will be made in the most significant fractures encountered. Attempts will be made to reconstitute the palaeocirculations in those fractures by analysing the geochemical characteristics of the infilling minerals and of the argilite at the fracture borders.

The hydraulic pressure field in the Callovo-Oxfordian formation is not well known and its causes are still unclear. The fact that a slight overpressure exists in comparison with adjacent carbonated formations is generally accepted, but it is still impossible to say if this field is homogeneous throughout the formation, and if it is effectively permanent or transient at the time scales of interest for disposal site evaluation. The architecture of the underground repository will provide a more complete vertical profile of the hydraulic pressure.

3.3 *To confirm diffusion coefficients measured on samples*

Preliminary studies have shown us that petro-physical variations of the Callovo-Oxfordian formation had little influence on the measured parameter values (measurement spectrum inferior by an order of magnitude). The research programme will confirm this fact by a series of samples and measurements made in experimental drifts situated in the mid-plane of the formation and in the upwards and downwards oriented drifts according to the different petro-physical variability orders observed.

The key question is the upscaling of these lab-scale parameters to ones suitable for describing host formation properties at relevant time and space scales. Two approaches will be used:

- On-site measurements at a pluri-decimetric scale using solute-migration experiments simulating radionuclide transfers. With a view to preparing those experiments, two methodological tests (DI and FM-C tests) were launched at Mont-Terri. The results of those tests will be made available next year.

- Looking for natural markers of diffusion processes, followed by inverse modelling to estimate global values for diffusion parameters. A reflection is underway on this question based on the ^{37}Cl measurements already made (see Figure 4), but they must take into account the chemical analysis of the interstitial water mentioned below and the detailed geological history of the site during the Cenozoic era. An extensive sampling programme during the sinking of the main access shaft is foreseen to aid this analysis.

In an other hand, researches are being carried out to verify whether or not Onsagger phenomena could modify the diffusion process. Preliminary results on Soret effect are available.

3.4 *To confirm the conservative delay coefficients used for actinides*

The important thing here is to confirm the geological media's physical and chemical conditions controlling the precipitation and sorption balances of actinide elements. From the preliminary characterisation of the chemistry of the argillites' interstitial water, the observed chloride content is in the order of 1-2 g/L. The pH, buffered by the presence of calcite, may present values included between 7 and 9 according to our current knowledge. We are faced with a reducing environment as evidenced by the presence of pyrite disseminated throughout the whole layer. The following points must be verified the pH and P_{CO_2} values, the major cation (Na+, Ca++, K+) content (the presence of sulphates will also need to be confirmed because of their aggressiveness towards concrete and, consequently, their possible impact on the source term of certain ILW packages).

The most delicate operations will involve interstitial-water sampling, while avoiding oxidation effects or bacteriological contamination. First samples will therefore be taken as soon as the underground Callovo-Oxfordian formation is accessible.

Moreover, the disturbance effects due to underground structures will be checked e.g. by experiments on the evolution of an oxido-reduction front under the influence of forced low-hygrometry ventilation and on the propagation of an alkaline plume from an injection borehole.

Figure 4. **Total chloride content and δ³⁷Cl log in the Callovo-Oxfordian**

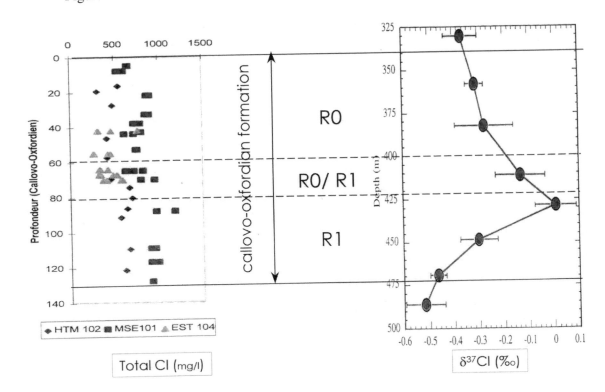

Total Cl (mg/l)

δ³⁷Cl (‰)

3.5 *To determine groundwater flow within carbonated formations around the host rock*

The purposes of the investigations are:

- To confirm that the more transmissive porous levels of the Oxfordian formation are not likely to be exploited;
- To determine the groundwater flow directions and velocities.

During the first phase, a local hydrogeological model of the site will be constructed based on hydraulic parameters measured in boreholes around the underground laboratory. This model will be used to predict the draw-down of groundwater levels caused by the sinking of the URL shafts.

During the second phase, these simulations will be compared to the variations of hydraulic charges in the different porous levels, measured at pluri-hectometric distances during shaft sinking.

During the third phase, the local model will be integrated within the regional hydrogeological model, in part aided by data on water origin and residence time provided by chemical and isotopic analysis of the waters sampled from shaft-sinking follow-up boreholes.

The modeled groundwater flow directions and velocities will be finally be confirmed by hydrogeological data gained from new measurement boreholes located in as yet unexplored areas.

4. Conclusions

The future Performance Assessment will include all the compartments of the repository and the geosphere up to the biosphere. The degree of complexity of the computational models used will be determined on the basis of an in-depth phenomenological analysis of the different situations through which the repository will go. The limitations and possible numerical artifacts of the used codes will be identified.

In order to support the Performance Assessment, we have to collect a range of data:

- The research programme in the underground laboratory will make it possible to determine the current characteristics of the local hydrogeological system, if all essential the mechanisms affecting the measurements can be identified and explained.

- Increasing confidence in the representation of its evolution over many hundreds of thousands of years requires deconvolution of this instaneous picture in order to know the functioning mode of the system throughout its geological history. This analysis can be supported by use of natural tracers as "self-analogs". However, since this later one requires use of conceptual transport models in the deconvolution process, we must be careful not to be trapped in a situation of circular reasoning.

- The host formation and its geologic surrounding cannot be considered separately given that they have the same history. The coherence of all the assumptions with the successive past geologic events must be verified.

References

[1]   B. Paris et R. Roberts, 1997
        "Mesures de transmissivité par essais de puits et analyse de sensibilité"
        Atlas des posters des Journées Scientifiques CNRS/ANDRA, Bar-le-Duc, 27 et 28 octobre 1997, p 50-51.

[2]   J.C. Robinet A. Jullien et A. Pasquiou, 1997
        "Mesures de perméabilité sur échantillons d'argilites de l'Est par analyse de profils hydriques et étude d'isothermes d'adsorption"; Atlas des posters des Journées Scientifiques CNRS/ANDRA, Bar-le-Duc, 27 et 28 octobre 1997, p 52-53.

[3]   X. Vitart G. Beaudoing et J. Pocachard, 1997
        "Mesures de perméabilité et de coefficient de diffusion des argilites par méthode de traçage"
        Atlas des posters des Journées Scientifiques CNRS/ANDRA, Bar-le-Duc, 27 et 28 octobre 1997, p 54-55.

[4]   Y. Babot, 1997
        "Hydrogéologie des formations carbonatées jurassiques dans l'Est du bassin de Paris"
        Actes des Journées Scientifiques CNRS/ANDRA, Bar-le-Duc, 27 et 28 octobre 1997, p 65-76, Éditions EDP sciences.

PART C:

POSTER SESSION

Use of a Matrix Diagram in Modelling Coupled Transport Processes in Performance Assessment

L.E.F. Bailey & S. Norris
Nirex Limited, UK

This paper describes how information on geosphere-transport processes is assimilated into the modelling of a disposal system for post-closure performance assessment. This description is based upon the overall iterative approach used by Nirex whereby performance assessments follow a cyclical development process. The production and use of a matrix diagram is an integral part of the Nirex model development programme, providing a clear audit trail from the identification of significant Features, Events and Processes (FEPs), to the models and modelling processes employed within a detailed safety assessment. As detailed in References [1-5], a five stage approach has been adopted, providing a systematic framework for addressing uncertainty and for documenting all modelling decisions and assumptions:

Stage 1:   FEP Analysis – compilation and structuring of a FEP database;
Stage 2:   Scenario and Conceptual Model Development;
Stage 3   Mathematical Model Development;
Stage 4:   Software Development;
Stage 5:   Confidence Building.

Confidence building, although referred to as Stage 5, occurs throughout the first four stages, by continual review of the process.

The scope of the FEP analysis work undertaken by Nirex to date has been as a demonstration of a methodology, rather than a complete implementation of the approach. It would therefore need to be reviewed in any future assessment cycle. Data from the Sellafield research and investigation programmes have been used, in conjunction with experience gained in Nirex 97 [6]. However, the approach would be applicable to any geological disposal facility.

A FEP database has been developed and structured as a Master Directed Diagram (MDD), showing elicited FEPs and the key relationships between them. The MDD was used to define a sufficient set of conceptual models that, together with their interactions, provides a full description of the disposal system and its future evolution. This set of conceptual models was then reviewed, to identify those providing a major influence on other areas of the disposal system, and a reduced set of conceptual models was derived. This formed the basis for the development of an influence matrix diagram.

A matrix diagram is used to examine all potential interactions between the conceptual models, and provides an indication of the interactions that have to be addressed in assessment models. Conceptual models are placed on the principal diagonal, and the interactions are considered in the off-diagonal elements. An interaction may be interpreted as one conceptual model supplying information as an input to another conceptual model. The relative strength of each interaction can be assigned a qualitative score. Only direct interactions are recorded on the matrix diagram, although indirect influences are easily identified. The position of the conceptual models on the matrix diagonal can usefully be ordered to highlight "clustering" of interactions and hence provide insights into appropriate approaches to modelling coupled processes.

As an example, a matrix diagram has been developed based on the understanding of the Sellafield site. Figure 1 shows part of this matrix diagram, identifying a number of conceptual models covering transport processes and other related conceptual models. For the transport processes considered of relevance at this site, namely *"Advection of Dissolved Species"*, *"Dispersion of Dissolved Species"*, *"Diffusion of Dissolved Species"*, *"RMD of Dissolved Species"* and *"Osmosis"*, it is seen that strong, direct two-way coupling is only present between *"Dispersion"* and *"Diffusion"*. Other direct interactions between these conceptual models are one-way only.

Indirect influences between these conceptual models occur via the conceptual model *"Properties of APL"* (aqueous phase liquid, taken to be the physical, chemical and thermal properties of groundwater, including any dissolved species present). This conceptual model is seen to have a strong influence on all of the above-mentioned transport processes. This suggests the most appropriate modelling approach for representing the effects of coupled transport processes would be to develop a conceptual model for the properties of the groundwater, which is then coupled to each of the transport processes, rather than necessarily coupling the transport processes directly to each other.

References

[1]   *Overview of the FEP Analysis Approach to Model Development*, Nirex Science Report S/98/009, 1998.

[2]   *Conceptual Basis of the Master Directed Diagram*, Nirex Science Report S/98/010, 1998.

[3]   *Overview Description of the Base Scenario Derived from FEP Analysis*, Nirex Science Report S/98/011, 1998.

[4]   *Modelling Requirements for Future Assessments based on FEP Analysis*, Nirex Science Report S/98/012, 1998.

[5]   *Development and Application of a Methodology for Identifying and Characterising Scenarios*, Nirex Science Report S/98/013, 1998.

[6]   *Nirex 97: An Assessment of the Post-closure Performance of a Deep Waste Repository at Sellafield* (5 Volumes), Nirex Science Report S/97/012, December 1997.

Figure 1.   Part of the matrix diagram developed based on the understanding of the Sellafield site, identifying a number conceptual models (CMs) covering transport processes and other related CMs. Influences between CMs are read in a clockwise manner.

Conceptual models (CMs), in order, form both the rows and the columns of the interaction matrix:

1. Properties of APL
2. Hydrogeological Properties of Rock
3. Transport Pathways in Geosphere
4. Dissolution/Precipitation
5. Mineralization: Recrystallization
6. Sorption
7. Groundwater Flow
8. Gas-driven Water Flow
9. Advection of Dissolved Species
10. Dispersion of Dissolved Species
11. Diffusion of Dissolved Species
12. Rock Matrix Diffusion of Dissolved Species
13. Osmosis

Interaction matrix (rows = influencing CM, columns = influenced CM; ■ = leading-diagonal CM cell):

CM	1	2	3	4	5	6	7	8	9	10	11	12	13
1 Properties of APL	■	0	10	0	10	0	0	0	10	10	10	10	6
2 Hydrogeological Properties of Rock	0	■	10	10	6	2	0	0	0	0	0	2	0
3 Transport Pathways in Geosphere	10	10	■	10	2	0	0	0	0	0	0	6	0
4 Dissolution/Precipitation	0	0	10	■	10	0	0	0	0	0	0	0	0
5 Mineralization: Recrystallization	10	10	2	10	■	10	0	0	0	0	0	0	0
6 Sorption	10	6	0	0	4	■	0	0	10	0	10	6	0
7 Groundwater Flow	0	2	0	0	0	0	■	10	10	10	0	0	0
8 Gas-driven Water Flow	0	2	0	0	0	0	10	■	10	0	0	0	0
9 Advection of Dissolved Species	0	0	0	0	0	0	0	0	■	10	0	0	0
10 Dispersion of Dissolved Species	10	0	0	0	0	0	0	0	10	■	10	0	0
11 Diffusion of Dissolved Species	10	0	0	0	0	0	0	0	10	10	■	10	0
12 Rock Matrix Diffusion of Dissolved Species	10	0	0	2	0	0	0	0	0	0	10	■	0
13 Osmosis	6	0	0	0	0	2	0	0	0	0	0	0	■

From Site Data to Models: Understanding the Influence of Fractures on Transport Properties through Natural and Artificial Field Tracer Experiments (the Tournemire Site)

G. Bruno, L. De Windt, Y. Moreau Le Golvan, A. Genty, J. Cabrera
IPSN, France

Abstract

Selected to develop *in situ* research programmes on confinement properties of potential geological formation for nuclear waste disposal, the argillaceous Tournemire site is characterised by the presence of fractures, some of them being hydraulically active. Even though flow rates are low in the site, these fractures might hamper the confinement properties of an argilite geological barrier. The hydraulic and transport properties being still unknown, the purpose of the present project is to gain a deeper knowledge on how radionuclides are transported through the fractures by means of tracer experiments. Such a programme will permit testing various methods for measuring transport characteristics but also evaluating various models of coupled chemical and transport processes.

The presented project aims at determining the influence of the presence of fractures on transport properties, through a multi-approach study, using natural and artificial tracers.

The main idea underlying the project is to perform *in situ* verifications of models and assumptions but also to develop a methodology to obtain site specific data and generic data, these latter's being potentially transferable to other sites.

Introduction

Due to the potential use of indurated clays as host rocks for radioactive waste disposal, numerous fundamental research programmes are developed on the confinement properties of argillaceous formations. In order to gain experience on site characterisation, The French Institute for Protection and Safety has selected the Tournemire site to develop research programmes.

As part of the characterisation programmes, the study and understanding of transport properties is of major interest. A special attention is given to fluid migrations because of their capability to damage the confinement structure and eventually to transport radionuclides towards the biosphere. Moreover different site characterisations in indurated clays have recently shown the presence of fractures either of tectonic or human origins. Hence, the characterisation of the experimental site concerned by the project has highlighted the juxtaposition of two vertical blocks, one being non-altered and the second one being highly affected by the presence of tectonic fractures. Hydraulically active fracture planes with carbonate fillings have also been discovered.

257

Studying an argilite with close characteristics to the clays of candidate sites for nuclear waste disposal will permit to extract the part of specific results from the part of generic results; these latter being transferable to future experimental sites or repositories.

Taking into account the above elements, the present paper details an international project project with several objectives:

1) To characterise and understand transport in indurated argillaceous media.

2) To gain a deeper knowledge on how radionuclides are transported through a fractured argilite.

3) To develop studying processes and tools to characterise transport.

4) To test various methods for measuring transport characteristics and evaluating various models of coupled chemical and transport processes.

To achieve the above objectives three main complementary approaches are envisaged, by using two types of tools, namely natural and artificial tracers:

(i) Distribution profiles of natural tracers perpendicular to a fracture.

(ii) Comparison of natural tracer distribution profiles in the matrix, between the fractured and the non-fractured zones.

(iii) Distribution of selected artificial tracers in a single fracture plane.

Geological setting

The Tournemire site is located in the South of France (Aveyron) in the Larzac Causse Area and consists of a three-layer model of sub-horizontal Jurassic sedimentary formations (Fig.1).

Figure 1. **Geological cross-section of the Tournemire site**

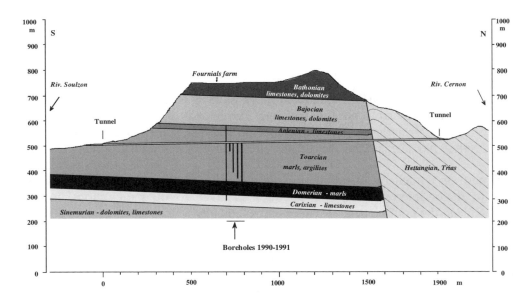

A 250 m thick indurated argillaceous layer (shales and marls) of the Toarcian and Domerian "sandwiched" between a 400-m thick layer of limestones and dolomites (Hettangian, Simurian and Carixian) at the lower side, and a 300 m limestones and dolomites layer (Aalenian, Bajocian &Bathonian) (Cabrera, 1991; Barbreau & Boisson, 1993; Boisson *et al.*, 1996). To the north the Cernon fault (E-W system) as uplifted the Triassic formations. A 100 year-old railway tunnel gives access to the argillaceous formations whereas two perpendicular galleries were drilled recently (3 year old).

The exploration of the site through the use of the tunnel and the galleries but also through several drilling campaigns has allowed to show the presence of fractures of tectonic origin in the argillaceous formations. The fracturing is mainly sub-vertical and includes microfissures (millimetre scale), fractures (centimetre scale) sealed by calcite and faults (decimetre to metre scale) including fault breccias (Cabrera, 1995). Most of this fracturing is associated to the N-S trending Pyrenean compressional event (50-60 million years ago). On the other hand, the different excavation phases (tunnel and galleries) have induced the development of a highly fractured disturbed zone (about 2 m thick in the main tunnel).

Beyond the observation of fractures in the clayey formations, the geological characterisation of the site, has shown the vertical juxtaposition of a densely fractured zone and a sparsely fractured zone. (Fig.2). Finally some of the encountered fractured are still hydraulically active such as the sub-vertical fractured plane discovered in the Eastern gallery. Such a fractured plane exhibits carbonate fillings as veins and geodes, and a weak but long-term water flow has been observed.

Figure 2. **Horizontal cross-section of the site**

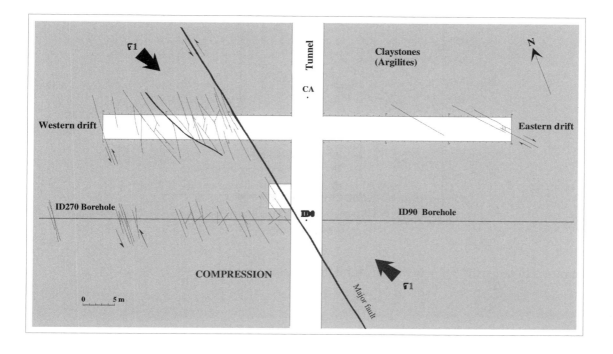

Natural tracers

Distribution profiles of natural tracers perpendicular to fractures

Previous isotopic studies on the transport phenomena in the Tournemire site (Moreau Le Golvan, 1997) have shown a clear link between the isotopic composition of porewater and the structural context : interstitial water of rock samples in densely fractured zones seems being affected by a secondary process of enrichment in heavy isotopes (Fig. 3). An hypothesis of water migration from the matrix to the faults during the compressive Pyrenean phase (ultrafiltration process) was proposed to explain the observed enrichment in heavy isotopes. However ultrafiltration processes are poorly documented and need to be confirmed. Thus, in order to understand the influence of fractures on fluid flows in the matrix, local tracer distribution profiles will be executed perpendicular to two kinds of fractures in the site:

- on a single fracture, water conducting in the Eastern gallery;

- in the western zone, where fractures are sealed by calcite and rather show local migrations between the matrix and the fractures.

This part of the project will focus on conservative tracers (Chloride and Deuterium) as well as the distribution of chemical compounds in the interstitial waters.

Figure 3. **Stable isotope content of interstitial water in the Tournemire site**

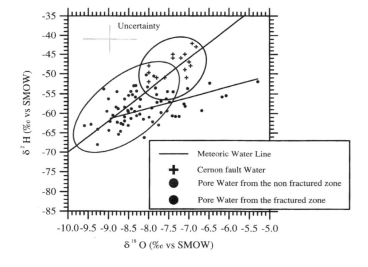

Influence of fracture density on porewater natural distribution profiles.

260

Figure 4. **Distribution of porewater deuterium content with depth: comparison between measured values and modelled diffusion profiles (METIS code – Ecole des Mines de Paris) Left: initial state, Formation water: dD=0‰, karstic water: dD=-70‰VSMOW**

Right: initial state, Formation water: dD=-52‰, karstic water: dD=-70‰VSMOW

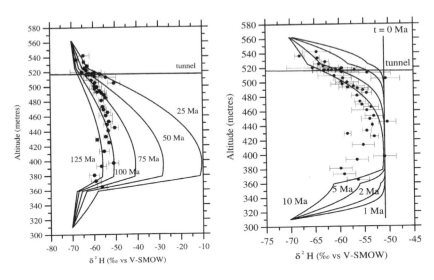

The interest of the site as support for the project, to characterise and understand large scale transport and the influence of fracture density, stands in the fact that the argillaceous formation is "sandwiched" between two limestone aquifers. Previous studies on the distribution of deuterium (Moreau Le Golvan, 1997) indicated a probable diffusive character of transport in the clayey matrix (Fig.4). However this preliminary study and another one concerning chloride distribution in a similar formation seem to indicate that water-rock interaction effects or analytical artefacts on isotopic and chemical composition could not be excluded. Given the above elements, this part of the project aims at following natural tracers in the overall extension of the formation, both in a densely fractured zone and in a sparsely fractured zone; detailed investigation will be carried out at the vicinity of the surrounding aquifers.

Two phases are considered to perform the analyses:

- Phase I: Full length distribution profiles in the sparsely and densely fractured zones for the following tracers: porewater chemical composition, stable isotopes of the water molecule (^2H, ^{18}O), chloride, $^{37}Cl/^{35}Cl$, noble gases. Determination of $^{36}Cl/^{35}Cl$ (P7), ^{14}C contents at the connections with aquifers.

- Phase II: Laboratory tracer behaviour study: determination of effective diffusion coefficient, effective porosity coefficient.

According to the results obtained in phases I and II, the influence of fracture density on the profiles will be assessed, leading to the modelling of large scale transport in the clayey formation. The obtaining of field parameters (e.g. effective diffusion coefficient) will be compared with laboratory data in order to assess eventual scale effects.

The interpretation of the analyses involves the integration of data on the burial and uplift history, the evaluation of the expected tracer concentrations in the formation water and in the aquifer (and their variations with time), the evaluation of the rock compaction and its consequences on the physico-chemical properties of the rocks.

Artificial tracing in a single fracture plane

As part of its contribution to the overall objective of the project, the artificial tracer experiment will help to understand and predict the transport and retention along water conducting features within the formations. This experiment will also permit to assess the transmissivity of the water conducting features, trying to determine its order of magnitude as well as its time dependence. Finally, information will be collected such as : the relative importance of matrix diffusion to advective transport, the importance of geochemical processes in transport (dissolution/precipitation, retention, chemical interactions ...), the channelling effects.

The realisation of the artificial tracer experiment is scheduled with several consecutive steps has described in figure 5.

Figure 5. **The different phases of the artificial tracer experiment**

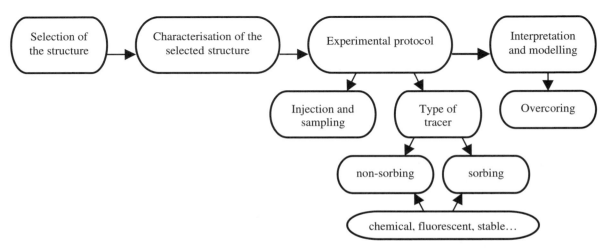

Figure 6. **Sub-vertical fracture plane**

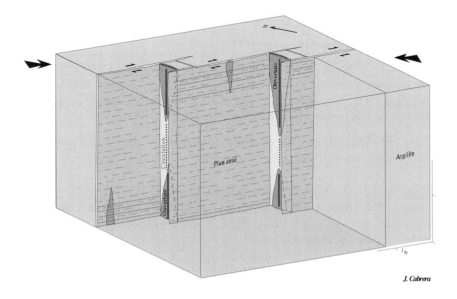

The fracture plane has already been chosen. This structure is located far enough from any perturbing elements like tunnel or gallery (development of an excavation disturbed zone). According to previous studies, the fracture plane is a decametre sub-vertical structure related to the Pyrenean compression (Fig.6). It presents carbonated fillings and geodes of calcite, whereas a weak water flow has been localised (De Windt et al., 1999; Cabrera, 1995).

The characterisation of the selected structure consists in the study of the geometry of the fracture as well as in the determination of dispersion parameters and the measurements of local transmissivity. The geometry of the fracture will be studied using different techniques such as seismic and electric resistivity tomographies, radar survey.

The characterisation of the selected fracture plane also involves the realisation of one or several hydraulic tests. Such tests are planned to provide quantitative estimates of hydraulic properties such as hydraulic conductivity, transmissivity, storativity as well as indications on the formation pore pressure and the radii of influence of the tests in order to define the scales at which the interpreted properties are representative. Precautions (scoping calculations) will have to be taken in the pulse phases or in the injections to avoid hydraulic fracturing of the structure. Finally, the characterisation of the selected fracture plane and especially the hydraulic tests will help dimensioning the tracer experiment.

As no radioactive tracers can be used in the experimental site, it is important to select the appropriate stable tracers. Two types of tracers can be considered depending on phenomena to focus on: non-sorbing tracers (transport processes) and sorbing tracers (coupled transport and retention). The use of ionic and non ionic species as well as the use of a cocktail of different tracers will also be considered.

The choice of the tracer has to be made keeping in mind that migration and transport must be large enough to be observable taking into account the detection means and will involve laboratory experiments (sorption isotherms, optimisation of the tracer analyses, diffusion and permeation experiments) performed on site samples, and followed by modelling exercises.

From the knowledge of the site and especially from the obtained experience on water sampling and on the fracture system, one can expect that the selected structure will be characterised by a very weak water flow and even no water. Hence, it will be important to work on the way of injecting the tracers in the structure; different types of water being conceivable such as water from the aquifers or synthesised water according to the knowledge on interstitial water chemistry.

The design modelling phase essentially concerns the development of scoping calculations in order to provide data on the performance of the experiment. Such calculations aim at adapting the experimental conditions to the optimisation of the measurement of the different parameters (either for the hydraulic tests or for the tracer experiments stricto sensu). The design modelling concerns different phases of the project :

- the modelling of the laboratory tracer experiments: optimisation of the choice of tracer, expected behaviour;
- the development of hydraulic tests either through the definition of the tests or their interpretation: e.g. value of the injected pressure, relaxation of the pressure with time, type of test (pressure-pulse, constant pressure flow, pressure build-up...);

- the tracer experiment *stricto sensu*: e.g. scale of the experiment, tracer profiles in the fracture and the rock matrix for different ranges of transport parameters, evaluation of the influence sphere of the selected tracer in the fracture plane, …

Taking into account the results from the previous phases, the design of the tracer experiment will be considered. The technical development of the experiment then concerns the choice of a type of protocol (unipolar, dipolar, multipolar), the manufacture of the injection/collection devices ("packers"), the selection of the type of injection (over-pressure, constant flow,…), the tracer collection and tracer analysis procedures.

The running of the experiment consists in several injections phases and associated tracer recoveries. The tracer will be monitored or analysed *in situ* or carefully sampled and conditioned for analysis in the laboratory. Specific techniques will of course depend on the nature of the chosen tracer.

Once the tracer test terminated it appears important to try to evaluate the influence sphere of the test . The position of the tracer, either in the fracture plane or in the matrix (diffusion) will be evaluated by different types of recognition boreholes. The position of the tracer in the fracture plane will be estimated by analysing filling materials sampled in different zones of the space between the injection and collection devices. The position and number of boreholes for sampling the fracture plane will be adapted using the results of the tracer test itself and the design modelling.

On the other hand, large diameter overcoring of the injection zone will permit to determine tracer concentration profiles and also to evaluate and analyse the diffusion of the tracer in the matrix.

Based on the results of the tracer experiment stricto sensu as well as on the hydraulic tests, hydraulic and geochemical behaviour of the water-conducting fracture will be provided. This will allow to interpret and model transport and retardation processes in the test fracture using the break-through curves obtained from the injection and collection phases.

The realisation of crosshole artificial tests in indurated argillaceous formations as well as the use of the information extracted for the global understanding of transport, will be a real innovation, transposable to the characterisation of other sites.

Conclusion

Once all the experimental data obtained a global interpretative and modelling phase, using different codes and models to first draw a picture of transport features over the whole studied formation and finally discriminate, among the used tools and the obtained results, which are transposable to other sedimentary sites.

The objective is to perform a general interpretation of transport in the site, taking into account the results of the preceding phases. This task is conceived as a stimulating reflection on the specific features of each model and code, their power and limits in accounting for the behaviour of a real system.

One way of quantifying the role of the fracture network would be to answer the following questions:

- What should be the density of these connected fracture pathways for advective transport to play an important role on the global flux of tracer towards the adjacent aquifers?

- Is this density compatible or realistic with the known structural characteristics of the fracture network in the formation?

The overall process described in this project sets the basis for a more general reflection on which are the theoretical, experimental and modelling tools which could be used to study flow and transport in any clayey massif, especially those foreseen for radioactive waste disposal or other dangerous materials. Hydrological results, such as regime change thresholds due to fracture networks, may be transposed to other site characterisation studies. To discriminate these generic results and assess their importance will be the object of this concluding task and involve all the participants.

The project is planned to be executed by an international consortium composed of: the French School of Mines, the University of Paris-Sud (Orsay), the University of Bern, the University of Heidelberg, SCK-CEN, CEA-DMT, GRS, EDF, the University of Turin and IPSN.

References

Barbreau, A. & Boisson, J.Y. 1993. *Caractérisation d'une formation argileuse. Synthèse des principaux résultats obtenus à partir du tunnel de Tournemire de janvier 1992 à juin 1993.* Rapport d'avancement n°1 du contrat CCE-CEA n° FI 2W CT91-0115, EUR 15736FR.

Boisson, J.Y., Cabrera, J. & De Windt, L., 1996. *Investigating faults and fractures in argillaceous Toarcian formation at the IPSN Tournemire research site.* Joint OECD/NEA-EC Workshop on "Fluid flow through faults and fractures in argillaceous formations", Bern, 207-222.

Cabrera, J. 1991. *Étude structurale dans le milieu argileux : LEMI du tunnel de Tournemire.* Rapport n°3 du contrat CEA-CABRERA.

Cabrera, J. 1995. *Site de Tournemire: programme Diffusion-Isotopes, synthèse géologique des sondages ID, campagne septembre-décembre 1994.* Rapport n°1 du contrat CEA-IPSN/ADREG.

De Windt, L., Cabrera, J., & Boisson, J.Y. 1999. *Radioactive waste containment in indurated shales: comparison between the chemical containment properties of matrix and fractures.* In Metcalfe, R. & Rochelle, C.A. (eds) Chemical Containment of Waste in the Geosphere. Geological Society, London, Special publications, 157, 167-181.

Moreau Le Golvan, Y., 1997. *Traçage isotopique naturel des transferts hydriques dans un milieu argileux de très faible porosité: les argilites de Tournemire (France).* Thèse Univ. Paris-Sud, Orsay, 177p.

Crushed-Rock Column Transport Experiments Using Sorbing and Nonsorbing Tracers to Determine Parameters for a Multirate Mass Transfer Model for the Culebra Dolomite at the WIPP Site, NM

C.R. Bryan and M.D. Siegel
Sandia National Laboratories, United States

Abstract

Site-characterization studies at the Waste Isolation Pilot Plant (WIPP) site in southeastern New Mexico, USA, have shown that, should radionuclides be released from the repository through inadvertent human intrusion, groundwater transport through the Culebra Dolomite would be the most significant pathway to the accessible environment (US DOE, 1996). The Culebra is a 7 m thick, variably fractured dolomite with massive and vuggy layers lying approximately 440 m above the WIPP repository. Field tracer tests and laboratory batch tests have been used to identify transport processes occurring within the Culebra and quantify relevant parameters for use in performance assessment of the WIPP. Interpretations of both single-well and multiple-well tracer tests suggest that a multirate mass transfer model best describes the transport of nonsorbing tracers in the Culebra Dolomite at scales of meters to tens of meters (Meigs *et al.*, 1997). Recent laboratory experiments using x-ray absorption imaging have confirmed that there are significant variations in diffusion rates within samples of the dolomite (Altman *et al.*, 1998). To constrain actinide diffusion rates and sorption parameters in the Culebra, and to build confidence in the multirate model and its applicability to sorbing species, crushed-rock column transport experiments with Culebra dolomite are currently being carried out at Sandia.

The multirate transport model assumes that mass transfer between advective porosity and diffusively-accessed porosity occurs at multiple rates (Haggerty and Gorelik, 1995). Previous studies have shown it's applicability to column-scale experiments (Haggerty and Gorelik, 1998). Columns of well-sieved clean, mineralogically similar particles of Culebra dolomite are expected to exhibit multirate behaviour – that is, a distribution of applicable diffusion rate coefficients – because for each grain, the matrix porosity is accessed by channels with a range of tortuosities and diameters. Also, although the crushed dolomite is well-sieved, a range of particle sizes is present and the shape of the grains varies from round, with a characteristic diffusive travel distance close to the radius of the sieve holes, to flake-like, with a much smaller average travel distance. The range in particle radii results in multirate behavior because smaller particles reach diffusional equilibrium faster than larger ones; net solute transport into the small grains approaches zero while larger grains continue to act as contaminant sinks.

Crushed-rock column experiments are currently being run at Sandia using columns of crushed, well-sorted Culebra dolomite and synthetic Culebra groundwater. Both nonsorbing and sorbing tracers (^{22}Na and ^{232}U, respectively) are being used. Tracer concentrations in the column effluent are measured by liquid scintillation, with a detection limit 4-5 orders of magnitude below the initial concentration. Breakthrough curves and eluted tails for square tracer pulses are fitted using STAMMT-L, a linear multirate transport code (Ver. 1.0; Haggerty, 1998), to yield the distribution (lognormal mean and standard deviation) of applicable diffusion rate coefficients (D_a/a^2, where D_a is the apparent diffusivity and a is the distance over which diffusion occurs). The storage capacity of the diffusive porosity is also estimated, and with measured porosity information, yields estimates of the sorption coefficient (K_d). As the column experiments investigate only one lithologic variant of the Culebra at a time, the K_d and the retardation are assumed not to be distributed. Contaminant retardations measured in the crushed-rock column experiments may provide more appropriate K_d data than batch tests by using more realistic solid/solution ratios and by accounting for slow diffusional uptake and slow desorption of radionuclides by the matrix material.

Five parameters were used to fit each data set; the mean and standard deviation of $\ln(D_a/a^2)$, the capacity coefficient (β_{tot}), the retardation factor in the mobile zone (R_m), and the Peclet number (Haggerty and Gorelick, 1998). To constrain the parameters, experiments were run at two flow rates, using two different sieve fractions of crushed Culebra dolomite (mean grain diameters, 0.23 and 0.65 mm).

Columns were run with nonsorbing tracers to estimate the Peclet number under the simplest conditions – when R_m is zero and β_{tot} is the ratio of the diffusive porosity to the advective porosity. Advective and diffusive porosities were estimated from the measured bulk density, the theoretical density for pure dolomite, and the estimated intergrain porosity.

Columns were then run under similar conditions with sorbing tracers. Fitting each breakthrough curve yields a mean and variance for $\ln(D_a/a^2)$ and a value for β_{tot}. Measured values of β_{tot}, and the K_ds derived from them, may be directly applicable to field-scale modeling. However, scaling the distribution of diffusion rate coefficients (D_a/a^2) derived from the crushed rock column experiments to larger dimensions may be difficult, as D_a and a may not be independent. If there is any lithologic variation with particle size, such as variations in tortuosity, restrictivity, porosity, or K_d, then scaling (D_a/a^2) as $1/a^2$ is not valid.

To test the robustness of the multirate model, and to provide confidence in the model parameters, diffusion rates and capacity coefficients derived from separate columns, each consisting of a single size fraction of dolomite particles, will be used to predict the behavior of a column containing a mixture of both size fractions. The distribution of (D_a/a^2) for the mixed column is the sum of the distributions determined for the single-grain size columns, weighted by capacity coefficient and the mass fraction of each in the mixture.

References

S.J. Altman, V.C. Tidwell, S.A. McKenna, and L.C. Meigs, 1998. "Use of X-ray absorption imaging to evaluate the effects of heterogeneity on matrix diffusion," in Characterization and Evaluation of Sites for Deep Geological Disposal of Radioactive Waste in Fractured Rocks, Proceedings from the Third ASPO International Seminar, Oskarshamm, June 10-12, SKB Technical Report, TR-98-10, 1998.

R. Haggerty, 1998. *STAMMT-L: Solute Transport and Multirate Mass Transfer – Linear Coordinates. User Manual Version 1.0.* Department of Geosciences, Oregon State University.

R. Haggerty, and S.M. Gorelick, 1995. "Multiple-Rate Mass Transfer for Modeling Diffusion and Surface Reactions in Media with Pore-Scale Heterogeneity." *Water Resources Research*, 31, 10, 2383-2400.

R. Haggerty, and S.M. Gorelick, 1998. "Modeling Mass Transfer Processes in Soil Columns with Pore-Scale Heterogeneity." *Soil Science Society of America Journal*, 62, 62-74.

L.C. Meigs, R.L. Beauheim, J.T. McCord, Y.W. Tsang, and R. Haggerty, 1997. "Design, Modelling, and Current Interpretations of the H-19 and H-11 Tracer Tests at the WIPP Site," in *OECD Proceedings: Field Tracer Experiments: Role in the Prediction of Radionuclide Migration*, Cologne, Germany, 28-30 August 1996.

US Department of Energy, 1996. Title 40 CFR Part 191 Compliance Certification Application for the Waste Isolation Pilot Plant, Vol. I, *DOE/CAO-1996-2184*. US DOE, Waste Isolation Pilot Plant, Carlsbad Area Office, Carlsbad, NM.

Sandia National Laboratories is a multiprogramme laboratory operated by Sandia Corporation, a Lockheed Martin Company, for the United States Department of Energy under contract DE-AC04-94AL8500.

Analyses Concerning the Estimation of the Release Behaviour of Heaps of the Former Mining Activities and Uranium Ore Mining

K. Fischer-Appelt & J. Larue
Gesellschaft für Anlagen- und Reaktorsicherheit (GRS) Köln, Germany

G. Henze
Bundesamt für Strahlenschutz (BfS) Berlin, Germany

J. Pinka
G.E.O.S. Ingeniergesellschaft Freiberg, Germany

In the German federal states Saxony and Thuringia there exist more than 3 600 heaps as remains of historical and uranium ore mining which partially show increased specific activities of radionuclides of the uranium-radium decay chain. Within the scope of the project "Radiological Registration, Investigation and Evaluation of Mining Deposits" (Altlastenkataster: BFS 1992, BFS 1997) a large number of data concerning the radioactive inventory of those heaps and their influence on surface- and groundwater were investigated. These informations were compiled and stored in a database (Technical Database for Environmental Radioactivity – FbU). Because of methodical and financial reasons, questions concerning the long term release of contained natural radionuclides as a result of weathering, leaching and groundwater transport were initially left aside. In addition to the radiological assessment of heaps carried out in the frame of the Federal project, it is planned to go ahead with a pre-selection of heap locations with regard to any possible need for remedial actions due to the protection of local inhabitants from increased exposure to radiation caused by radionuclide transport via the aquatic pathway.

In order to obtain decision criteria to enable this pre-selection of heap locations *Gesellschaft für Anlagen- und Reaktorsicherheit (GRS)* – commissioned by the *Federal Office for Radiation Protection (BfS)* – is currently developing a set of instruments with which an estimation of the radionuclide release rate (source term) can be made on a simplified basis.

For this purpose, three selected heaps that are characteristic of the former mining activities and uranium ore mining in Saxony are currently subjected to a thorough investigation of their geochemical, mineralogical and radioactive inventory and their hydrochemical composition of the leachate. At the same time, the elution behaviour of the heap material is simulated in columnar laboratory experiments. This is done on the one hand to generate input data for a definition of the source term on the basis of geochemical model calculations and on the other hand to compile such data for checking the calculated radionuclide release.

At each of the three locations drillings and exploratory excavations holes were performed. The heap materials and the underlying natural soils have been analysed mineralogically and chemically. Furthermore 50 kg of a mixed sample material has been filled in a column and than sprinkled with water under defined conditions. Conductivity, pH-value and redox potential of the circulating water were analysed permanently to obtain a trend of the ongoing hydrochemical development. Additionally samples were collected at regular intervals and analysed concerning their hydrochemical composition. The results of the analyses programme are depicted in the diagram in Fig.1 (material of the heap at the Haberlandmühle near the city Johanngeorgenstadt, Saxon Ore Mountains, Fig. 2).

Figure 1. **Time-dependent developement of the parameters Eh, pH and conductivity**

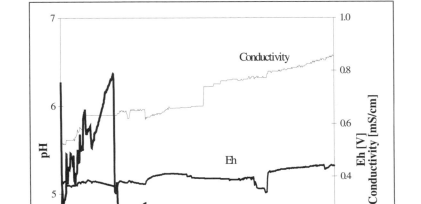

Figure 2 **View of the heap at the Haberlandmühle**

It is obvious that former carbonate which was originally distributed in the heap material is consumed due to the oxidation of sulphides while reacting with acid water. After completion of this reaction a constant pH-Eh-environment established. This buffer-effect is attributed to the weathering of the mica and feldspar which are the main minerals of the rocks in the heap material, and to their interaction with the sulphate-rich acid water. In other words the adjusted buffer system is characterised by a particular pH-value of the known aluminium buffer range as well as oxidizing redox conditions. Both the seepage water taken from the foot of the heap and the water of the column show that especially the main components aluminium, calcium, potassium, sulphate and free silica are leaving the heap material.

The next step of the work is the creation of a model of the geochemical processes developing within the heaps that is as realistic as possible. The associated model calculations needed for this purpose are performed with the PHREEQC code (PARKHURST 1995).

In this context, special attention needs to be paid to geochemical subprocesses like

- oxidation and reduction processes;
- equilibrium reactions between solution and solid phase (solution/precipitation);
- sorption (ion exchange, surface complexation);
- poss. colloid-borne contaminant transport.

Here, the identification of typical processes that have a substantial influence on the radionuclide release from heaps is particularly important.

As far as it is possible to obtain a sufficiently exact understanding of the geochemical process sequences going on within the investigated heaps, in another step it has to be checked in what way it is possible to apply the geochemical source term to other and in less detail analysed heaps in the sense of a simplified estimation of the release behaviour. If possible a simplified tool will be developed for modelling the geochemical source term of similar heaps. The reliability of such prognoses has to be checked by comparing the analysis results of the three selected heaps, by statistical correlation calculations, and by the comparison of research results on heaps that were obtained by other institutions.

The results of these investigations are to contribute to a guideline of requirements for the site specific investigation of heaps from former mining activities and uranium ore mining, too.

References

BfS (1992):
    Radiologische Erfasssung, Untersuchung und Bewertung bergbaulicher Altlasten, Abschlußbericht zum ersten Teilprojekt, BfS- Schriften 8/92.

BfS (1997):
    Radiologische Erfassung, Untersuchung und Bewertung bergbaulicher Altlasten, Abschlußbericht zum Teilprojekt 2: Altlasten Bergbau, Verifikation (TP 2), BfS-IB-7; Berlin.

PARKHURST, D. L. (1995):
    Users Guide to PHREEQC – A Computer Programme for Speciation, Reaction-Path, Advective-Transport, and Inverse Geochemical Calculations.
    U.S. Geol. Survey Water Resour. Invest. Rept., 95-4227: 143 S.; Lakewood, Colorado.

Dual Porosity Approaches for Transport in Fractured Media: A Numerical Investigation at the Block Scale

C. Grenier
CEA, Centre d'Études de Saclay, France

E. Mouche
CEA, Centre d'Études de Saclay, France

E. Tevissen
ANDRA, France

1.   Introduction

One of the major issues in assessing the performance of a deep repository in crystalline geologic media concerns the ability of the rock matrix to delay the transport of radionuclides because of its storage capacity. As a matter of fact, transport in the matrix is much slower than in the fractures (generally assumed purely diffusive) and matrix porosity is much greater than fracture porosity, providing large storage volumes accessible by diffusion process (see for instance ***Bear et al. 93***). Furthermore, for given species, retardation coefficient in the matrix is much greater than on fracture walls.

From the modelling point of view, the strong contrast in the values of flow and transport parameters between the fracture and the rock matrix leads to classical numerical difficulties and awkward computer simulation time for the transport problem: typically, a highly refined spatial discretization is necessary for the parts of the matrix blocks close to the fracture in order to guarantee that the concentration gradients are well accounted for. This requirement is indeed expensive to meet in most cases encountered since matrix blocks have large extensions compared to fracture apertures. As a consequence, the number of meshes required quickly turns inadequate and the fractured network geometry can't be modelled explicitly.

In order to face situations involving a dense and well connected network of fractures, dual porosity, and dual permeability approaches have been developed in the hydrogeological field during the last two decades. In these studies, the flow problem has been extensively studied (see the petroleum engineering literature for instance) but the transport problem has only been treated for a limited range of geometries.

This observation led us to study the different "operational" dual porosity models proposed in the literature, i.e. which can be implemented in a 3D performance model. The approaches developed by *Bibby 81*, *Huyakorn et al. 83* serve here as a basis and are further extended to 2D and 3D matrix blocks geometries. These are called here "sugar box" geometries as in the petroleum engineering literature, and correspond to a set of 2D or 3D orthogonal fractures. These simple geometries are not supposed to correspond to the real geometrical features but to an equivalent mean geometry.

These approaches are implemented and compared to reference calculations for which the explicit geometry is taken into account. The size of the domain is chosen in order to meet two opposite requirements: the reference calculation remains tractable, the simulation domain includes a sufficient number of matrix blocks cells in order to apply an homogenisation approach (the size of a REV is roughly the size of a matrix block). The simulation of mass transfers in the matrix blocks within the dual porosity approach is done by means of numerical simulation or analytically.

2.   Reference calculations for explicit *sugar box* geometries

The dual porosity approach is compared in the following to reference simulations involving an explicit 2D sugar box geometry. For examples of 3D sugar box computations see *Genty et al. 98*. The work is performed within the numerical finite element code CASTEM2000, currently developed at the CEA. A mixed and hybrid finite element formulation is used (*Dabbene 98*); it guaranties a good quality of the velocity, concentration fields and fluxes for the cases of high fracture-matrix contrasts encountered here.

The flow problem is permanent, limited to the fracture network, and corresponds to a mean hydraulic gradient along the diagonal. The transport problem involves the release of an initial concentration plume close to the upstream corner of the domain as can be seen on Figure 1(a). Three matrix diffusion coefficient values are considered in this study, corresponding to a fracture dominated transport case (lowest matrix diffusion value), a matrix dominated transport situation (largest matrix diffusion value) and a median case. The parameter set considered is given on Table 1. The different transport regimes can be clearly visualized on the corresponding concentration fields provided on Figure 1(b), 1(c) and 1(d) for a 5x5 matrix blocks geometry.

Table 1. **Parameter values for the reference calculation (explicit geometry). A set of three different values is considered for effective diffusion coefficient**

Parameters	Fractures	Matrix blocks
Porosity	$\omega_{fr} = 0.1$	$\omega_{ma} = 0.001;0.005;0.01$
Fracture Aperture	$e = 0.1m$	–
Transmissivity	$T_{fr} = 10^{-8} m^2/s$	–
Hydraulic Gradient	$j = 5.10^{-4} m/m$	$j = 5.10^{-4} m/m$
Darcy velocity	$U_{fr} = T/ej = 5.10^{-11} m/s$	$U_{ma} \approx 0$
Effective Diffusion Coefficient	$D_{fr} = 10^{-11} m^2/s$	$D_{ma} = 10^{-14};10^{-13};10^{-12} m^2/s$
Dispersion coefficient	$\alpha_L = 4m; \alpha_T = 0.4m$	–

Figure 1. **Reference calculation (for sake of illustration on a 5 x 5 matrix blocks geometry):**

(a)   initial condition and concentration fields for three matrix diffusion data sets.

(b)   Concentration field for $D_{ma} = 10^{-14} \, m^2/s$, $time = 2.510^{11} \, s$, $C_{max} = 2.610^{-4} \, U.A.$

(c)   Concentration field for $D_{ma} = 10^{-13} \, m^2/s$, $time = 10^{12} \, s$, $C_{max} = 5.710^{-5} \, U.A.$

(d)   Concentration field for $D_{ma} = 10^{-12} \, m^2/s$, $time = 2.510^{12} \, s$, $C_{max} = 2.410^{-5} \, U.A.$

(a)

(b)

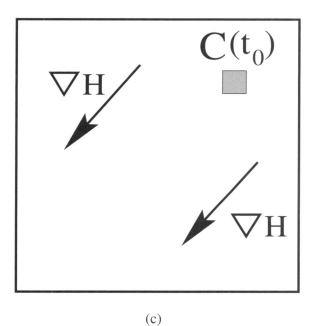

(c)

(d)

The first step in the study relates to the size of the domain to be considered to get an averaging effect. The curves corresponding to total output flux proved symmetrical for domains larger than 10x10 matrix blocks. We keep this size in the following.

The quality of the simulation for the transport problem was studied considering different types of spatial discretization as well as refinement strategies. The regular type of geometry provided on Figure 2(a) proved better for transport dominated by matrix diffusion whereas the nested geometry on Figure 2(b) showed necessary for transport dominated by fracture advection. For rough levels of discretization at the matrix blocks interface towards the fracture planes, matrix storage effects are over estimated whereas peak arrival time is underestimated. The median matrix diffusion data set proved the most sensitive to the spatial discretization. These conclusions are fully in line with former results obtained for single fracture geometries.

Figure 2. **Two spatial discretization strategies shown on a 5 x 5 matrix blocks sugar box**

(a) Regular spatial discretization.   (b) Adaptative spatial discretization.

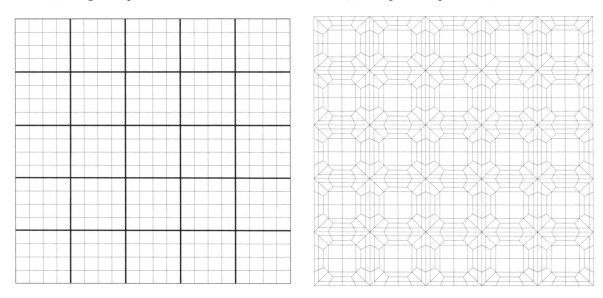

3.   Dual porosity models studied

The dual porosity approach consists in (see ***Bibby 81, Huyakorn et al. 83***): i) associate an equivalent porous medium to the fracture network with equivalent flow and transport parameters; ii) attach to each node describing this porous medium a virtual matrix block in which matrix diffusion is solved analytically or numerically. This is schematically summed up on Figure 3. The coupling between both equivalent media is provided by a source term in the fracture transport equation and by an imposed concentration condition for the matrix block picked up on the corresponding equivalent fracture mesh. The size of the REV is theoretically equal to a multiple of the size of the matrix block. We consider here the size of the matrix block. The equations for the dual porosity transport problem are:

$$\omega_f \frac{\partial C_f}{\partial t} = \vec{\nabla}(\overline{D}_f \vec{\nabla} C_f - \vec{U} C_f) + SourceTerm,$$

for the equivalent fracture. The source term is the opposite of the variation of mass in corresponding matrix block. For the equivalent matrix cell,

$$\omega_m \frac{\partial C_m}{\partial t} = \vec{\nabla}(D_m \vec{\nabla} C_m),$$

with boundary conditions $C_m^{BC} = C_f$.

C stands for concentrations in the equivalent media associated to the fractures (f) or matrix blocks (m). U for Darcy velocity in the equivalent fracture, D for equivalent diffusion coefficient or dispersion tensor, ω for equivalent porosity.

Two matrix diffusion simulation strategies were tested. The numerical simulation of the source term (followed by **Huyakorn et al. 83**) towards the matrix proved too expensive in terms of computational costs. Furthermore, the extension of the method to 3D problems is not straightforward. The analytical solution (followed by **Bibby 81** and extended by the authors to 2D and 3D problems) was preferred. The analytical solution is classical (**Carslaw and Jaeger 59**) for the situation considered (see Figure 3(b)). The solution is expressed as a time convolution product of the concentration evolution in the fracture and the Green function for the diffusion problem in the matrix block (see **Grenier 99** for more details).

Figure 3. **Dual Porosity model used**

(a) Principle of Dual Porosity approach.

(b) Diffusive transport problem solved in matrix blocks.

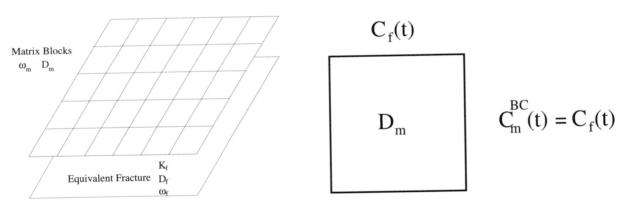

The comparison between the reference and dual porosity simulation is conducted as follows:

(i) Derive and validate flow properties for the equivalent fracture medium (permeability and porosity);

(ii) Derive and validate transport properties for the equivalent fracture medium without matrix diffusion effects (longitudinal and transverse dispersion coefficients);

(iii) Add matrix diffusion mechanism and compare with reference calculation including matrix blocks.

The second point is achieved based on ***Schwarz and Smith 88***. These results, based on pure advection in a regular fracture network are then extended to advective and dispersive transport in the same fracture geometry. The dispersion coefficients are close to 1/3 of the size of the regular matrix blocks. This results leads to strong numerical constraints: in order to keep mesh Peclet numbers around 1, it is necessary to discretize the equivalent fracture medium in 9 elements (2D). This requisite proved expensive in terms of calculation costs but can not be circumvented for situations where fracture transport is dominant. Using a coarser grid for instance would result in an over estimation of dispersion mechanism in the equivalent fracture.

4.   Results and conclusions

The validation was performed for three values of matrix diffusion coefficient (see data set in Table 1) considering the breakthrough curve at the outlet of the medium, the total mass in the domain as well as mass fractions in both sub domains and the concentration field in the equivalent fracture. The results for the median case for matrix diffusion coefficient are provided on Figure 4. This case is representative of the quality of the results for the other matrix diffusion coefficients values. Over all results obtained, the dual porosity approach leads to very good results compared to the reference simulation. This result is not surprising for the lower matrix diffusion simulation case since the transport process is dominated by fracture transport and the assumption of constant concentration imposed at the boundaries of the matrix block cell (see figure 1(b)) is met. For larger values of matrix diffusion coefficient the approach proved nevertheless robust (error less than 5%), providing a versatile framework for situations varying from dominant fracture transport to dominant matrix transport.

Figure 4.   **Reference calculation vs dual porosity model simulation for the 10x10 matrix blocks domain and the median matrix diffusion coefficient value $D_{ma} = 10^{-13} m^2 / s$:**

(a) Total mass flux leaving the domain.

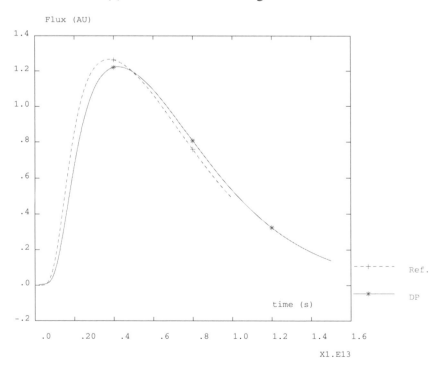

(b) Total mass variations within the domain (fractions of total mass corresponding to the fracture and matrix sub domains are provided). DP stands for Dual Porosity and Ref. for reference simulation.

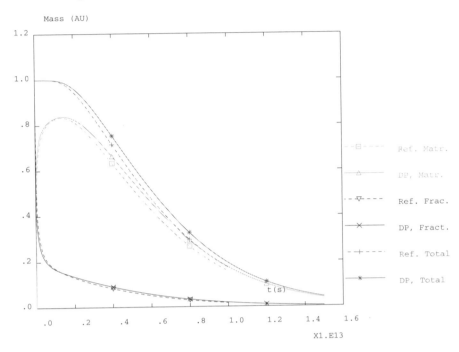

The CPU time is improved by a factor of five for the dual porosity approach based on analytical simulation of matrix transport (CPU time is not significantly improved when using numerical simulations). This result is satisfactory although a stronger improvement was expected. The cost in terms of CPU remains here high due to the fact that dispersion produced by advection in a regular fracture network provides small values of dispersion coefficients (roughly one third of the block size) imposing a relatively fine discretization for the equivalent fracture medium.

Work on 3D sugar box geometry is ongoing, proving efficient in terms of CPU time. An extension of the method towards heterogeneous matrix blocks is straightforward and will be considered in the future.

References

J. Bear, C.F. Tsang, and G. de Marsily. Flow and Contaminant Transport in Fractured Rock, ACADEMIC PRESS, 1993.

R. Bibby. Mass transport of solutes in dual porosity media. Water Resources Research, Vol. 17, No. 4, Pages 1075-1081, August 1981.

H. S. Carslaw and J. C. Jaeger. Conduction of heat in solids. OXFORD UNIVERSITY PRESS, 1959.

F. Dabbene. Mixed Hybrid Finite Elements for Transport of Pollutants by Underground Water. Proceedings of the 10th International Conference on Finite Elements in Fluids. Tucson, Arizona, USA. January 5-8 1998.

A. Genty, C. Grenier and E. Mouche. Influence of matrix diffusion on solute transport in a 3D fracture network. Proceedings of Computational Methods in Water Resources XII, 15-20 june 1998. COMPUTATIONAL MECHANICS PUBLICATIONS.

C. Grenier. Modeling of transport in a sugar box type fractured block : Dual porosity approaches. Internal Report (in French). CEA 1999.

P. Huyakorn, B. H. Lester and J. W. Mercer. An efficient finite element technique for modelling transport in fractured porous media. Single species transport. Water Resources Research, Vol. 19, No. 3, Pages 841-854, june 1983.

F. Schwarz, and L. Smith. A continuum approach for modelling mass transport in fractured media. Chemical Engineering Science, Vol. 48, No. 14, Pages 2537-2564, 1993.

Anisotropy of Migration in Soft Clays

M. J. Put, G. Volckaert and J. Marivoet
SCK•CEN, Belgium

Abstract

Anisotropy of the migration properties is demonstrated for a sedimentary soft clay formation (Boom Clay). The measured values for the hydraulic conductivity and for the apparent diffusion constant in the direction parallel with the stratification of sedimentation are a factor of two higher than the values perpendicular to the stratification.

The influence of the anisotropy due to the stratification of the clay formation is not negligible for the calculation of the migration of radionuclides and hence also included in the performance assessment modelling.

Introduction

In Belgium, a sedimentary plastic clay formation (Boom Clay) is studied as a potential host for a deep geological repository for spent fuel and high level radioactive waste. This clay formation is chosen for its favourable barrier properties:

- low hydraulic conductivity;
- essentially diffusive transport;
- high sorption capacity for cations;
- slightly alkaline and strongly reducing conditions limiting the solubility and the mobility of many radionuclides;
- high plasticity resulting in the self healing of possible fissures.

The Boom Clay is a very low-permeable stratum situated, at the Mol site, at the depth of 160 to 270 metres, and is the uppermost clay formation of an alternating sequence of clays and sands. The clay formation was deposited about 30 million years ago. It is plastic clay with water content of 40% in volume. For the assessment of the safety of a possible repository, reliable values of the large-scale migration parameters of the formation are needed. Due to the layering of the sedimentary formation anisotropy of migration was expected. Therefore, clay samples cored perpendicular and parallel to the stratification, and 3-dimensional *in situ* tests were used to assess the anisotropy of the migration of radionuclides.

All this research is performed in close co-operation between the experimentalists and the performance assessment specialists. The values of the parameters obtained on clay cores, and the anisotropy is confirmed by large-scale *in situ* experiments. The anisotropy is included in the performance assessment models.

Methodology

The hydraulic properties are determined by percolation experiments on clay cores, by *in situ* single point hydraulic tests, by *in situ* 3-dimensional interference tests, and by an *in situ* macro permeameter test [Monsecour (1990), Put (1992), Volckaert (1995), Ortiz (1996), Wemaere (1997)]. Figure 1 shows an *in situ* set-up for hydraulic and migration testing. In 1982 SCK•CEN has built an Underground Research Laboratory (URL) at a depth of 220 metres. Due to the plasticity of the clay, a patented technique without packers can be used to install single and multiple filters in the formation. From the URL, vertical and horizontal multiple piezometers were installed in the clay formation. These multiple piezometers were used to perform large-scale *in situ* hydraulic interference tests. Due to the pressure difference between the URL and a filter installed in the clay formation, interstitial water can be drawn from this filter. For the interference test, the water flow of this filter is monitored as a function of time, and the pressure evolution in the surrounding filters is also measured. The test allows for the determination of the three components of the hydraulic conductivity and of the storage coefficient.

Figure 1: **Example of a large-scale *in situ* piezometre for hydraulic and migration   experiments from the HADES underground research facility**

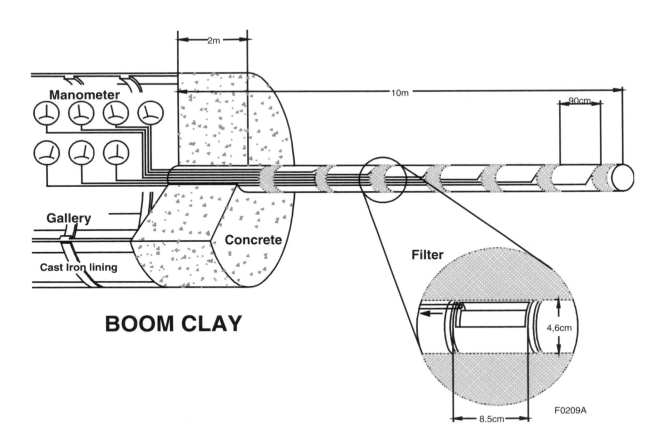

The diffusion properties are measured by diffusion and by percolation experiments on small clay cores drilled perpendicular and parallel to the stratification and by large-scale *in situ* experiments (Put (1988, 1990, 1992), Henrion (1991)).

To assess the vertical homogeneity of the Boom Clay, clay cores are sampled over the thickness of the formation. The vertical components of the migration parameters measured for the two main tracers (HTO and iodide) only show minor variations, indicating a rather homogeneous behaviour for vertical transport in spite of the layered structure of the formation.

Results and discussion

The results of the hydraulic tests clearly show the anisotropy of the sedimentary Boom Clay formation. Figure 2 gives a comparison of the values for the two main components of the hydraulic conductivity. Values obtained for the hydraulic conductivity are 1.7×10^{-12} m s^{-1} for the vertical component and 4.1×10^{-12} m s^{-1} for the horizontal component. These values are in good agreement with the values obtained from percolation experiments on clay samples cored perpendicular and parallel to the stratification. The storage coefficient does not show an isotropic behaviour, a value of 1.3×10^{-5} m^{-1} was obtained.

Figure 2: **Comparison of the values of the main components (vertical and horizontal) of the hydraulic conductivity measured on clay cores (filled square markers) and *in situ* (open square markers)**

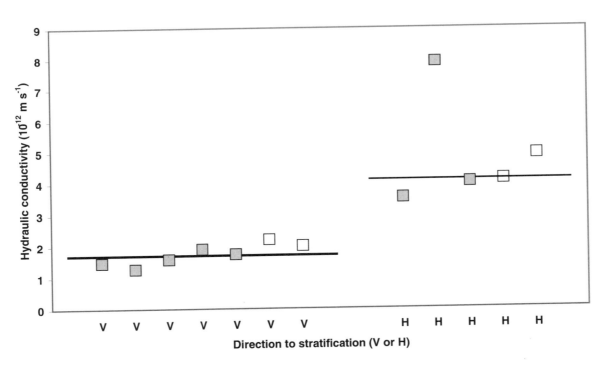

The results of diffusion tests clearly show the anisotropy for diffusion. The key parameters for diffusion in plastic clay are the "diffusion accessible porosity", the "retardation factor", and the "apparent diffusion coefficient". Only the diffusion coefficient shows anisotropy behaviour. The obtained values of the apparent diffusion coefficient for water labelled with tritium (HTO) are 4×10^{-10} $m^2 s^{-1}$ for the horizontal component and 2×10^{-10} $m^2 s^{-1}$ for the vertical component. For the diffusion accessible porosity a value of 0.35 was obtained, and this value can differ for different species depending on the size and on the electric charge (ion exclusion) of the species. For example, for the iodine anion a value of 0.12 is obtained. The parameter values obtained on small clay cores (centimetre scale) and their anisotropy are also validated on a larger scale (metre scale) by 3-dimensional *in situ* migration experiments. Figure 3 compares the results of the model simulation, considering anisotropy and using the values of migration parameters measured on clay cores, with the experimental concentration evolution for eleven years since the injection. This experiment shows a good agreement between the predictive simulation and the measured concentration profiles.

Figure 3: **Model simulation (curves) and measured tritium concentrations (symbols) in the interstitial liquid for a large-scale *in situ* migration experiment with tritiated water (HTO). Results for eleven years are shown for the injection filter (5), and for the sampling filters at distances of 1 m (4 and 6) and 2 m (3 and 7) from the injection filter.**

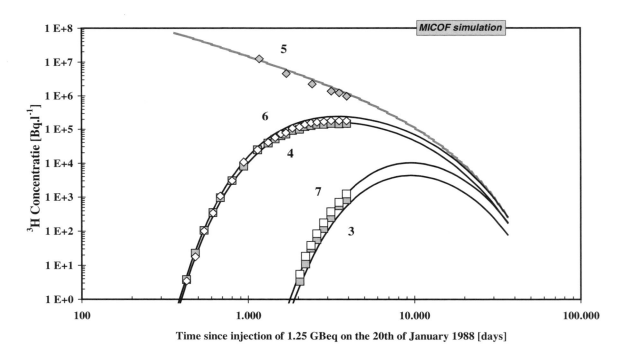

The results of the percolation experiments at different advection velocities also allowed assessing of the influence of dispersion on a centimetre scale. For the dispersion length a value of 10^{-3} m was obtained (Put 1990).

Conclusions

Due to the stratification of a sedimentary plastic clay formation (Boom Clay), anisotropy was expected. The experimental results of the migration properties indeed showed an anisotropy for the hydraulic conductivity and for the apparent diffusion coefficient. The values obtained for migration parallel with the stratification of the formation are a factor of two higher than vertical on the stratification. Thus the migration parallel with the stratification is faster than perpendicular. The effect is not negligible and has to be included in the performance assessment calculations. As expected, the other migration properties show no isotropic behaviour.

Acknowledgements

Part of this study is financially supported by NIRAS/ONDRAF (the National Agency for Radioactive Waste and Nuclear Fuels – Belgium) and by the European Community through its R&D programme on "Nuclear Fission and Safety".

References

M. PUT, P. HENRION (1988)
An improved method to evaluate radionuclide migration model parameters from flow-through diffusion tests in reconsolidated clay plugs
Radiochimica Acta, 44/45, 343-347 (1988).

M. PUT, M. MONSECOUR, A. FONTEYNE, H. YOSHIDA (1990)
Estimation of the migration parameters for the Boom clay formation by percolation experiments on undisturbed clay cores
MRS, Scientific Basis for Nuclear Waste Management XIV, Vol. 212, 1990, pp. 823-82.

M. MONSECOUR, M. PUT, A. FONTEYNE, H. YOSHIDA (1990)
Migration experiments in the underground facility at Mol to validate safety assessment model
OECD/NEA Conf. "Geoval'90"
Stockholm, 14-17 May, 1990.

P. N. HENRION, M. J. PUT, M. VAN GOMPEL (1991)
The influence of compaction on the diffusion of non-sorbed species in Boom clay
Radioactive Waste Management and the Nuclear Fuel Cycle, 1991, Vol. 16(1), pp. 1-14.

M. PUT, M. MONSECOUR, A. FONTEYNE (1992)
Mobility of the dissolved organic material in the interstitial Boom clay water
Radiochim. Acta, 58/59, 315-317, (1992).

M. PUT, P. DE CANNIERE, H. MOORS, A. FONTEYNE, P. DE PRETER (1992)
Validation of performance assessment model by large scale in situ migration experiments
Proc. International symposium on "Geological Disposal of Spent Fuel, High-Level and Alpha-Bearing Wastes", pp 319-326, IAEA-SM-326/37
Antwerpen, 19-23 October 1992
Ed. IAEA, Vienna, 1993.

G. VOLCKAERT, L. ORTIZ, M. PUT (1995)
In situ water and gas injection experiments performed in the HADES underground research facility.
Proc. Of the "5th International Conference on Radioactive Waste Management and Environmental Remediation", Berlin, 3-7 September 1995, pp. 743-750.
Ed.: S. Slate; The American Society of Mechanical Engineers, 1995, ISBN 07918 1219 7.

L. ORTIZ, G. VOLCKAERT, M. PUT (1996)
Characterisation of the hydraulic properties of the Boom Clay formation around the HADES underground research facility. Poster.
NEA/EC Workshop on "Fluid flow through faults and features in argilaceous formations".
Bern, 10-12 June 1996.

I. WEMAERE, J. MARIVOET (1997)
Can we extrapolate hydraulic parameters of the Boom Clay from the local to the regional scale. The Boom Clay Seminar
Alden Biesen, Bilzen, Belgium, 8-9 December 1997
Ed. SCK/CEN, BLG-758, 1997.

Evaluation of Transport Model Uncertainties using Geostatistical Methods: A Case Study Based on Borehole Data from the Gorleben Site

Klaus-Jürgen Röhlig
GRS, Köln, Germany

Brigitta Pöltl
GRS, Köln, Germany

1. The Gorleben site and the data set

The Gorleben salt dome is located in the north-eastern part of Lower Saxony. It is currently under investigation as a potential site for the final disposal of all kinds of solidified, especially heat-producing radioactive waste. The planned repository will be located at a depth of about 850 m (approximately 600 m below the top of the salt dome).

The evaluation of the retention potential of the sediments above the salt dome will be an essential part of a long-term safety analysis for the repository. Such an evaluation should be based on the performance of groundwater and nuclide transport model studies. Furthermore, it should include an assessment of the impact of the uncertainties arising from spatial variability and imprecise knowledge concerning the features of the site. The work presented here is intended to be a first step towards an integrated approach for such an assessment.

The Gorleben salt dome extends from a depth of about 3 500 m up to 260 m below the surface. Its tertiary clay cover has been partially removed by subglacial erosion forming a system of channels, one of which is the "Gorleben Erosion Channel". This channel has a length of more than 16 km and its width ranges between 1 and 2 km. Its erosional features extend to the caprock at a depth of about 250 m and in some locations they are in contact with the rock salt. The channel is filled with sandy and gravely sediments forming a system of two aquifers separated by clay layers, the so-called Lauenburg clay complex. This complex includes clay layers as well as sandy and silty lenses. The spatial distribution of the low-permeable clay layers which has a significant influence on the groundwater flow regime and the contaminant transport is the main subject of the study presented here.

Hydrogeological investigations are performed in an area of more than 300 km² around the salt dome. A selection of 340 borehole logs are compiled and partially re-interpreted in Annex 6 of LUDWIG (1994). 331 of them are situated south of the river Elbe. Much less information was available about the northern part (right of the river Elbe in Figure 1) because it formerly belonged to the German Democratic Republic. Due to the significantly different database concerning the southern and northern areas, only the southern part has been taken into account in the work presented here.

2. Geostatistical analyses and concept for the uncertainty and sensitivity calculations

Based upon the data described above, GRS had already performed preliminary two-dimensional geostatistical analyses of stratigraphic and petrographic features. Using the results of these preliminary studies, a three-dimensional analysis of the spatial distribution of the low-permeable clay layers mentioned above had been carried out (RÖHLIG 1998). In order to take into account secondary information concerning the stratigraphy of the site, the utilisation of a curvi-linear co-ordinate system based on stratigraphic information had been proposed by PORTER and HARTLEY (1997). The approach presented there has been modified, extended, and used for a part of the 3D analysis.

The 3D analysis had lead to a reference distribution for the clay layers obtained by ordinary kriging and to a series of equally probable conditional sequential indicator simulation results both for a domain with a horizontal extension of 16 km x 16 km and a depth of 400 m (RÖHLIG 1998). For the study presented here, a vertical cross section through this domain which follows the channel axis (roughly south-north, Figure 1) has been chosen for the performance of 2D groundwater, pathway, and contaminant transport calculations. This cross section is following the main groundwater flow direction and is regarded as representative for this area.

Figure 1. **Generalised hydrogeological cross section (based on Vogel and Schelkes, 1996); model area for flow and transport calculations**

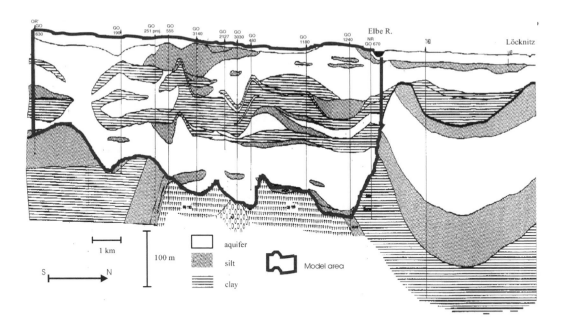

In order to assess the impact of the uncertainty concerning the spatial clay distribution as well as of parameter uncertainty, three series of calculations have been performed: For the first series the sequence of indicator simulation results has been used whereas the parameters (conductivities, porosities, dispersion lengths) have been left constant (reference values). The second series was based upon one single spatial clay distribution model (the one obtained using ordinary kriging). For each calculation of this series, Monte Carlo realisations have been used as input parameters. For the third series, both indicator simulation results for the spatial clay distribution and Monte Carlo realisations for the input parameter sets have been used. Additionally, a reference calculation has been performed using the ordinary kriging results for the facies distribution and the reference values for the parameters. Each of the series consisted of 10 000 steady-state freshwater and pathline calculations respectively of 1 000 transient contaminant transport calculations for a non-sorbing, non-decaying tracer.

3. Calculation results

Figure 2 shows some realisations including the calculated pathways. In Figures 3 and 4, the results of the contaminant transport calculations are presented for the reference case and for one indicator simulation realisation, respectively.

Figure 2. Cross-sections through several indicator simulation results, pathways (darker areas = clay, groundwater flow roughly from south to north) (from RÖHLIG, 1998)

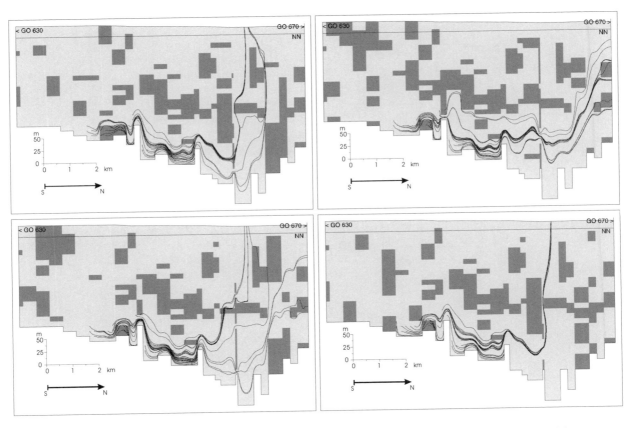

The calculation results have been used for uncertainty and sensitivity analyses with respect to *advective groundwater travel time* (calculated for the pathlines), the *time evolutions* of the integrated contaminant fluxes across the model boundaries, the *maxima* of these fluxes and the *arrival times* of these maxima. The uncertainty analyses have been carried out using simple and time-dependent univariate statistics, respectively. As an example, Figure 5 shows a comparison for the percentiles of the flux time evolutions at the upper boundary for the three series. The sensitivity calculations are based on scatterplots, correlation and rank correlation coefficients for the entities in question or for their logarithms.

Figure 3. **Nuclide concentrations after 100, 800, and 1 400 years for the reference case**

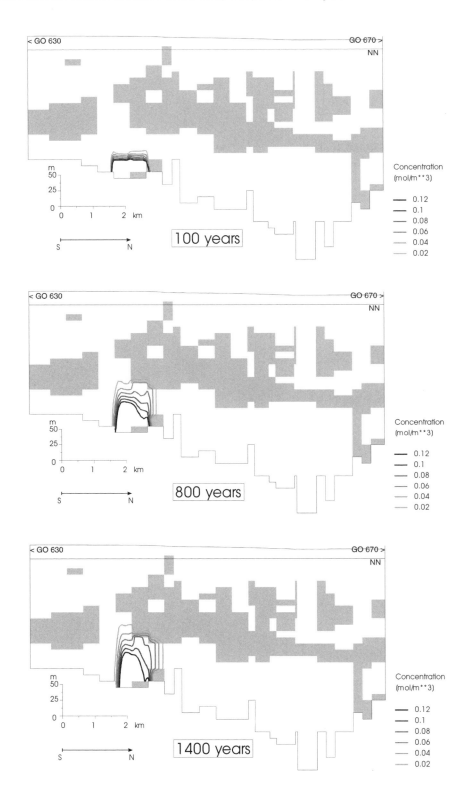

Figure 4. **Nuclide concentrations after 100, 800, and 1 400 years for one indicator simulation realisation**

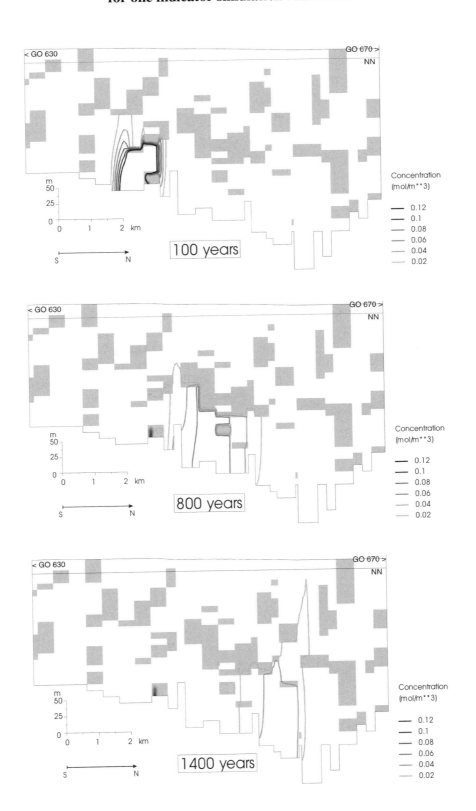

The uncertainties of the *advective groundwater travel time*, the *maximum* of the flux at the upper model boundary and the *arrival time* of the flux maximum caused by parameter variations (second series) are larger than the ones caused by facies distribution uncertainty (first series). During the first 20 000 years problem time the uncertainty of the *time evolution* of the flux is for the second series larger than for the first series, whereas for later problem times the uncertainty for the second series is decreasing (cp. the difference between high and low percentiles in Figure 5). As expected, the uncertainties for the third series (both parameter and facies distribution variation) are greater than the ones for the first and for the second one. Especially the third series is showing significantly greater probabilities for more critical values (earlier arrival times, greater maxima).

The conductivity of the clay layers, the ratio of the conductivities between clay and higher-conducting layers, and to some extent the conductivity of the higher-conducting layers have been identified as most sensitive with respect to *advective groundwater travel times*. Figure 6 shows scatterplots for the minima per realisation of the advective groundwater travel times versus conductivities.

In contrast, the *maxima* of the fluxes as well as the *arrival times* of these maxima are mainly determined by the conductivity of the higher-conducting layers and to some extent by the conductivity of the clay. Figure 7 shows scatterplots for flux maxima at the upper boundary versus conductivities.

In general, for the second series the sensitivities for the "clay" conductivities are larger, whereas higher sensitivities for the "non-clay" conductivities have been calculated for the third series. This indicates that for the kriging results of the second series the compact clay layer covering the lower aquifer is of significant influence on the flow and transport behaviour. For the indicator simulation results of the third series, the clay cover is much more heterogeneous and less compact. Therefore, the calculation results depend much more on the conductivities of the "non-clay" areas where in principal the flow and the transport takes place.

Figure 5. Percentiles of the flux time evolutions at the upper boundary for the three series

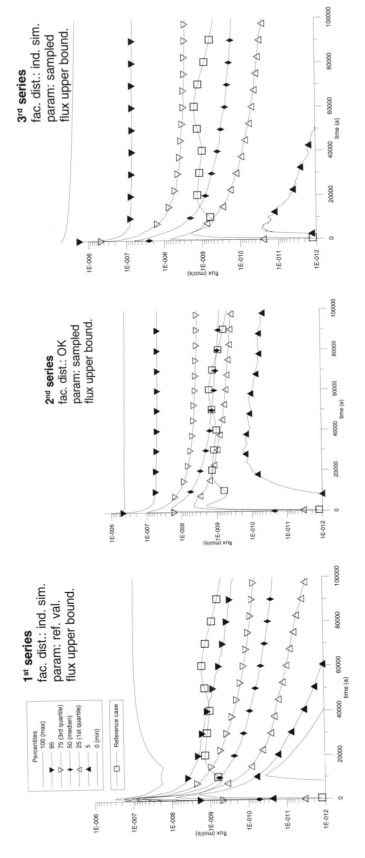

Figure 6. **Scatterplots for minima of advective groundwater travel times versus "clay" conductivities (above) and "non-clay" conductivities (below)**

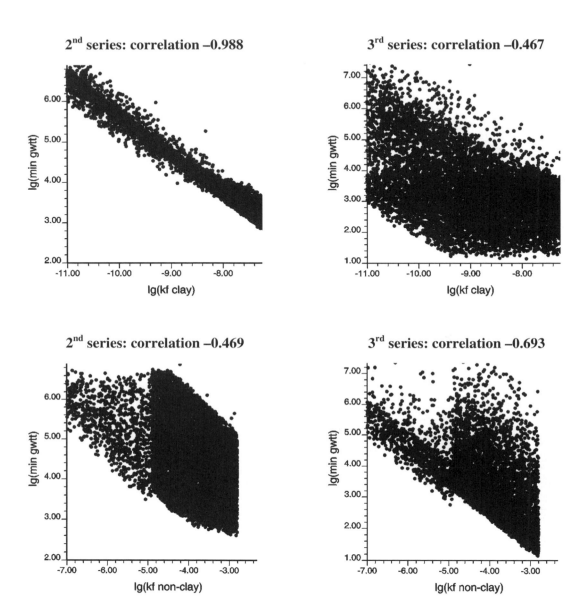

Figure 7. **Scatterplots for flux maxima at the upper boundary versus "clay" conductivities (above) and "non-clay" conductivities (below)**

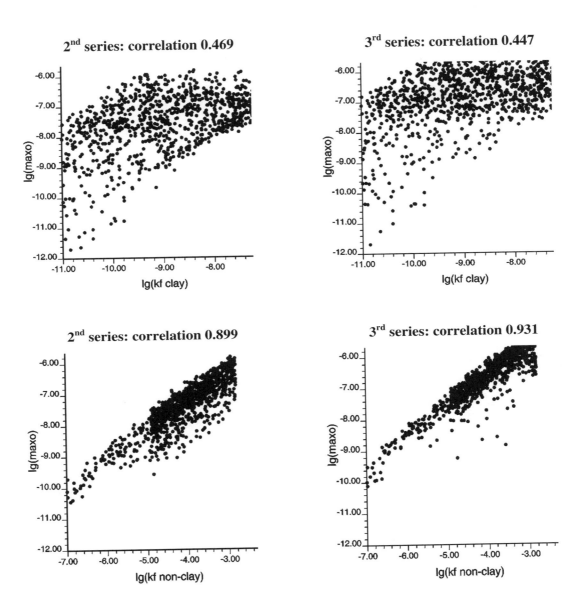

Conclusions and future work

The study has demonstrated that the Gorleben data set can be used as a base for geostatistical analyses which might be a first step towards an integrated assessment of the impact of spatial variability of the geological features of the site on safety indicators.

In the future, the methodology for the representation of spatial variability as well as the groundwater flow and nuclide transport modelling has to be improved.

For an improved representation of spatial variability,

- more than two hydrogeological units should be considered;
- spatial variability of hydrogeological parameters should be taken into account;
- the area north of the river Elbe should be taken into consideration;
- the consideration of additional (especially soft) information has to be developed.

A more adequate modelling of flow and transport processes would require

- the use of models taking into account the influence of the salinity-dependent density on the flow;
- the consideration of 3D groundwater flow and nuclide transport models.

References

Ludwig, R. (1994):
Projekt Gorleben. Hydrogeologische Grundlagen für Modellrechnungen. Kenntnisstand 1994. BGR-Archiv No. 112 002 (unpublished report). BGR Hannover.

Porter, J. D. and Hartley, L. J. (1997):
The Treatment of Uncertainty in Groundwater Flow and Solute Transport Modelling. Application of Indicator Kriging to Stratigraphic and Petrographic Data from the Gorleben Site. EUR 17829 EN, Luxembourg.

Vogel, P. and Schelkes, K. (1996):
Influence of initial conditions and hydrogeological setting on variable density flow in an aquifer above a salt dome. Calibration and Reliability in Groundwater Modelling, IAHS Publ. no. 237, 373-381.

Röhlig, K.-J. (1998):
Geostatistical Analysis of the Gorleben Channel. GeoENV98. Second European Conference on Geostatistics for Environmental Applications. November 18-20, 1998, Valencia. Kluwer Academic publishers, Dordrecht, Boston, London (in press).

Coupled Phenomena Potentially Affecting Transport by Diffusion in Silt Rocks: Evaluation Approach and First Results Regarding Thermal Diffusion

E. Tevissen (1), M. Rosanne (2), N. Koudina (2), B. Prunet-Foch (2), P.M. Adler (3)
French Radioactive Waste Management Agency (ANDRA)
LPTM – CNRS Meudon
Institut de Physique du Globe Paris (IPG)

1.   Introduction

The French Radioactive Waste Management Agency (ANDRA) has been investigating an argillite formation at Bure (France). The Meuse/Haute-Marne site, on the eastern margin of the Paris basin, presents a simple geological context of alternating carbonate and argilaceous formations. Some transport properties of the host formation (Callovo-Oxfordian argilites) have been characterised by hydraulic tests in deep boreholes and by specific measurements of permeability and diffusion on core samples.

The results indicate that the permeability of the rock is very low, ranging from 10^{-11} to 10^{-13} m/s provided by hydraulic tests to $3 \ 10^{-14}$ m/s obtained on core samples. Considering the natural hydraulic gradients, the conceptual model for solute transport in the Callovo-Oxfordian layer can be summarised as follows. The solute transport throughout the formation to the surrounding carbonate formations will be due to fickian diffusion i.e. the flux of solutes is proportional to the concentration gradient.

Therefore, diffusion coefficients were measured on argilite core samples. These coefficients associated with a pure fickian diffusion model, are used in preliminary safety assessment exercises to evaluate the potential migration of radionuclide in the Callovo-Oxfordian formation.

However, other causes of transport exists in nature and should not be forgotten (de Marsily G. 1987). Coupled phenomena (thermal osmosis, thermodiffusion, chemical osmosis…) may play a role in fluid, solute and heat transport in rocks such as Callovo-Oxfordian argilites and then may affect the diffusive transport of radionuclides.

2.   Approach

These coupled phenomena are described in the reference frame of the thermodynamics of irreversible processes. In general terms a coupled process is one in which flow of any kind (e.g. fluid, solute…) is driven by the gradient of potential not usually associated with that flow. Regarding solute transport in the argillites, the issue is to know if other gradients than concentration gradients could affect the diffusive process at a pluridecametric scale.

In order to do this, a general approach based on both experimental investigation and numerical simulation is defined:

- transport experiments are performed on argilite core samples of less than 1 cm thick, first with a simple salinity gradient in order to provide a precise quantification of fickian diffusion and then, coupling a salinity concentration gradient and another gradient (thermal, electric or osmotic);

- numerical models based on Onsager's reciprocal relations are used first to estimate the coupled coefficients from the experiments at a centimetric scale and then to evaluate the effect of coupled phenomena at the formation scale, considering real potential gradients.

The approach is illustrated with the results obtained on thermal diffusion of NaCl in Callovo-Oxfordian rocks: first evaluation of the Soret coefficient (S_T) and the potential effect at a decametric scale.

3.   First Results Regarding the Influence of Thermal Gradients on Diffussion

Evaluation of diffusion properties and the influence of thermal gradients on argilite samples

If advection, chemical interaction and coupled processes are neglected, the transient transport is driven by Fick's first law:

$$J_1 = -De\nabla C$$

Where J_1 is the mass flux of solute, C is the concentration in the porous medium, t is the time, De is the effective diffusion coefficient.

The effect of temperature on fickian diffusion is quantified in order to have the ability to distinguish the effect of temperature gradient on diffusion (thermal diffusion) from the effect of temperature on the purely diffusive uncoupled process. This effect is calculated considering the Nernst equation.

The heat and solute flux affected by thermal diffusion is given by the following equations:

$$J_1 = -De\nabla C - CS_T De\nabla T$$

$$J_q = -\lambda\nabla T - \rho RT^2 S_T De\nabla C$$

Through diffusion experiments of NaCl on a 0.5 cm thick argillite sample are performed without thermal gradients and under two thermal gradients of 130K/cm and 50 K/cm successively. The schema of the experimental device is given below (fig. 1).

The experimental results, concentration of NaCl in the inlet reservoir are shown Fig 2.

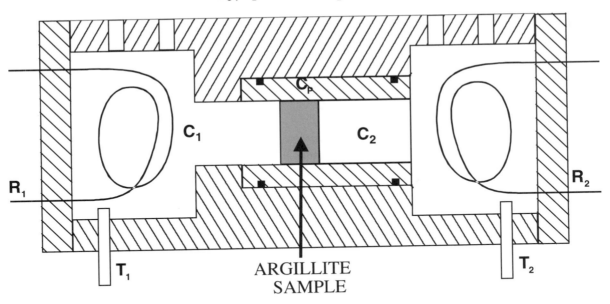

Figure1: **Experimental set up and design of migration experiment on a 0.5 cm argillite sample. R_1, R_2 : heating resistances ; C_1 , C_2 : Fluid chamber concentrations; T_1 , T_2 : thermocouples**

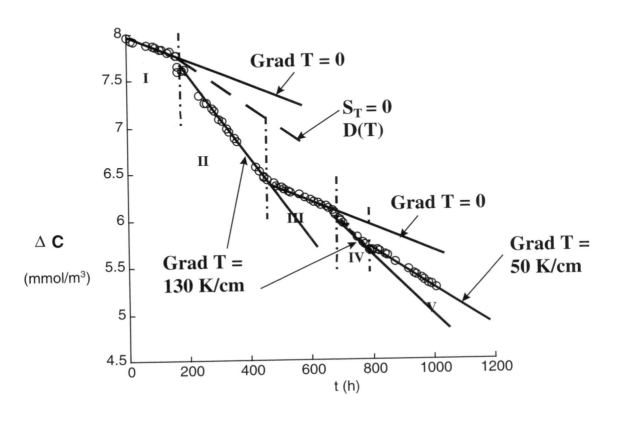

Figure 2 : **Experimental results obtained on argillite core sample. L = 0.5 cm   At t = 0 : $C(NaCl)_1$ = 9 10^{-3} M ; $C(NaCl)_2$ = 1 10^{-3} M. Phase I and phase III : pure diffusion-Phase II and phase IV: $\nabla T = 130 K/cm$ - Phase V : $\nabla T = 50 K/cm$**

The flux of NaCl is increased by a temperature gradient. As shown in Fig. 2 ($S_T = 0$, dotted line) this increase could be explained partially by the effect of temperature on fickian diffusion. The effect of temperature gradient is then established. Assuming an effect of thermal diffusion, it provides an estimation of Soret coefficient of $S_T = 0.1$ to 0.3 K^{-1}. From this evaluation of S_T, and solving equation (2) and (3), assuming constant concentrations at boundary conditions, the effect of a 0.1 K/m temperature gradient on the monodimensionnal diffusion of NaCl at a decametric scale in the argillites may be simulated. Results expressed in terms of flux at the upper limit are shown in Fig.4.

4.   Discussion

The estimation of the Soret coefficient determined in this study is significantly higher than those given in the literature (Schott J. 1973, Jamet P.V. 1991, Horsemann S.T. 1996) which usually range between 10^{-4} and $2\ 10^{-2}$ K^{-1}. As a consequence, the order of magnitude of the Soret Coefficients has to be verified. In particular, the potential occurrence of thermo-osmosis during the experiments must be checked.

Nevertheless, the simulations show that even for such a high Soret coefficient, the potential influence of thermo-diffusion on the migration of solutes in the argillites remains limited under natural thermal gradient conditions.

This type of approach could be applied to other coupled phenomena.

Figure 4:   **Shema of the simulated system Tp-Tn = 1K.**

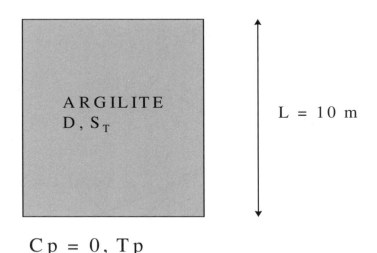

302

Figure 4: **Comparison of simulated NaCl flux at the upper limit of the system under pure diffusion (ST = 0) and under diffusion coupled with thermal diffusion.**

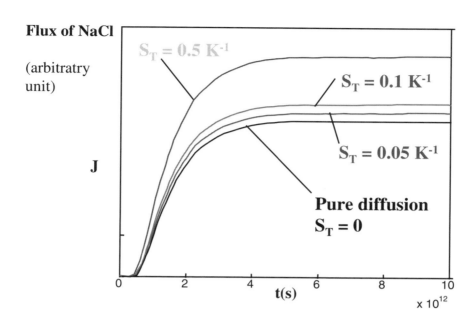

References

Horsemann S.T., Higgo J.J.W. Alexander J., and Harrington J.F., 1996
Water, Gas and Solute Movement Through Argillaceous Media.
NEA-OCDE Report CC-96/1

Jamet P.V., 1991
Sur certains aspects du couplage en milieux poreux entre champs de température et de concentration. Approche par la thermodynamique des processus irréversibles et modélisation de l'effet thermo-gravitationnel dans les solutions binaires
École des Mines de Paris. Doctoral thesis.

De Marsily G. Fargue D. et Goblet P., 1987
How much do we know about coupled transport processes in the geosphere and their relevance to performance assesment?
GEOVAL 87 Proc pp 475 – 491.

Schott J., 1973
Contribution à l'étude de la thermodiffusion dans les milieux poreux. Application aux possibilités de concentrations naturelles.
University of Toulouse. Doctoral thesis.

PARTICIPANTS LIST

BELGIUM

PUT, Martin
SCK/CEN
Geological Waste Disposal Project
Boeretang, 200
B-2400 Mol

Tel: +32 (14) 333 221
Fax: +32 (14) 323 553
Eml: mput@sckcen.be

CANADA

FLAVELLE, Peter
Senior Specialist, Geoscience and Waste A
Wastes and Decommissioning Division
Canadian Nuclear Safety Commission
P.O. Box 1046, Station B, 280 Slater Stre
Ottawa, Ontario K1P 5S9

Tel: +1 613 995-3816
Fax: +1 613 995-5086
Eml: flavellep@cnsc-ccsn.gc.ca

JENSEN, Mark
Ontario Power Generation
700 University Ave, H16-E27
Toronto, Ontario, M5G 1X6

Tel: +1 416 592 8672
Fax: +1 416 592 4485
Eml: mark.jensen@
              ontariopowergeneration.com

FINLAND

BLOMQVIST, Runar
Geological Survey of Finland - GTK
Betronimiehenkuja 4
FIN-02150 Espoo

Tel: +358 205 502 469
Fax: +358 205 5012
Eml: runar.blomqvist@gsf.fi

JAKOBSSON, Kai
STUK
Nuclear Waste and Materials Regulation
P.O. Box 14
FIN-00881 Helsinki

Tel: +358 (9) 759 88 308
Fax: +358 (9) 759 88 670
Eml: kai.jakobsson@stuk.fi

POTERI, Antti
Technical Research Centre of Finland
Tekniikantie 4C
Espoo PO Box 1604
FIN-02044 VTT

Tel: +358 0 456 5059
Fax: +358 0 456 5000
Eml: antti.poteri@vtt.fi

VIENO, Timo
VTT Energy
P.O. Box 1604
FIN-02044 VTT

Tel: +358 9 456 5066
Fax: +358 9 456 5000
Eml: timo.vieno@vtt.fi

VIRA, Juhani
Posiva Oy
Töölönkatu 4
FIN-00100 Helsinki

Tel: +358(0)9 2280 3740
Fax: +358(0)9 2280 3719
Eml: juhani.vira@posiva.fi

FRANCE

BRUNO, Gérard
IPSN
IPSN/DPRE/SERGD - BP6
F-92265 Fontenay-aux-Roses Cedex

Tel: +33 (0)1 46 54 77 55
Fax: +33 (0)1 46 54 71 35
Eml: gerard.bruno@ipsn.fr

DE MARSILY, Ghislain
Laboratoire de géologie appliquée
B 123
Univ. Pierre et Marie Curie
4 Place Jussieu
F-75252 Paris Cedex 05

Tel: +33 44 27 44 27
Fax: +33 44 27 51 25
Eml: gdm@ccr.jussieu.fr

LEBON, Patrick
ANDRA
1-7 rue Jean Monnet
F-92298 Chatenay-Malabry Cedex

Tel: +33 (0)1 46 11 80 82
Fax: +33 (0)1 46 11 84 10
Eml: patrick.lebon@andra.fr

MOUCHE, Emmanuel
Commissariat à l'énergie atomique
(CEA)
Bâtiment 454
Centre d'études nucléaires de Saclay
F-91191 Gif-sur-Yvette Cedex

Tel: +33 (0)1 69 08 22 54
Fax: +33 (0)1 69 08 82 29
Eml: emouche@cea.fr

OUZOUNIAN, Gérald
ANDRA
Parc de la Croix Blanche
1-7, rue Jean Monnet
F-92298 Chatenay-Malabry Cedex

Tel: +33 (0)1 46 11 83 90
Fax: +33 (0)1 46 11 84 10
Eml: gerald.ouzounian@andra.fr

GERMANY

FEIN, Eckhard
GRS
Theodor-Heuss Strasse 4
D-38122 Braunschweig

Tel: +49 (531) 8012 292
Fax: +49 531 8012 200
Eml: fei@grs.de

ROEHLIG, Klaus-Juergen
Gesellschaft für Anlagen- und
Reaktorsicherheit (GRS) mbH
Schwertnergasse 1
D-50667 Köln

Tel: +49(0)221 2068 796
Fax: +49(0)221 2068 939
Eml: rkj@grs.de

RÖTHEMEYER, Helmut
Bundesamt für Strahlenschutz
Willy-Brandt-Str.5
Postfach 10 01 49
D-38206 Salzgitter

Tel: +49 (5341) 885 600
Fax: +49 (5341) 885 605
Eml: hpiepenburg@bfs.de

SCHELKES, Klaus
Bundesanstalt für Geowissen-
schaft und Rohstoffe (BGR)
Stilleweg 2
D-30655 Hannover

Tel: +49 511 643 2616
Fax: +49 511 643 2304
Eml: k.schelkes@bgr.de

STORCK, Richard
Gesellschaft fuer Anlagen und
Reaktorsicherheit (GRS) mbH
Theodor-Heuss-Strasse 4
D-38122 Braunschweig

Tel: +49 (0)531 8012 205
Fax: +49 (531) 8012 211
Eml: sto@grs.de

WOLLRATH, Jürgen
Bundesamt für Strahlenschutz
BfS
Postfach 10 01 49
D-38201 Salzgitter

Tel: +49 (0)5341 885 642
Fax: +49 (0)5341 885 605
Eml: JWollrath@BfS.de

JAPAN

IJIRI, Yuji
Tokai Works, JNC
Tokai-mura
Ibaraki 319-1194

Tel: +81 29 287 0928
Fax: +81 29 287 3258
Eml: ijiri@tokai.jnc.go.jp

KOREA (REPUBLIC OF)

KANG, Chul-Hyung
KAERI
Head
Geological Disposal System Dev.
P. O. Box 105, Yusong
Taejon 305-600

Tel: +82 (0)42 868 2632
Fax: +82 (0)42 868-2035
Eml: chkang@kaeri.re.kr

SPAIN

ASTUDILLO, Julio
ENRESA
Emilio Vargas 7
E-28043 Madrid

Tel: +34 91 566 8120
Fax: +34 91 566 8165
Eml: jasp@enresa.es

GOMEZ-HERNANDEZ, Jaime
Dept. Hidraulica,
Escuela de Caminos
Universidad Politecnica
de Valencia (UPV)
E-46071 Valencia

Tel: +34 96 387 96 14
Fax: +34 96 387 76 18
Eml: jaime@dihma.upv.es

RODRIGUEZ AREVALO, Javier
Consejo de Seguridad Nuclear
CSN
C/Justo Dorado 11
E-28040 Madrid

Tel: +34 (91) 3460282
Fax: +34 (91) 3460588
Eml: jra@csn.es

SANCHEZ-VILA, Xavier
Technical University of Catalona
C/Gran Capitan, s/n
E-08034 Barcelona

Tel: +34 93 401 1698
Fax: +34 93 401 7251
Eml: dsanchez@etseccpb.upc.es

SWEDEN

DVERSTORP, Björn
Swedish Nuclear Power Inspectorate
SKI
S-106 58 Stockholm

Tel: +46 (0) 8 6988486
Fax: +46 (0) 8 6619086
Eml: bjorn.dverstorp@ski.se

STROMBERG, Bo
Office of Nuclear Waste
Swedish Nuclear Power Inspectorate (SKI)
S-10658 Stockholm

Tel: +46 8 698 8485
Fax: +46 8 661 9086
Eml: bo.stromberg@ski.se

WIKBERG, Peter
Swedish Nuclear Fuel and
Waste Management Co. (SKB)
Box 5864
S-102 40 Stockholm

Tel: +46 (8) 459 8563
Fax: +46 (8) 661 5719
Eml: skbpw@skb.se

SWITZERLAND

FRANK, Erik
HSK
Swiss Federal Nuclear Safety Inspectorate
CH-5232 Villigen-HSK

Tel: +41 (56) 310 39 45
Fax: +41 (56) 310 39 07
Eml: frank@hsk.psi.ch

HADERMANN, Jörg
Paul Scherrer Institute
Waste Management Laboratory
CH-5232 Villigen PSI

Tel: +41 56 310 2415
Fax: +41 56 310 2821
Eml: joerg.hadermann@psi.ch

MAZUREK, Martin
Rock/Water Interaction Group
University of Bern
GGWW, Institutes of Geology
and of Mineralogy and Petrology
Baltzerstr.1
CH-3012 Bern

Tel: +41 (0)31 631 87 81
Fax: +41 (0)31 631 48 43
Eml: mazurek@mpi.unibe.ch

SOLER, Josef
Paul Scherrer Institut
CH-5232 Villigen PSI

Tel: +41 56 310 2390
Fax: +41 56 310 2821
Eml: josef.soler@psi.ch

UNITED KINGDOM

NORRIS, Simon
Nirex UK Ltd
Curie Avenue,
Harwell, Didcot
Oxfordshire OX11 ORH

Tel: +44 (0) 1235 825310
Fax: +44 (0) 1235 820560
Eml: simon.norris@nirex.co.uk

WATTS, Len
BNFL
Room 2 Rutherford House
Warrington WA3 6AS

Tel: +44 1925 83 43 44
Fax: +44 1925 83 20 16
Eml: len.watts@bnfl.com

UNITED STATES OF AMERICA

ALTMAN, Susan
Sandia National Laboratories
P.O. Box 5800 MS 0735
Albuquerque, NM 87185-0735

Tel: +1 505 844 2397
Fax: +1 505 844 4426
Eml: sjaltma@sandia.gov

BEAUHEIM, Richard L.
Sandia National Laboratories
13255 W. Center Dr.
Lakewood, CO 80228-2428

Tel: +1 (303) 984 4192
Fax: +1 (303)
Eml: rlbeauh@sandia.gov

CODELL, Richard
U.S. Nuclear Regulatory Commission
Mail Stop T7D13
Washington, DC 20555

Tel: +1 (301) 415 8167
Fax: +1 (301) 415 5399
Eml: rbc@nrc.gov

EISENBERG, Norman
Consultant
1208 Harding Lane
Silver Spring MD 20905-4004

Tel: +1 301 384 6507
Fax:
Eml: nae@bellatlantic.net

LEVICH, Robert A.
International Programme Manager,
USDOE/YMP
1551 Hillshire Drive,
Las Vegas NV 89134

Tel: +1 (702) 794 5449
   Fax: +1 (702) 794 5431
Eml: bob_levich@ymp.gov

MATTHEWS, Mark
USDOE/CAO/WIPP
PO Box 3090
Carlsbad, NM 88221

Tel: +1 505 234 7467
Fax: +1 505 234 7061
Eml: matthem@wipp.carlsbad.nm.us

PATTERSON, Russ
USDOE/YMP
1551 Hillshire Drive
Las Vegas, NV 89134

Tel: +1 702 794 5469
Fax: +1 702 794 5559
Eml: russ_patterson@notes.ymp.gov

WENDELL, Weart
Sandia National Laboratories
Albuquerque,
New Mexico 87185-0771

Tel: +1 505 844 4855
Fax: +1 505 844 0591
Eml: wdweart@sandia.gov

CONSULTANT

SMITH, Paul
(Consultant to NEA/OECD)
Safety Assessment Management Ltd.
9 Penny Green
Settle
North Yorkshire BD24 9BT

Tel: +44 1729 823 839
Fax: +44 1729 824893
Eml: paul@samltd.demon.co.uk

NEA SCIENTIFIC SECRETARIAT

LALIEUX Philippe
Presently at:
ONDRAF/NIRAS
Avenue des Arts, 14
B-1210 Bruxelles

Tel: +32 2 212 10 82
Fax: +32 2 212 51 65
Eml: p.lalieux@nirond.be

NEA Publications of General Interest

1999 Annual Report (2000)   *Free: available on Web.*

NEA News
ISSN 1605-9581   Yearly subscription: FF 240 US$ 45 DM 75 £ 26 ¥ 4 800

Geologic Disposal of Radioactive Waste in Perspective (2000)
ISBN 92-64-18425-2   Price: FF 130 US$ 20 DM 39 £ 12 ¥ 2 050

Radiation in Perspective – Applications, Risks and Protection (1997)
ISBN 92-64-15483-3   Price: FF 135 US$ 27 DM 40 £ 17 ¥ 2 850

Radioactive Waste Management in Perspective (1996)
ISBN 92-64-14692-X   Price: FF 310 US$ 63 DM 89 £ 44

Radioactive Waste Management

Geological Disposal of Radioactive Waste – Review of Developments in the Last Decade (1999)
ISBN 92-64-17194-0   Price: FF 190 US$ 31 DM 57 £ 19 ¥ 3 300

Water-conducting Features in Radionuclide Migration (1999)
ISBN 92-64-17124-X   Price: FF 600 US$ 96 DM 180 £ 60 ¥ 11 600

Features, Events and Processes (FEPs) for Geologic Disposal of Radioactive Waste
An International Database (2000)
ISBN 92-64-18514-3   Price: FF 150 US$ 24 DM 45 £ 15 ¥ 2 900

Porewater Extraction from Argillaceous Rocks for Geochemical Characterisation (2000)
ISBN 92-64-17181-9   Price: FF 380 US$ 60 DM 113 £ 37 ¥ 6 350

Regulatory Reviews of Assessments of Deep Geological Repositories – Lessons Learnt (2000)
ISBN 92-64-05886-9   Price: FF 210 US$ 32 DM 63 £ 20 ¥ 3 400

Strategic Areas in Radioactive Waste Management – The Viewpoint and Work Orientations
of the NEA Radioactive Waste Management Committee (2000)   *Free: paper or Web.*

Stakeholder Confidence and Radioactive Waste Disposal (2000)
ISBN 92-64618277-2   *Free: paper or Web.*

Progress Towards Geologic Disposal of Radioactive Waste: Where Do We Stand? (1999)
Free: paper or Web.

Confidence in the Long-term Safety of Deep Geological Repositories – Its Development
and Communication (1999)   *Free: paper or Web.*

Order form on reverse side.

ORDER FORM

OECD Nuclear Energy Agency, 12 boulevard des Iles, F-92130 Issy-les-Moulineaux, France
Tel. 33 (0)1 45 24 10 10, Fax 33 (0)1 45 24 11 10, E-mail: nea@nea.fr, Internet: www.nea.fr

Qty	Title	ISBN	Price	Amount
			Postage fees*	
			Total	

*European Union: FF 15 – Other countries: FF 20

❏ Payment enclosed (cheque or money order payable to OECD Publications).

Charge my credit card   ❏ VISA   ❏ Mastercard   ❏ Eurocard   ❏ American Express

(N.B.: You will be charged in French francs).

Card No.	Expiration date	Signature
Name		
Address	Country	
Telephone	Fax	
E-mail		

OECD PUBLICATIONS, 2, rue André-Pascal, 75775 PARIS CEDEX 16
PRINTED IN FRANCE
(66 2001 01 1 P 1) ISBN 92-64-18620-4 – No. 51685 2001